Lecture Notes in Applied Mechanics

Volume 1

Series Editor

Prof. Dr.-Ing. Friedrich Pfeiffer

Springer

Berlin
Heidelberg
New York
Barcelona
Hong Kong
London
Milan
Paris
Singapore
Tokyo

Set-Valued Force Laws

Dynamics of Non-Smooth Systems

Christoph Glocker

Springer

Professor Dr.-Ing. Christoph Glocker
Institute of Mechanical Systems
Dept. of Mechanical and Process Engineering
ETH Zentrum - CLA J21.1
CH-8092 Zürich
SWITZERLAND
email: christoph.glocker@imes.mavt.ethz.ch

With 65 Figures

ISBN 3-540-41436-3 Springer Verlag Berlin Heidelberg New York

Cataloging-in-Publication Data applied for

Glocker, Christoph:
Set valued force laws : dynamics of non-smooth systems /
Christoph Glocker. - Berlin ; Heidelberg ; New York ; Barcelona ;
Hong Kong ; London ; Milan ; Paris ; Singapore ; Tokyo : Springer, 2001
(Lecture notes in applied mechanics ; Vol. 1)
ISBN 3-540-41436-3

Springer-Verlag Berlin Heidelberg New York
a member of BertelsmannSpringer Science+Business Media GmbH

http://www.springer.de

Printed in Germany

Cover design: Design & Production GmbH, Heidelberg
Typesetting: Digital data supplied by author

SPIN: 10792269 62/3020xv-5 4 3 2 1 0 – printed on acid-free paper

Preface

As one of the oldest natural sciences, mechanics occupies a certain pioneering role in determining the development of exact sciences through its interaction with mathematics. As a matter of fact, there is hardly an area in mathematics that hasn't found an application of some form in mechanics. It is thus almost inevitable that theoretical methods in mechanics are highly developed and laid out on different levels of abstraction. With the spread of digital processors this goes as far as the implementation in commercial computer codes, where the user is merely confronted on the surface with the processes that run in the background, i.e. mechanics as such: in teaching and research, as well as in the context of industry, mechanics is much more, and must remain much more than the mere production of data with the help of a processor. Mechanics, as it is talked about here, traditionally includes a wide spectrum, ranging from applied mechanics, analytical and technical mechanics to modeling and experimental mechanics, as well as technical realization. It also includes the subdisciplines of rigid body mechanics, continuum mechanics, or fluid mechanics, to mention only a few.

One of the fundamental and most important concepts used by nearly all natural sciences is the concept of linearization, which assumes the differentiability of mappings. As a matter of fact, all of classical mechanics is based on the availability of this quality. The most simple examples, however, such as the absolute value at the origin show that linearization may fail. The next possible structural level is convexification, where linear spaces are replaced by convex cones. Inequality problems, as they occur in mechanics with dry friction or problems of contact, can be dealt with in this way, while they have to be excluded in classical mechanics. In good textbooks on the dynamics of rigid bodies one can thus find a general treatment of multibody systems which enables the student to deal with complex mechanical systems. Yet questions of dry friction and unilateral contact, along with the problem of impacts are merely addressed in specialized, simple examples. The reason for this lies in the fact that the mathematical methods for the description and treatment of inequalities have only been developed during the last decades and are still in a state of flux. Classical mechanics, on the other hand, had already reached one of its primes towards the end of the 18th century. It is therefore not surprising that commercially available computer software only allows the treatment of inequalities in a rudimentary fashion, quasi as an appendage. A general implementation would require deep structural changes, i.e. changes in the core of extensive computer programs. It is however foreseeable that this situation will change within the next decade. Today, the attempt has been made to use smooth functions obtained from regularization. This approximate description is unsatisfying, both from a theoretical as well as from a physical point of view, because it does not produce the desired but merely a similar behavior, not to mention the numerical difficulties which then express themselves in the form of stiff differential equations.

One thing is true in any case: We cannot think of the inclusion of inequalities in mechanics as a small supplemental step. We must rather speak of a basic extension of the classical theories, even of a completely new direction in mechanics, a new area of extremely fertile basic research that is developing and spreading at high speed. Used correctly, inequalities are an incredibly powerful tool. They further complete the range of what we are able to model and justice can be done to minimal models, i.e. models that portray, in a condensed form, the essential dependencies on parameters in relation to a given set of questions in accordance with Hertz's third requirement. In this context, we will have to rethink and reinterpret the physical content of non-uniqueness and non-existence of solutions, as well as in dynamics the lost property of continuous dependence of the trajectories on the initial conditions.

Two names especially should be mentioned in connection with inequality problems in mechanics: Professor Jean Jacques Moreau from Montpellier, who likely made the most important contribution to finite freedom dynamics – and much more – in his already 1988 published CISM-course "Unilateral Contact and Dry Friction in Finite Freedom Dynamics", and Professor Panagiotis D. Panagiotopoulos, who passed away much too early. Professor Panagiotopoulos' work on nonconvex variational inequalities in the context of his 1982 theory on "hemivariational inequalities" was widely circulated in mechanics and mathematics. Almost all the ideas developed in the present treatise on "Multibody Dynamics with Unilateral Contacts", started in the extremely creative environment at his institute in Thessaloniki, where I was a guest for a year as fellow of the Feodor-Lynen-Program supported by the Alexander von Humboldt-Stiftung. Last but not least, I would like to thank my doctoral supervisor, Proffessor Friederich Pfeiffer, who provided me with the opportunity to complete this work at the Institute B of Mechanics at the Technical University of Munich, and who over the years, with his extensive experience in the area of machine dynamics, was able to convey the necessary practical background.

Munich, January 2001 *Christoph Glocker*

Contents

1. Introduction

When dealing with problems in multibody systems, there is a point where questions about the influence of Coulomb friction occur. This is — unfortunately — the common situation where non-smooth dynamics is accessed, and worst, this is also the point where one remembers each word heard about friction in school in order to implement it immediately in the most general way in multibody theory: There is a coefficient μ for sliding friction and another coefficient $\mu_0 > \mu$ for stiction. In the case of sliding the friction force F_T is computed to be μ times the normal force F_N, opposing the direction of the relative velocity v, whereas for stiction one has the restriction $|F_T| \leq \mu_0 F_N$. With that friction law, one has chosen one of the most complicated force laws that occur in application problems. It seems to be so easy and so clear at a first view, however, when trying to apply it, or even when just trying to write it down as a mathematical expression, one immediately encounters a lot of serious and not expected problems of very different nature.

1.1 Friction Laws

When we draw the graph of this friction law for the one-dimensional case (Figure 1.1, left diagram) we must accept its set-valuedness at $v = 0$. The vertical line at this point is not drawn by mistake, but expresses real physical behavior: Any force within this set may occur, without changes in the velocity $v = 0$. The friction characteristic is therefore not expressed by a function like classical force laws describing springs or dashpots, but by a set-valued map, similar to the representation of bilateral constraints. We also recognize that the friction characteristic in the left diagram of Figure 1.1 does not express the behavior of the model that we have in mind. There is something missing. We intuitively understand that, for a sticking contact, the friction force must reach the point $\pm\mu_0 F_N$ before sliding may start. On the other hand, F_T may jump to any point within the set $[-\mu_0 F_N, +\mu_0 F_N]$ when transitions to stiction occur. There is no necessity for the friction force to take values $\pm\mu_0 F_N$ in this situation. The friction model therefore contains some hysteresis behavior (Figure 1.1, right diagram), where the area of differences between "loading" and "unloading" has shrunk to one point, $v = 0$. This behavior might be described by the additional condition $|F_T| \to \max$ on the tangential force,

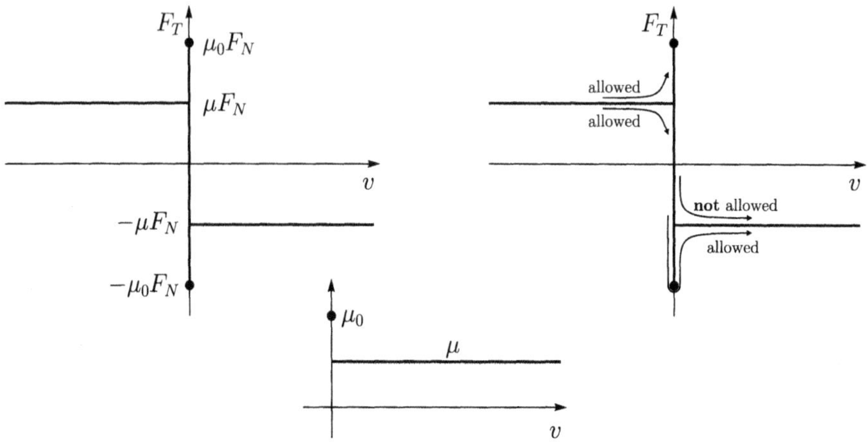

Figure 1.1. Friction characteristics without and with hysteresis

finally leading to a complete description of the friction law that we have learned in school. Except for the approach proposed in [91] there is not a single attempt to apply this friction law to multiple contact situations because of its difficult hysteresis behavior at $v = 0$. Instead, force laws with friction coefficients depending smoothly on v are usually used, such as the Stribeck curves in Figure 1.2. For practical applications it is often sufficient to use the characteristic from the left diagram in Figure 1.2. Hysteresis-caused effects, however, are then finally excluded. A theoretical framework on how to deal with force elements with hysteresis may be found in the book [101].

By restricting ourselves to hysteresis-free force characteristics we have already bypassed some of the problems contained in the force law in Figure 1.1. There are, however, other structural difficulties resulting mainly from

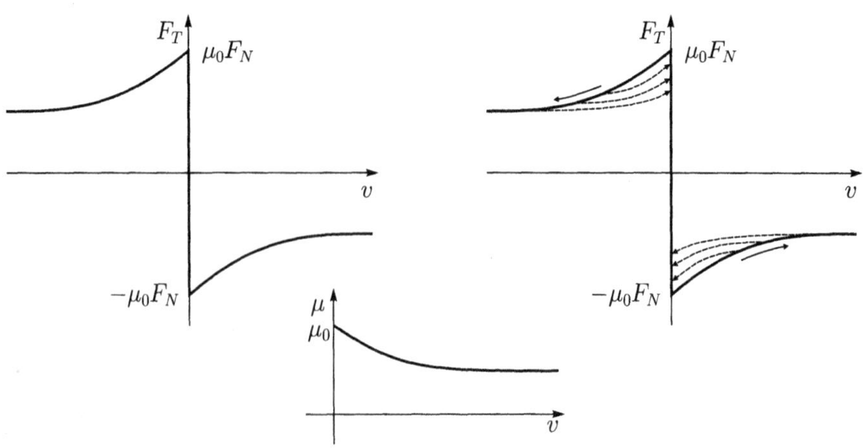

Figure 1.2. Coulomb-type Stribeck curves without and with hysteresis

the dependence of F_T on the normal force F_N. Thinking about a system that is unilaterally or bilaterally constrained in the normal direction, it becomes immediately clear that the normal force F_N is unknown for general dynamic situations. The overall contact law therefore denies a splitting into independent normal and tangential portions. Moreover, it acts on different kinematic levels since unilateral constraints are expressed by the complementarity of positive normal forces $F_N \geq 0$ and displacements $d \geq 0$, see Figure 1.3, whereas the friction law is stated in terms of velocities.

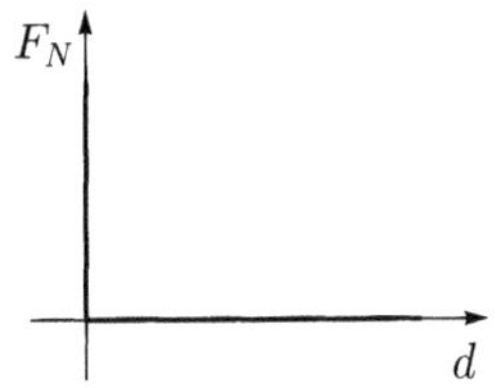

Figure 1.3. Unilateral constraint

Even if it would be possible to state these two portions of the contact law on a common kinematic level (and indeed, this can be done as we will see) there is a final hurdle to get over: Coulomb friction does not fit in potential theory. There is no potential function, neither on the displacement nor on the velocity level, that allows to express the Coulomb friction law in terms of generalized gradients. A similar property is known from some classical force elements such as oil bearings, where the force vector field is composed of both, gradient fields and twisted fields. For given normal forces, however, the Coulomb friction law splits into two independent parts, both being derived from potential functions. These are the unilateral constraint from Figure 1.3 and the friction characteristic in Figure 1.4, which we call dry friction.

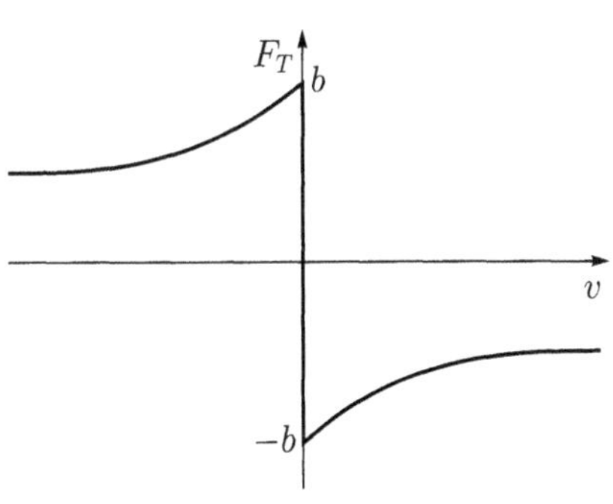

Figure 1.4. Dry friction Stribeck curve (given bounds of normal force)

Note that the force laws from Figures 1.3 and 1.4 are still multivalued, a fact that seems to contradict a unique determination of the forces at the points of set-valuedness. However, one must not forget that force laws always occur in combination with some equations of motion of a dynamical system, which guarantee uniqueness in many cases. Consider, for example, the equation of motion $m\ddot{x} + c = F$ $(m > 0)$ together with a force law on the acceleration level stated as $F = +b$ for $\ddot{x} < 0$, $F = -b$ for $\ddot{x} > 0$, and $F \in [-b, +b]$ for $\ddot{x} = 0$ $(b > 0)$. It is immediately clear that the solution of this equilibrium problem is unique, i.e. $F = +b$ for $c > b$, $F = -b$ for $c < -b$, or $F = c$ if $-b \leq c \leq +b$, because the graphs of the equation of motion and the force law intersect at exactly one point for a positive mass m and any choice of the external load c, see Figure 1.5.

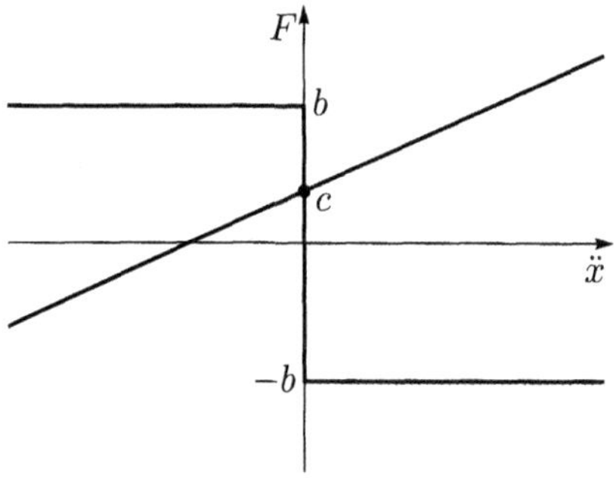

Figure 1.5. Equilibrium problem on acceleration level

A force law stated on the acceleration level might look strange since forces in dynamics depend on velocities and displacements only. However, exactly the same procedure is observed, for example, in the treatment of bilateral holonomic constraints: When the constraint is expressed in its original form on the displacement level, the corresponding differential equation is said to be of index 3. After index reduction by differentiation of the constraint it is said to be of index 2 and of index 1, if the constraint is additionally stated on the velocity and on the acceleration level, respectively. Since bilateral constraints are nothing else than set-valued force laws, the motivation for stating the friction law (Figure 1.5) on the acceleration level becomes clear: To regard dry friction as a force element where nonholonomic constraints are switched on and off, depending on the system's dynamics.

Once again, let us come back to the friction law depicted in Figure 1.4. Suppose a motion that starts with sliding and that reaches, after some time, a state where the relative velocity v vanishes. At this instance of time the set of values that the tangential force F_T may take suddenly increases from a single number to the entire interval $[-b, b]$. The force F_T may therefore jump to any value within this set, depending on the overall external loads as well as on the dynamics. Since forces and accelerations are connected by the Newton-Euler equations, a discontinuity in the accelerations as functions of time might occur, caused by the jump of F_T. Similar discontinuity events result also from the unilateral constraint in Figure 1.3. Moreover, since the set of the admissible forces is unbounded for distances $d = 0$, we may even observe jumps in the velocities, as a result of contact impulsions.

Accelerations play a fundamental role for the comprehension of dynamic systems. In contrast to classical mechanics where these magnitudes are always well defined, one has to consider certain restrictions for non-smooth motion. As we have learned from the discussion above, there are events causing discontinuities in the sense that only some left or right limit of the accelerations exists, and that the accelerations become useless, or they are even undefined, when impacts are treated. On the other hand, accelerations are extensively used in classical analytical dynamics, leading to fundamental and famous principles that should be carried over, at least by parts, to non-smooth motion.

1.2 Literature Survey

Classical mechanics involves a lot of different mathematical and physical disciplines. Obviously the range of scientific areas gets even bigger when non-smooth dynamics is accessed. The literature is thus huge, and the following summary cannot provide a complete overview, since this would fill an entire book which has already been written by BROGLIATO [21] with its 1015 references. Instead, only a relatively small number of contributions is mentioned, just to indicate some of the directions in which today non-smooth dynamics is developing. Only a few comments are made on the contents, because nearly all of the material cited below is fully contained and extensively discussed in the excellent review [22] written by BROGLIATO, TEN DAM, PAOLI, GÉNOT and ABADIE, which should be consulted for more information on this subject. It contains another 240 references and will hopefully appear soon as an ASME AMR paper. Further references related to oscillations of non-smooth systems may be found in the overview [119] by POPP, and the status of multibody dynamics in mechanics is addressed in [130] by SCHIEHLEN. In the short survey presented below, mainly people working on finite-dimensional dynamic problems are mentioned. The big group of researchers in continuum mechanics, in which non-smooth problems have been investigated for a much longer time, has only been considered partially, as they are too many. Currently, a lot of concepts that are known in continuum mechanics is re-invented in some modified form for the area of dynamics. This, however, seems to be a natural process as it happens not for the first time, and is probably necessary when adopting new ideas from other areas of research.

Rigid body dynamics: The classical book on mechanics is "Theoretische Mechanik" [52] written in 1949 by HAMEL. In the author's opinion it would be worth translating this book into English to make it accessible for a bigger class of readers. With respect to classical mechanics in terms of tensor coordinate notation we refer to the recently published book [107] by PAPASTAVRIDIS and to the articles [76], [77], [75] of MAISSER. Modern books on classical dynamics are ARNOLD [7] and, in the abstract mathematical language, the work of ABRAHAM, MARSDEN and RATIU, see e.g. [1], [2], [78]. Methods and principles used in classical mechanics are discussed in the books [20] by BREMER and [34] by FISCHER and STEPHAN. Rigid body systems are treated in WITTENBURG [145] and, with respect to screw theory, in the famous book [11] of BALL. Kinematics may be found in ANGELES [5].

Nonsmooth analysis: The standard reference on convex analysis and optimization theory is the book [121] by ROCKAFELLAR. In the book [24] by CLARKE the generalized gradient is introduced for Lipschitz continuous functions, a tool necessary to treat non-convex optimization problems. This work is continued in [25] by CLARKE, LEDYAEV, STERN and WOLENSKI. The booklet [123] by ROCKAFELLAR extends the generalized gradient to the non-Lipschitzian case. The book [9] by AUBIN and EKELAND provides a comprehensive source for non-convex and non-smooth analysis, in particular containing set-valued maps. Set-valued analysis is continued in the book [10] by AUBIN and FRANKOWSKA. Finally, there is the recently published book [124] by ROCKAFELLAR and WETS which presents a unified framework of most of this material.

Numerical integration: One of the main topics of present research in non-smooth dynamics is the development of reliable numerical integration algorithms for measure differential inclusions from both, the practical and the theoretical point of view. This implies the treatment of differential algebraic equations, see in particular volume II of the two books [45], [46] on ordinary differential equations by HAIRER, NORSETT and WANNER, and the computer code MEXX [71] especially designed for multibody systems by LUBICH, NOWAK, PÖHLE and ENGSTLER. One of the main problems of differential algebraic equations is the stabilization of the constraints. For measure differential inclusions this problem remains, but it is accompanied by much more, such as complementarity conditions which switch off and on the constraints, and additional impacts which result from dirac point measures. Measure differential inclusions looked upon as sequences of differential algebraic equations which are initialized by the solutions of complementarity problems lead to numerical schemes which are called in [22] the event based algorithms. Although these algorithms can not handle the general case, they have been proven to be suitable for certain classes of problems with not too many contact points, say up to 100, such as the vibration feeder treated in [146]. Some basic ideas concerning the mechanical interpretation of d'Alembert/Lagrange's principle in inequality form and on how to combine the different branches of the solution may already be found in the book [18] of BOLTZMANN. In rough outline, two complementarity problems, one for the velocities and another one for the accelerations have to be solved to complete the initialization process as described in [37], [116]. Event based numerical schemes are presented by BARAFF in [14] as well as by TEN DAM in [29] with particular emphasis on the computation of the contact and release sets for linear systems. Event based algorithms become more and more impracticable when the number of discontinuity events increases. This happens for problems with very many unilateral constraints or for problems at which accumulation points occur in the sequences of the time instances at which constraints have to be switched on and off. In these cases, time-stepping methods have to be preferred. These are difference schemes including fully the complementarity conditions and the impact rules, allowing a simultaneous treatment of impulsive and non-impulsive forces together with the involved inequalities; in the author's opinion the only way to deal with really big application problems. First schemes of BDF-type methods were provided by LÖTSTEDT, see e.g. [67], [68], [70], [69], although the impact laws have still been treated separately. The algorithm used by MOREAU and JEAN [91], [87], [88], [89], [94], [58], [59], a discretization of MOREAU's sweeping process based on a midpoint rule, provides an extremely powerful discretization scheme with which systems containing up to thousands of unilateral frictional contacts have been computed successfully. The related questions on the convergence, existence and uniqueness of solutions are treated by MABROUK in [74] and by MONTEIRO MARQUES in [84]. There is much progress in this direction. Recently, different discretization schemes have been provided, such as an algorithm based on displacements by PAOLI and SCHATZMAN [104], [105], [106] with proven convergence, or several other well-developed codes by the authors STEWARD, TRINKLE, ANITESCU and POTRA, see e.g. [136], [135], [6] and also [22] for further references. From the mechanical point of view, non-uniqueness of solutions and non-continuous dependence on the initial values have to be accepted. They often describe the dynamical behavior of physical devices in a very reasonable manner. Thus, it is not surprising that these questions occur also in a mathematical framework as addressed by BALLARD in [12] and SCHATZMAN in [128], [129].

Numerical treatment of inequality laws: Today, it is more and more accepted to use tools from convex analysis for the formulation and mathematical treatment of set-valued force laws. Variational formulations, projections, proximations and optimization problems associated with the inclusions to be dealt with are immediately recognized. They turn up in the upmost natural manner which is in perfect accordance with the spirit of classical mechanics. In the following we give a brief overview of methods today used to solve inclusions numerically. As a matter of course, MOREAU applies tools from convex analysis to formulate the unilateral contact problem with Coulomb friction. The numerical evaluation of the resulting multi-contact problem is performed by a Gauß-Seidel iteration [93], [94] in which the contacts are solved cyclically independent of each other. The iterative algorithm [96] introduced by PANAGIOTOPOULOS splits the Coulomb friction multi-contact problem into one overall normal and one overall tangential subproblem which are then alternately solved until the solution converges. The variational calculus associated with convex and non-convex minimization problems of this sort can be found in great detail in [101], [97], [100], [99] for continuum mechanics, whereas finite dimensional dynamic systems are treated in [98]. Another powerful method for frictional contacts is the augmented Lagrangian approach which was introduced into mechanics by ALART and CURNIER [3], see e.g. KLARBRING [65], SIMO and LAURSEN [133] for a review, and CURNIER, HE, KLARBRING [28] for the kinematical and mechanical formulation of the contact laws. It has been successfully applied in dynamics, for example, by the simulation of the three-dimensional vibration feeding process [146]. A formulation of the frictional contact problem in terms of the contact forces leads to linear and nonlinear complementarity problems, respectively, depending on the choice of the numerical discretization scheme (explicit versus implicit, for example). This has been attempted by nearly every author working on multi-contact problems, and we will only give a small selection of references. Such formulations together with the associated variational inequalities and minimization problems may be found in [62], [63], [64] by KLARBRING and BJÖRKMAN, but also in the papers [136], [103], [142], [102] by STEWART, TRINKLE, PANG, SUDARSKY and LO along with proofs on existence and uniqueness of solutions. The theory of linear complementarity, linear and nonlinear programming together with available algorithms is found in the paper [26] by COTTLE and DANTZIG, and in the books [27], [95] by COTTLE, PANG, STONE and by MURTY. Solution methods for nonlinear complementarity problems based on NCP-functions and a semi-smooth Newton algorithm are treated by DE LUCA, FACCHINEI, KANZOW, FISCHER, see e.g. [73], [72], [32].

Modeling and applications: Coulomb's law was vehemently discussed in Germany in 1910 because of the inconsistencies exposed by Painlevé. In the famous four articles [66], [83], [51] and [120] KLEIN, V. MISES, HAMEL and PRANDTL tried to find a reasonable resolution of this problem, as compatible as possible with the view of mechanics at that time. As arguments related to physics on the one hand and to the mathematical model on the other hand were mixed up, no general consensus was found. Nevertheless, the today's view was already expressed and also confirmed by PFEIFFER's calculation [109] on the elastic model: The assumption of a rigid body is reasonable. Coulomb's friction law is also reasonable. The putative inconsistencies can be resolved when allowing a velocity jump without collision. There is no reason why motion should not start with a velocity jump. An even more striking example of an impact without collision is the sliding rod, see MOREAU [91] for computational results, and GÉNOT and BROGLIATO [36] for a detailed analysis. This system is probably best suited to dwell on Coulomb friction, rigid bodies and dynamics. In non-smooth mechanics there is, however, more than frictional contacts. In continuum mechanics, for instance, one-dimensional multifunctions with and without hysteresis are used to model the stress-strain laws responsible for the breaking of fibers and delamination processes, see PANAGIOTOPOULOS [101], [100] and STAVROULAKIS [134]. In rigid body dynamics, set-valued force laws occur naturally as parallel and serial connections of unilateral primitives and have been studied and classified in [40]. Impacts require sophisticated and individual modeling. Standard models are impacts obeying the maximum dissipation principle of MOREAU [91], or Poisson-type impact laws [37], [116] based on a unilateral formulation. Multiple impacts in rigid body systems, however, must be treated with care. The superposition of simple impact laws by addition may fail due to complicated mechanical impulse transfer which, of course, can not be *calculated* by the rigid body approach. A theoretical framework, however, is still available as described by FRÉMOND in [35]. The most famous example of such sequential impact behavior is Newton's cradle which has been investigated by a number of authors, see e.g. HURMUZLU and MARGHITU [53], IVANOV [56], [55], STRONGE [140] and CHATTERJEE [23] for general comments, related problems and discussions on the rigid body assumption. Nevertheless, a large number of essentially different applications in dynamics has already been treated successfully by using non-smooth elements. All the ingredients mentioned above are indeed needed to develop reasonable mechanical models for practical problems. The dynamics of granular materials has been studied by MOREAU [92], [93] and by JEAN [57]. Deep drawing processes were investigated by JOURDAN, JEAN and ALART [60]. Both problems have a whole bunch of contact points and are excellently suited for applying time-stepping methods. Various applications in machine dynamics, computed by event driven algorithms, are collected in the review papers [115], [114], [113], [112] of PFEIFFER. They contain vibratory feeders, frictional chimney dampers, chains and gears, assembly tasks as peg-in-hole problems and snap joints, frictional turbine blade dampers, unilaterally constrained elastic robots, and gear rattling problems. Some of them may also be found in the book [116]; others are analyzed in accompanying papers, such as the chimney dampers together with STIEGELMEYR in [137], or gear shift process and models of wet clutches together with HAJ-FRAJ in [49] and [50]. The starting point of all these investigations was the original woodpecker paper [110], followed by an impact theory [118] for gear rattling problems [111].

1.3 Subjects and Contents

Nearly all publications touching non-smooth dynamics are concentrated on the unilateral contact problem with Coulomb friction and/or impacts. Non-smooth dynamics, however, does not only consist of unilateral constraints and dry friction laws, although these are indeed the most significant elements. In order to access a bigger class of set-valued interactions, we allow forces that may be derived from scalar potential functions. One-way clutches that are unilateral constraints on the velocity level, or pre-stressed springs may serve as examples. All of the following is done within this framework. Of course, this relatively simple class covers only a small number of imaginable force laws; on the other hand it is very well suited to understand the connections to classical mechanics. The contact laws in the frictional contact problem are usually stated on different kinematic levels, mainly influenced by the discretization schemes used for the numerical evaluation: Formulations on displacement, velocity, and acceleration level are found in the literature. Thus, the aim of this book is twofold. Besides the main purpose of determining the accelerations of a dynamic system that is subjected to scalar force potentials, we will give some analytical expressions that connect the three different levels of representation, i.e. the formulations using the displacement, the velocity and the acceleration level. Secondly, we try to discuss how methods used in classical mechanics fit into this setting, and how classical terminology might be understood when applied to non-smooth systems. We do not treat the impact itself. Impacts and impact equations are beyond the scope of this book. We are only interested in the accelerations of the system. Pre- and post-impact velocities, however, enter fully our description. As already pointed out, we restrict ourselves to dry friction, that is Coulomb friction with a given normal force. This restriction is necessary because all of the following is based on a one-dimensional potential theory. We allow, however, arbitrary friction laws like Stribeck-curves and also multivalued displacement dependent interactions. Nevertheless, we will discuss at least some mathematical formulations of the spatial Coulomb friction law in Chapter 10 of this book, because of its enormous importance for practical applications.

The book is organized as follows: In Chapter 2 we review some fundamental concepts in mechanics that hold beyond the rigid body assumption. Within the framework of the extended interaction law and the linear and angular momentum principle as general axioms in mechanics and dynamics, respectively, definitions of the most basic terms are put together. In particular, mechanical systems and subsystems together with their internal and external forces are introduced in order to formulate the extended law of interaction, to define the dynamic equilibrium, and to proceed to the virtual work of a system as one of the key points in all of mechanics. In the last section most of the material presented so far is restated in terms of resultants, a step necessary to pave the way to rigid body systems.

Chapter 3 contains classical multibody theory. By starting with the definition of a rigid body, the rigid body kinematical equations are derived and then combined with the principle of linear and angular momentum from Chapter 2 to give the classical Newton-Euler equations. Propositions on the dynamic equilibrium of a rigid body are stated in order to carefully investigate the role played by the internal and external forces. The main matter of this chapter, however, is the classical mechanical theory of perfect holonomic rheonomic constraints needed to join together different bodies to a rigid multibody system. It is intended to give precise definitions of the constraint forces of perfect constraints on the one hand, and to introduce the quite different concept of active and passive forces on the other hand, both necessary to approach the classical principle of d'Alembert/Lagrange. With respect to the constraints, physically compatible and, as their orthogonal counterparts, incompatible virtual displacements are needed to solve the dynamic equilibrium problem of a multibody system when starting from the virtual work principle. The chapter will close with the classical result that only the set of equations associated with the compatible virtual displacements, called the projected Newton-Euler equations, is needed to determine the generalized accelerations of the system.

In Chapter 4 we turn to non-smooth dynamics. We discuss possible discontinuity events in the velocities and in the accelerations of the system as functions of time that might occur in non-smooth dynamics. Here, the velocities are assumed to be functions of bounded variations, admitting a countable number of finite jumps that express the discontinuous changes due to impacts. In the integration, this leads to absolutely continuous displacements. In smooth dynamics the velocities of the system may be obtained by integration of the accelerations. This, however, already fails in the presence of velocity jumps. Although the accelerations, as the time derivatives of the velocities, are even defined in this case up to a countable number of points, their integration would not yield the overall velocity function, but only their absolutely continuous portion. In order to overcome this situation one has to introduce, instead of the accelerations, the differential measure of the velocities. This measure, which is also capable of describing impacts, may then be split into three parts, that is the Lebesgue part, an atomic measure, and a singular measure. The accelerations are then the density functions of the Lebesgue part and may be replaced by the left and the right derivatives of the left-continuous and the right-continuous regularization of the velocities, without any changes in the values of that measure. This step allows us to reduce the number of points where the accelerations are not defined and leads, together with the results of Chapter 3, to some directional Newton-Euler equations.

The topic of Chapter 5 is to introduce set-valued force laws which can be expressed with the help of potential functions. We start with the projected Newton-Euler equations of Chapter 3 and allow kinematic discontinuities as discussed in Chapter 4. Due to these discontinuities, the Newton-Euler equations hold for almost every time point but not for the entire time interval that we are interested in. This is an unsatisfactory fact, because exactly the points of interest are left out. In order to get rid of that restriction, we replace these equations by a corresponding equality of measures, which was introduced by Moreau and which also covers the impulsive motion by the use of the differential measure from Chapter 4. In this context, the generalized forces acting on the system have also to be understood as measures, where the same decomposition as for the differential measure of the velocities applies in a natural manner. Under the assumption of vanishing singular measures, this decomposition yields finally three sets of equations: The impact equations, and two directional Newton-Euler equations. The impact equations result from the atomic measure and connect the (probably) different values of the right and the left velocities at the countable time points of impact by some impact impulsions. The first set of directional Newton-Euler equations describe the impact-free evolution of the system with respect to future events by using the right accelerations and the right limit of the applied forces as functions of time, whereas the second set does the same job for past events. The impact and directional Newton-Euler equations, however, do not provide a complete description of the dynamics of the system, since up to now neither force laws nor impact laws have been introduced. These laws are also the only source for the occurrence of discontinuities in the velocities and in the accelerations, as all the terms investigated so far depend smoothly on the displacements and the velocities, as known from classical mechanics. Depending on their particular choice, one can finally obtain a differential equation for classical force laws, a differential inclusion if the force laws are expressed by bounded multifunctions, or even a measure differential inclusion if the forces are set-valued and unbounded. All cases are included in the presented description; we will, however, concentrate on the impact-free motion and no longer consider the impact equations. Starting out from a rather general set-valued map for hysteresis-free force laws, we will narrow down the problem towards one-dimensional force characteristics, and we will point out, step by step, what assumptions have to be made and what forces drop out in the chosen approach. As a result, the directional Newton-Euler equations become directional differential inclusions, with generalized forces being expressed by generalized force directions and scalar force values. The latter have to be taken from a convex set which is expressed by the generalized gradient of some scalar displacement or velocity potentials. In this setting, especially the connection between Lagrange multipliers and scalar forces from single-valued force laws becomes clear. With the help of an example, we can finally show how this description may be obtained for a given model of a mechanical system.

Chapter 6 is exclusively devoted to the representation and mathematical formulation of set-valued one-dimensional force laws as introduced in Chapter 5. In the first section we discuss a reasonable decomposition of the force characteristics for practical problems, leading to a continuous part, a set-valued step function, and an indicator-type force law. We show that the latter two may be further decomposed into the most basic set-valued force laws known, the unilateral primitives, that may be represented by scalar complementarity conditions of Signorini-Fichera type. As a result, this decomposition may be interpreted as a connection of corresponding force elements parallel to each other or in series, just as it is done by designing classical force laws at the stage of modeling. The second part of Chapter 6 deals with different mathematical formulations of the force laws in the framework of convex analysis. As just indicated, we focus here on force laws that are derived from scalar convex potentials. Our main goal is to provide five equivalent representations that are later used in Chapters 8 and 9 to state again the complete evolution problem in some modified but equivalent form, and to find some connections to classical mechanics. The five representations are obtained by expressing the force laws in terms of the subdifferential of convex analysis, in terms of a global variational inequality involving the values of the force potential at different points, in terms of the inverse of the subdifferential mapping that requires the introduction of the conjugate potential, in terms of the corresponding conjugate variational inequality, and finally in terms of Fenchel's equality that connects the potentials and their conjugates when it is written as an inequality. At the end of Chapter 6 some combinations of unilateral primitives are discussed with respect to their relevance for practical applications. As a surprising fact we will recognize that each arrangement of unilateral primitives, both on the displacement and on the velocity level, leads to force elements well-known from machine dynamics.

Chapter 7 deals with a quite unaccustomed step, the re-formulation of the force laws on the acceleration level. Force laws as introduced in Chapter 5 depend on the system's displacements and velocities only. As long as the force laws are single-valued, the forces may be computed directly, if the displacements and the velocities are known. This situation changes in the set-valued case that already occurs in classical mechanics in the presence of bilateral constraints: The values of the constraint forces are not known a priori but they adjust themselves such that a motion of the system may be realized on the corresponding constrained manifold. This is expressed by an equilibrium problem of the accelerations and the constraint forces, finally leading to the famous principle of least constraints. This concept may also be used for set-valued potential force laws, but requires some continuity assumptions on the trajectories. We recall that we are interested in computing the left and the right acceleration of the system, involving also the left and the right limit of the applied forces as functions of time. The re-formulation of the force laws on the acceleration level requires the evaluation of these force limits, which is performed exemplary for the motion leading to future events, applied to the unilateral and the single-step type force elements. We present a formulation of the acceleration force laws in terms of a potential function that we call the acceleration force potential. This potential will smooth the way to the principle of least constraints and its dual, being discussed in Chapters 9 and 8, respectively.

In Chapter 8 we come back to the directional Newton-Euler equations of Chapter 5 in order to solve them for the unknown scalar force values. By introducing certain index sets and by using the representation of the force laws on the acceleration level according to Chapter 7, we can derive three equivalent formulations of the problem: An LCP formulation with positive semidefinite matrix and a dimension equal to the number of unilateral primitives in use, according to the decomposition presented in the first section of Chapter 6. A variational inequality involving force variations that is the dual representation of Gauß' principle, and that is obtained from the results presented in the second section of Chapter 6. Finally, a convex quadratic program on the force values with affine inequality constraints that makes use of the conjugate force potentials introduced in Chapter 6, and that is recognized to be the dual problem to the principle of least constraints. From the latter formulation one can immediately obtain the following well-known result: In general, there is no unique solution for the scalar force values. They may be obtained uniquely, however, if the associated generalized force directions are linearly independent.

Chapter 9 is the last part that contributes to potential theory. By the use of the index sets of Chapter 8, the formulation of the scalar force laws on the different kinematic levels of Chapter 7, the corresponding variational expressions from Chapter 6, and the representation of the scalar contact laws by potentials in the configuration space as introduced in Chapter 5, we finally return to the investigation of the differential inclusions that are based on the directional Newton-Euler equations in Chapter 5. In this chapter we try to put all the results obtained so far into the framework of classical analytical mechanics. We discuss extensions of the famous three classical variational principles, i.e. the principle of d'Alembert-Lagrange as long as displacement variations are considered, the principle of Jourdain that involves variations of the velocities, and the principle of Gauß that deals with the variations of the generalized accelerations. We also formulate a strictly convex minimization problem on the generalized accelerations, corresponding to the classical principle of least constraints of Gauß. In this case the cost function is composed of a positive definite quadratic form, sometimes called the "Zwang" of the system, and a polyhedral convex function regarding the non-smooth force potentials in which also unilateral and bilateral constraints are included. As a consequence, the generalized accelerations obtained as the optimal solutions to this problem are always unique, no matter whether the associated Lagrange multipliers of Chapter 8 have been uniquely determined or not.

In Chapter 10 potential theory is left. As the most important force law that goes beyond potential theory we investigate spatial Coulomb friction. In the first two sections the kinematic equations of the spatial contact problem are developed for smooth surfaces. Section three shows how to consider the additional contact forces in the equations of motion. As already pointed out, Coulomb's frictional law is much more complicated as all the examples presented up to Chapter 9. We recall that one peculiarity of Coulomb friction is based on the different kinematic levels in use: The normal portion that takes into account impenetrability is originally stated on the displacement level, whereas the tangential behavior is expressed in terms of velocities. Here, the methods presented in Chapter 7 come in handy which allow us to reformulate the contact law on the acceleration level. This is done in the fourth section of Chapter 10. The fifth section then treats sliding contacts. The formulation of a linear complementarity problem is straightforward in this case, but it shows that the LCP matrix may become indefinite as a consequence of the absence of potential theory. Rolling spatial contacts are treated in sections six to eight. Even for given normal forces which change the Coulomb friction law into a dry friction problem, simple complementarity formulations as in Chapter 8 are not available, since the problem under consideration is no longer one-dimensional. Basically there are two ways of treating spatial force laws in the framework of complementarity formulations: If the sets of admissible forces are defined or approximated by polytopes, one can expect linear complementarity to be available in a slightly more complicated form than in Chapter 8. This approach will be used in the sixth section by a polytopic approximation of the friction disc. The second way is to treat directly sets with curved boundaries. Linear complementarity formulations are in general not capable of describing such situations, and one must try to put them into the framework of nonlinear problems. The formulation of the corresponding complementary inequalities, however, is not as straightforward as for the linear case, especially for sets where the inequalities depend on additional parameters that have to be determined by solving nonlinear equations. In the seventh and eighth section we will show how standard NCP formulations can be obtained, using the example of the conic friction law. Hereby, a formulation requiring as less equations as possible is presented in section seven, whereas a differentiable version with one additional equation is derived in section eight. In the ninth section some comments are made on how to treat numerically nonlinear complementarity problems. We shortly outline the way from an NCP via NCP-functions to a set of semi-smooth nonlinear equations, which might then be solved by using a specially designed Newton algorithm. Finally, some numerical simulations on a point mass moving on a horizontal plane are presented in order to show first computational results with the formulation developed in section eight being used.

Chapter 11 treats impacts by an example. Although the impact equations have been already stated in Chapter 5, the impact itself was not further discussed. It is also not intended to present a framework of a general impact theory in this chapter, as the topic is too extensive. We just want to give one example of a quite unexpected velocity jump that occurs due to a non-differentiable bilateral constraint. After a short review of impacts we present the mechanical model at which the problem of a C^0-constraint turned up. The impact-free motion is treated by Lagrange's equations of first kind in order to show that the velocity jump is caused by an instantaneous change in the gradients of the constraints. This is in sharp contrast to normal collisions, where the gradients of the inequality constraints are usually assumed not to change during the impact. The non-dissipative impact law describing the velocity jump is then developed by two different methods: Time-scaling with the displacements being fixed provides some analogies to classical constraints and may hence be used to keep the concept of perfect constraint forces also for the case of impact. Reflections give more insight into the geometry behind the impact and provide some links to energy-preserving collisions: It will be shown that any reflection may be interpreted as an abstract collision, i.e. an event at which some relative velocity is inverted.

Chapters 12 to 14 show typical applications from machine dynamics, in which set-valued force laws have been used with great success. In Chapter 12 we present an outline of the modeling of an electropneumatic drilling machine, starting with its operating principle and closing with a comparison of measured and simulated time signals that confirm the validity of the chosen model. In order to give a first idea of this relatively simple device, we anticipate some data of the model: The overall system is described by eight generalized coordinates, five of which are used for the machine's parts, while the remaining three are used for an operator's hand-arm-model. Altogether we have used 15 one-dimensional force elements, seven of them being set-valued: One adiabatic air cushion, three classical linear spring-damper-combinations, three elastic joints with backlashes, one elastic joint with backlash and hysteresis, five Stribeck friction characteristics and two unilateral (contact-impact) constraints.

Chapter 13 treats another type of percussion drills at which the hammering process is not generated by a crank-slider-mechanism as in Chapter 12, but by a pair of gearing cam disks. In contrast to the electropneumatic drilling machine with only potential force laws used in the model, this drill requires Coulomb friction characteristics in order to properly take into account the force regime at the contacting cams. We have used the same hand-arm-model as in Chapter 12 together with three linear and two angular displacement coordinates for the machine's parts, resulting in a total of eight degrees of freedom. The force elements were modeled as five linear spring-damper-combinations, one Coulomb friction characteristic, and two unilateral constraints, completed by corresponding frictional impact laws.

In Chapter 14 a model of a parabolic friction damper used in airborne gas turbines is presented, which may serve as an example on how to evaluate contact kinematics in the planar case, according to the methods presented in Chapter 10. The entire model consists of five rigid bodies, representing two neighboring turbine blade-platform-configurations and the damper held between them by two frictional contacts. The kinematics of the problem is a bit complicated insofar as the contact points at the platforms and the damper are not spatially fixed, but moving on the contours, even in the case of stiction by some rolling motion of the damper. We use seven coordinates to model this system, accompanied by four spring-damper-combinations, three external loads, and two frictional contacts. Obviously, modeling multibody systems is a very "individual" process that always, as a general rule, should be accompanied by measurements.

2. Fundamental Concepts

In this chapter, we briefly review some basic concepts in dynamics that hold beyond the rigid body assumption. We start with the classification of forces into internal and external forces as one of the key points in all of mechanics, followed by the introduction of the law of interaction as a general axiom. We then define the dynamic equilibrium via the Newton-Euler equations, and state the virtual work of a dynamic system together with some informal comments on the classical principle of d'Alembert/Lagrange.

2.1 Internal and External Forces

By a *mechanical system* S we understand a Lebesgue measurable set of points x in the three-dimensional oriented space which interacts with its environment by forces and moments. We identify the points $x \in S$ as the endpoints of a radius vector $\mathbf{r}$ starting at some inertially fixed point O, and we denote by

$$\mathrm{d}\mathbf{F}(x) \quad \text{and} \quad \boldsymbol{\xi}(x) \times \mathrm{d}\mathbf{F}(x) + \mathrm{d}\mathbf{M}(x) \tag{2.1}$$

distributions of forces and moments on S. Here, the moment distribution is composed of a free portion $\mathrm{d}\mathbf{M}$ and a portion induced by forces $\boldsymbol{\xi} \times \mathrm{d}\mathbf{F}$. The latter thus requires some point of reference which is chosen according to Figure 2.1 to be any moving or non-moving point A with $\boldsymbol{\xi} = \mathbf{r}_{AO} + \mathbf{r}$ the radius vector from A to x.

For S a mechanical system and K any Lebesgue measurable subset of S, we call K a *mechanical subsystem* of S, and we denote the set of all subsystems of S inclusively S by $\mathcal{S}$. Two subsystems $H, \bar{H} \in \mathcal{S}$ are called a pair of *complementary subsystems* of $K \in \mathcal{S}$, if $H \cup \bar{H} = K$ and $H \cap \bar{H} = \emptyset$. Note that any two disjoint subsystems $H, G \in \mathcal{S}$ are complementary with respect to $(H \cup G) \in \mathcal{S}$. A subsystem $K \in \mathcal{S}$ may interact by forces with its environment, that is the environment of S together with the remainder $S \backslash K$ of S, which leads in a natural way to the following classification of forces:

Definition 2.1.1. *Let $K \in \mathcal{S}$, and let* $(\mathrm{d}\mathbf{F}, \mathrm{d}\mathbf{M})$ *as defined in (2.1) be distributions of forces and moments on K, short a force on K.*

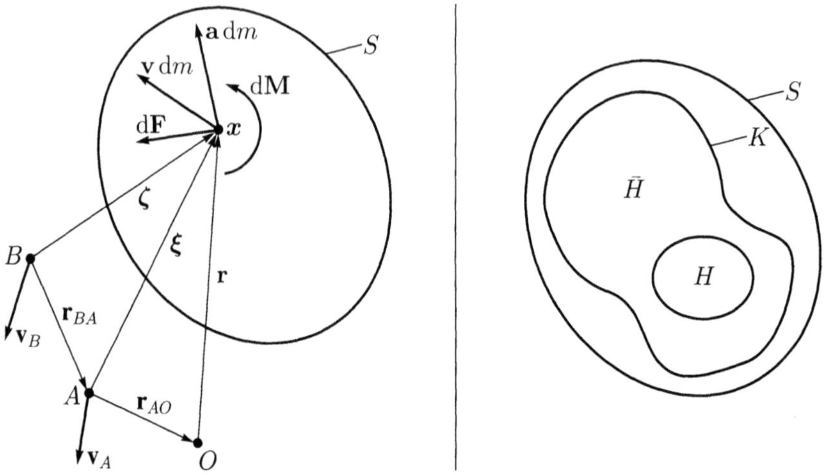

Figure 2.1. Mechanical system and subsystems

(i) A force on K is called an external force *of K, if it acts from the environment of K on K. It is denoted by* $(\mathrm{d}\mathbf{F}^{(e,K)}, \mathrm{d}\mathbf{M}^{(e,K)})$.

(ii) A force is called an internal observable force *of K, if it is not an external force of K, and if there exists a pair of complementary subsystems H, $\bar{H}$ of K such that the force describes the action of $\bar{H}$ on H. Notation:* $(\mathrm{d}\mathbf{F}^{(o,K|H)}, \mathrm{d}\mathbf{M}^{(o,K|H)})$.

(iii) Any other forces are called internal non-observable forces *of K and are denoted by* $(\mathrm{d}\mathbf{F}^{(n,K)}, \mathrm{d}\mathbf{M}^{(n,K)})$.

Convention: Note that the external and the internal non-observable forces of K are only defined on K. We extend these domains to S by setting $(\mathrm{d}\mathbf{F}, \mathrm{d}\mathbf{M}) = (0, 0)$ when outside of K. This provides integration on S with the same value of the integral as if integration would have been performed on K. In the same fashion we extend the domains of the internal observable forces to S.

In the literature on classical dynamics, see e.g. [52], [107], [34], [20], only the term internal force is used and has to be understood as the sum of internal observable and non-observable forces, where usually distributions of free moments $\mathrm{d}\mathbf{M}$ are not considered. Note that the internal observable forces become part of the external forces of some subsystems of K. This fact is understood by the *cutting principle*: To make the internal observable forces of K visible by taking a suitable division of K into subsystems, and to isolate each of the subsystems in the sense that it is treated as an independent mechanical system. One simple consequence on the definition of the internal non-observable forces consists in the fact that they cannot be turned into external forces with respect to any subsystem H of K, which ever it would be. Internal non-observable forces are thus beyond any way of detection, and thus quite pathologically for practical applications. We therefore claim that the effect of the non-observable forces on the environment vanishes:

Postulate *on internal non-observable forces: For every* $K \in \mathcal{S}$ *we demand*

$$\int_K \mathrm{d}\mathbf{F}^{(n,K)} \equiv 0 \quad \text{and} \quad \int_K \boldsymbol{\xi} \times \mathrm{d}\mathbf{F}^{(n,K)} + \mathrm{d}\mathbf{M}^{(n,K)} \equiv 0 \tag{2.2}$$

with $(\mathrm{d}\mathbf{F}^{(n,K)}, \mathrm{d}\mathbf{M}^{(n,K)})$ *being the internal non-observable forces of* K.

Equation (2.2) ensures that every force which *may* have effect on the environment is observable and may thus be turned into an external force by applying the cutting principle.

2.2 The Law of Interaction

The classification into external and internal forces is one of the key points in all of mechanics. Without this classification, none of the fundamental laws in mechanics could be stated, such as the law of interaction which we are going to formulate now.

Definition 2.2.1. *Let* K_1 *and* K_2 *be two disjoint subsystems of* S, *and let* $(\mathrm{d}\mathbf{F}_1, \mathrm{d}\mathbf{M}_1)$ *and* $(\mathrm{d}\mathbf{F}_2, \mathrm{d}\mathbf{M}_2)$ *be given force distributions on* K_1 *and* K_2. *We say that* $(\mathrm{d}\mathbf{F}_1, \mathrm{d}\mathbf{M}_1)$ *and* $(\mathrm{d}\mathbf{F}_2, \mathrm{d}\mathbf{M}_2)$ *satisfy the* interaction formula, *if*

$$\begin{aligned} &\int_{K_1} \mathrm{d}\mathbf{F}_1 = -\int_{K_2} \mathrm{d}\mathbf{F}_2 \quad \textit{and} \\ &\int_{K_1} \boldsymbol{\xi} \times \mathrm{d}\mathbf{F}_1 + \mathrm{d}\mathbf{M}_1 = -\int_{K_2} \boldsymbol{\xi} \times \mathrm{d}\mathbf{F}_2 + \mathrm{d}\mathbf{M}_2. \end{aligned} \tag{2.3}$$

The law of interaction: *Internal observable forces occur as pairs and satisfy the interaction formula: For any* $K \in \mathcal{S}$ *and any pair of complementary subsystems* H, $\bar{H}$ *of* K, *one has*

$$\begin{aligned} &\int_H \mathrm{d}\mathbf{F}^{(o,K|H)} = -\int_{\bar{H}} \mathrm{d}\mathbf{F}^{(o,K|\bar{H})} \quad \text{and} \\ &\int_H \boldsymbol{\xi} \times \mathrm{d}\mathbf{F}^{(o,K|H)} + \mathrm{d}\mathbf{M}^{(o,K|H)} = -\int_{\bar{H}} \boldsymbol{\xi} \times \mathrm{d}\mathbf{F}^{(o,K|\bar{H})} + \mathrm{d}\mathbf{M}^{(o,K|\bar{H})}, \end{aligned} \tag{2.4}$$

i.e. the effect of the forces $(\mathrm{d}\mathbf{F}^{(o,K|H)}, \mathrm{d}\mathbf{M}^{(o,K|H)})$ *acting from* $\bar{H}$ *on* H *equals the effect of the forces* $(\mathrm{d}\mathbf{F}^{(o,K|\bar{H})}, \mathrm{d}\mathbf{M}^{(o,K|\bar{H})})$ *acting from* H *on* $\bar{H}$.

The *law of interaction* stated in the form (2.4) is an axiom in any field of mechanics. It fully incorporates Newton's third law (actio = reactio) for particles which reads: Forces occur as pairs, are anti-parallel and of the same size, and lie on a common line of interaction. To see this, take $H = \{x_1\}$, $\bar{H} = \{x_2\}$, $K = \{x_1, x_2\}$ and perform integration. This yields $\mathbf{F}_1 = -\mathbf{F}_2$ and

$\boldsymbol{\xi}_1 \times \mathbf{F}_1 + \mathbf{M}_1 = -\boldsymbol{\xi}_2 \times \mathbf{F}_2 - \mathbf{M}_2$, and Newton's third law applies iff $\mathbf{M}_1 = -\mathbf{M}_2$, in particular if $\mathbf{M}_1 = -\mathbf{M}_2 = 0$. The interaction formula for the moments then reduces to $(\boldsymbol{\xi}_1 - \boldsymbol{\xi}_2) \times \mathbf{F}_1 = 0$, hence $\mathbf{F}_1 = -\mathbf{F}_2 = \lambda(\boldsymbol{\xi}_1 - \boldsymbol{\xi}_2)$ for some $\lambda \in \mathbb{R}$. In order to apply the interaction formula one needs forces occurring as pairs. It is therefore senseless to speak about the law of interaction of just a single force. In particular, the law of interaction can not be immediately extended to external forces, since it is not known whether they have partners or not.

2.3 The Dynamic Equilibrium

The terms "internal" and "external" depend on the subsystem under consideration but not on the nature of the forces themselves: A force $(\mathrm{d}\mathbf{F}^{(o,K|H)}, \mathrm{d}\mathbf{M}^{(o,K|H)})$ acting from a subsystem $\bar{H}$ of K on the subsystem H of K is an external force with respect to H, i.e. it belongs to $(\mathrm{d}\mathbf{F}^{(e,H)}, \mathrm{d}\mathbf{M}^{(e,H)})$, whereas it is an internal observable force with respect to K. The external forces of H are thus composed of two portions: The force which acts from $\bar{H}$ on H on the one hand, and the part of the external forces of K which acts on H on the other hand. The latter may be obtained as follows: Let $H, \bar{H} \in \mathcal{S}$ be a pair of complementary subsystems of K. Since $H \cap \bar{H} = \emptyset$ and $H \cup \bar{H} = K$, we divide up the external forces of K into

$$\begin{aligned}\int_K (\mathrm{d}\mathbf{F}^{(e,K)}, \mathrm{d}\mathbf{M}^{(e,K)}) &= \int_K \chi_H \, (\mathrm{d}\mathbf{F}^{(e,K)}, \mathrm{d}\mathbf{M}^{(e,K)}) \\ &\quad + \int_K \chi_{\bar{H}} \, (\mathrm{d}\mathbf{F}^{(e,K)}, \mathrm{d}\mathbf{M}^{(e,K)}) \\ &=: \int_K (\mathrm{d}\mathbf{F}^{(e,K|H)}, \mathrm{d}\mathbf{M}^{(e,K|H)}) \\ &\quad + \int_K (\mathrm{d}\mathbf{F}^{(e,K|\bar{H})}, \mathrm{d}\mathbf{M}^{(e,K|\bar{H})}),\end{aligned} \tag{2.5}$$

where χ_A denotes the characteristic function of A, i.e. $\chi_A(x) = 1$ for $x \in A$ and $\chi_A(x) = 0$ for $x \notin A$. In order to state now the external forces of H, we need the portion of the external forces of K according to the first term in the second line of (2.5) together with the action of $\bar{H}$ on H. This yields

$$\begin{aligned}(\mathrm{d}\mathbf{F}^{(e,H)}, \mathrm{d}\mathbf{M}^{(e,H)}) =& (\mathrm{d}\mathbf{F}^{(e,K|H)}, \mathrm{d}\mathbf{M}^{(e,K|H)}) \\ &+ (\mathrm{d}\mathbf{F}^{(o,K|H)}, \mathrm{d}\mathbf{M}^{(o,K|H)}),\end{aligned} \tag{2.6}$$

an expression that is needed in the Newton-Euler equations and also in the definition of the dynamic equilibrium.

Definition 2.3.1. *We say the Newton-Euler equations apply for a subsystem* $K \in \mathcal{S}$, *if*

$$\int_K \mathbf{a}\,dm - d\mathbf{F}^{(e,K)} = 0 \quad \textit{and}$$
$$\int_K \boldsymbol{\xi} \times (\mathbf{a}\,dm - d\mathbf{F}^{(e,K)}) - d\mathbf{M}^{(e,K)} = 0 \tag{2.7}$$

holds with **a** *being the acceleration of the points* $x \in K$ *as indicated in Figure 2.1,* dm *the mass distribution on* K*, and* $(d\mathbf{F}^{(e,K)}, d\mathbf{M}^{(e,K)})$ *the external forces of* K*.*

The first equation in (2.7) is known as the *principle of linear momentum*, that is the mathematical expression of Newton's second law, and the second equation is the *principle of angular momentum*, also known as Euler's axiom. Both equations together, here called the *Newton-Euler equations*, are the axioms of classical dynamics.

The dynamic equilibrium: *A subsystem* $K \in S$ *is called to be in the* dynamic equilibrium, *if the Newton-Euler equations (2.7) apply for* K *itself and for any of its subsystems* H*.*

Beneath the dynamic equilibrium there are other equilibria, such as the static equilibrium that is obtained from (2.7) by setting $\mathbf{a} = 0$. Thus, the dynamic equilibrium is more general than the static, and we see that every system in the static equilibrium is also in the dynamic equilibrium. For a system in the dynamic equilibrium, however, the static equilibrium may no longer apply. For the same reason we must not expect equations (2.7) to hold generally, since there are equilibria that go beyond classical dynamics. For example, relativistic dynamics is not contained in (2.7), but we can say that every system in the dynamic equilibrium is also in the equilibrium in the sense of relativistic mechanics.

Proposition 2.3.1. *The dynamic equilibrium is not influenced by the choice of the point of reference in Euler's axiom: If both equations in (2.7) hold for some point of reference* A*, then Euler's axiom holds for any other point* B*.*

Proof. Let B be any point different from A, and denote by $\boldsymbol{\zeta}$ the position of x relative to B as indicated in Figure 2.1. Then $\boldsymbol{\zeta} = \mathbf{r}_{BA} + \boldsymbol{\xi}$, and Euler's axiom for B becomes $\int_K \boldsymbol{\zeta} \times (\mathbf{a}\,dm - d\mathbf{F}^{(e,K)}) - d\mathbf{M}^{(e,K)} = \mathbf{r}_{BA} \times \int_K (\mathbf{a}\,dm - d\mathbf{F}^{(e,K)}) + \int_K \boldsymbol{\xi} \times (\mathbf{a}\,dm - d\mathbf{F}^{(e,K)}) - d\mathbf{M}^{(e,K)} \equiv 0$, since (2.7) holds by assumption. □

We will now show that the definition of the dynamic equilibrium is compatible with the law of interaction.

Proposition 2.3.2. *For* $K \in S$ *in the dynamic equilibrium, the internal observable forces of* K *do not violate the law of interaction, i.e. for any pair of complementary subsystems* H*,* $\bar{H}$ *of* K *equation (2.4) holds.*

Proof. The proof follows the one given in [34]. Since K is in the dynamic equilibrium, the Newton-Euler equations (2.7) hold for K, for H, and for $\bar{H}$. First, we apply (2.7) on H. With $(\mathrm{d}\mathbf{F}^{(o,K|H)}, \mathrm{d}\mathbf{M}^{(o,K|H)})$ in (2.4) being the internal observable forces of K which act from $\bar{H}$ on H, the external forces of H can be stated according to (2.6), and the Newton-Euler equations for H become

$$\begin{aligned}\int_H \mathbf{a}\,\mathrm{d}m &= \int_H \mathrm{d}\mathbf{F}^{(e,K|H)} + \mathrm{d}\mathbf{F}^{(o,K|H)}, \\ \int_H \boldsymbol{\xi} \times \mathbf{a}\,\mathrm{d}m &= \int_H \boldsymbol{\xi} \times (\mathrm{d}\mathbf{F}^{(e,K|H)} + \mathrm{d}\mathbf{F}^{(o,K|H)}) + \mathrm{d}\mathbf{M}^{(e,K|H)} + \mathrm{d}\mathbf{M}^{(o,K|H)}.\end{aligned} \tag{2.8}$$

Applying (2.7) on $\bar{H}$ we obtain analogously

$$\begin{aligned}\int_{\bar{H}} \mathbf{a}\,\mathrm{d}m &= \int_{\bar{H}} \mathrm{d}\mathbf{F}^{(e,K|\bar{H})} + \mathrm{d}\mathbf{F}^{(o,K|\bar{H})}, \\ \int_{\bar{H}} \boldsymbol{\xi} \times \mathbf{a}\,\mathrm{d}m &= \int_{\bar{H}} \boldsymbol{\xi} \times (\mathrm{d}\mathbf{F}^{(e,K|\bar{H})} + \mathrm{d}\mathbf{F}^{(o,K|\bar{H})}) + \mathrm{d}\mathbf{M}^{(e,K|\bar{H})} + \mathrm{d}\mathbf{M}^{(o,K|\bar{H})},\end{aligned} \tag{2.9}$$

where the external forces of $\bar{H}$ are composed of the portion $(e, K|\bar{H})$ of the external forces of K and the terms indexed by $(o, K|\bar{H})$ which reflect the action of H on $\bar{H}$. Since H, $\bar{H}$ disjoint and $H \cup \bar{H} = K$, addition of (2.8) and (2.9) yields with the help of (2.5)

$$\begin{aligned}\int_K \mathbf{a}\,\mathrm{d}m &= \int_K \mathrm{d}\mathbf{F}^{(e,K)} + \int_H \mathrm{d}\mathbf{F}^{(o,K|H)} + \int_{\bar{H}} \mathrm{d}\mathbf{F}^{(o,K|\bar{H})}, \\ \int_K \boldsymbol{\xi} \times \mathbf{a}\,\mathrm{d}m &= \int_K \boldsymbol{\xi} \times \mathrm{d}\mathbf{F}^{(e,K)} + \mathrm{d}\mathbf{M}^{(e,K)} \\ &\quad + \int_H \boldsymbol{\xi} \times \mathrm{d}\mathbf{F}^{(o,K|H)} + \mathrm{d}\mathbf{M}^{(o,K|H)} \\ &\quad + \int_{\bar{H}} \boldsymbol{\xi} \times \mathrm{d}\mathbf{F}^{(o,K|\bar{H})} + \mathrm{d}\mathbf{M}^{(o,K|\bar{H})}.\end{aligned} \tag{2.10}$$

Finally, we state the Newton-Euler equations (2.7) for K,

$$\begin{aligned}\int_K \mathbf{a}\,\mathrm{d}m &= \int_K \mathrm{d}\mathbf{F}^{(e,K)}, \\ \int_K \boldsymbol{\xi} \times \mathbf{a}\,\mathrm{d}m &= \int_K \boldsymbol{\xi} \times \mathrm{d}\mathbf{F}^{(e,K)} + \mathrm{d}\mathbf{M}^{(e,K)},\end{aligned} \tag{2.11}$$

and compare these expressions with (2.10). As a result, we obtain (2.4) which completes the proof. □

The following relations between the law of interaction, the dynamic equilibrium and the internal and external forces should be noted: By the above calculations we have not proven the law of interaction, but we have only shown that there is no conflict with the dynamic equilibrium. Note that the law of interaction holds generally, that means it is also true for systems beyond the dynamic equilibrium. The same can be done for static systems, for example, where the Newton-Euler equations have to be replaced by the equilibrium of forces and moments.

For the internal non-observable forces we have postulated in (2.2) that their effect on the environment vanishes. This is in any case weaker than proposing the law of interaction for this class of forces to hold, since the occurrence as pairs is not needed.

We emphasize again that the interaction formula (2.3) is only meaningful for forces occurring as pairs, and so is the law of interaction. The question whether the law of interaction may be extended to the external forces may be easily answered: For the external forces of a mechanical system S there are only two possibilities. Either they are given by god, and we don't have to worry about the question whether they have partners or not, or one may find a mechanical system $S^\star$ containing S and being free of any external forces. In the latter case any external force of any subsystem of $S^\star$ is an internal force of $S^\star$ and has thus a partner fulfilling the law of interaction.

Corollary 2.3.1. *In order to decide whether a system is in the dynamic equilibrium, the following is useful: Let K be a subsystem of S.*

(i) If H, $\bar{H}$ is a pair of complementary subsystems of K, each of them being in the dynamic equilibrium, then K is in the dynamic equilibrium.

(ii) Let G, H be two disjoint but not complementary subsystems of K in the dynamic equilibrium. Then $G \cup H$ is in the dynamic equilibrium.

(iii) Let $\mathcal{H} := \{H_i\}$ be a countable disjoint division of K into subsystems H_i, i.e. $H_i \cap H_j = \emptyset$ for any i, j and $\bigcup_i H_i = K$, and suppose that every H_i is in the dynamic equilibrium. Then $\bigcup_i H_i = K$ is also in the dynamic equilibrium.

Proof. All three propositions are straight-forward consequences on the definition of the dynamic equilibrium and the law of interaction for the internal observable forces:

(i) In order to prove the assertion we have to show that the Newton-Euler equations (2.7) hold for *any* subsystem G of K. There are two possible cases: Either, G is contained in H (or in $\bar{H}$). As a subsystem of H (or $\bar{H}$, respectively) the Newton-Euler equations apply for G, because H (and $\bar{H}$) is in the dynamic equilibrium by assumption. Or, G is not contained in H or $\bar{H}$. Then we may split G into $G_H := G \cap H$ and $G_{\bar{H}} := G \cap \bar{H}$, where $G = G_H \cup G_{\bar{H}}$ and $\emptyset = G_H \cap G_{\bar{H}}$. As subsystems of H and $\bar{H}$, respectively, the Newton-Euler equations apply for G_H and $G_{\bar{H}}$ in a form according to (2.8) and (2.9), and after summation in the form according to (2.10). Since G_H and $G_{\bar{H}}$ are complementary subsystems of G, the internal forces $(o, G|G_H)$ and $(o, G|G_{\bar{H}})$ with respect to G drop out by the law of interaction (2.4), and one obtains from (2.10) the Newton-Euler equations for G in the form according to (2.11). Thus, the Newton-Euler equations hold for G in both cases and, since G was arbitrary, for any subsystem of K which proves the assertion.

(ii) Consider the subsystem $L := H \cup G$ of K. Then H and G is a pair of complementary subsystems of L, and one may proceed as in (i).

(iii) Set $L_1 := H_1$, $L_{i+1} := L_i \cup H_{i+1}$ for $i = 1, 2, \ldots$. The sets L_i and H_{i+1} are complementary subsystems of L_{i+1} with H_{i+1} being in the dynamic equilibrium by assumption. For L_i in the dynamic equilibrium we thus obtain from (ii) that L_{i+1} is in the dynamic equilibrium, and we know that the dynamic equilibrium holds for $L_1 = H_1$. □

2.4 The Virtual Work of a Dynamic System

The virtual work done by a force is the product of the force and some linear virtual displacement. The virtual work done by a free moment is the product of the free moment and some angular virtual displacement. Since we have both, forces and free moments, we need two sets of *independent* virtual displacements, the *linear* which we denote by $\delta\mathbf{s}$ and the the *angular* which we denote by $\delta\boldsymbol{\varphi}$. We consider a mechanical system S as in Figure 2.1 together with virtual displacements $\delta\mathbf{s}(x)$, $\delta\boldsymbol{\varphi}(x)$ on S. The overall linear virtual displacement $\delta\mathbf{r}$ at a point x becomes

$$\delta\mathbf{r} = \delta\mathbf{s} + \delta\boldsymbol{\varphi} \times \boldsymbol{\xi} \tag{2.12}$$

when A is chosen as the point of reference, see Figure 2.2. Note that $\delta\mathbf{r}$ is composed of two parts, the linear virtual displacement $\delta\mathbf{s}$ and a linear virtual displacement $\delta\boldsymbol{\varphi} \times \boldsymbol{\xi}$ induced by $\delta\boldsymbol{\varphi}$, which may be regarded as the effect of a virtual rotation $\boldsymbol{\xi} \to \boldsymbol{\xi} + \delta\boldsymbol{\varphi} \times \boldsymbol{\xi}$ with angle $|\delta\boldsymbol{\varphi}|$ and axis $\lambda\, \delta\boldsymbol{\varphi}$ passing through A. By definition, virtual displacements are small and independent of time. When a system S is slightly moved to end up in a modified configuration $S^\star$, the points $x \in S$ are shifted to the new points $x^\star$ in $S^\star$. The small deviation in the displacement of point x may then be expressed by $\mathbf{r}^\star - \mathbf{r}$ which is the $\delta\mathbf{r}$ in (2.12).

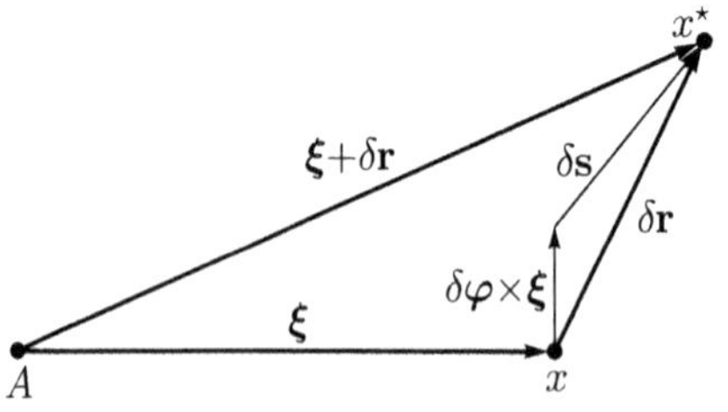

Figure 2.2. Virtual displacements

Before stating the principle of virtual work in dynamics we shorten the notation of the internal observable forces by setting

$$\begin{aligned}(\mathrm{d}\mathbf{F}^{(o,K|H\bar{H})}, \mathrm{d}\mathbf{M}^{(o,K|H\bar{H})}) :=& (\mathrm{d}\mathbf{F}^{(o,K|H)}, \mathrm{d}\mathbf{M}^{(o,K|H)}) \\ &+ (\mathrm{d}\mathbf{F}^{(o,K|\bar{H})}, \mathrm{d}\mathbf{M}^{(o,K|\bar{H})}),\end{aligned} \tag{2.13}$$

when H, $\bar{H}$ is a pair of complementary subsystems of K. With this abbreviation, the law of interaction (2.4) reads

$$\int_K \mathrm{d}\mathbf{F}^{(o,K|H\bar{H})} = 0 \quad \text{and} \quad \int_K \boldsymbol{\xi} \times \mathrm{d}\mathbf{F}^{(o,K|H\bar{H})} + \mathrm{d}\mathbf{M}^{(o,K|H\bar{H})} = 0. \tag{2.14}$$

Further, we drop from now the term "observable", since internal non-observable forces are nowhere considered from now.

For every fixed time t, we build the *virtual work* $\delta W^{(K)}$ of the mechanical subsystem K with the inertia terms, the external forces of K, and pairs of internal forces acting between subsystems H, $\bar{H}$ that are complementary with respect to K. To get the physically correct work expression, every free moment has to be multiplied with the angular virtual displacement $\delta\boldsymbol{\varphi}$, and every force with the linear virtual displacement at its point of application, that is with $\delta\mathbf{r}$ from (2.12). The inertia term as a translational magnitude has to be treated like the forces, but with reversed sign. After "summing up" all virtual work terms one obtains

$$\begin{aligned}\delta W^{(K)} := &\int_K \left(\delta\mathbf{s} + \delta\boldsymbol{\varphi} \times \boldsymbol{\xi}\right)^T \left(\mathbf{a}\,\mathrm{d}m - \mathrm{d}\mathbf{F}^{(e,K)} - \mathrm{d}\mathbf{F}^{(o,K|H\bar{H})}\right) \\ &+ \int_K \delta\boldsymbol{\varphi}^T \left(-\mathrm{d}\mathbf{M}^{(e,K)} - \mathrm{d}\mathbf{M}^{(o,K|H\bar{H})}\right),\end{aligned} \tag{2.15}$$

which puts the virtual displacements and the forces in duality. Due to $(\delta\boldsymbol{\varphi} \times \boldsymbol{\xi})^T\mathbf{u} = \delta\boldsymbol{\varphi}^T(\boldsymbol{\xi} \times \mathbf{u})$ this equation may be equivalently rewritten as

$$\begin{aligned}\delta W^{(K)} = &\int_K \delta\mathbf{s}^T \left(\mathbf{a}\,\mathrm{d}m - \mathrm{d}\mathbf{F}^{(e,K)} - \mathrm{d}\mathbf{F}^{(o,K|H\bar{H})}\right) \\ &+ \int_K \delta\boldsymbol{\varphi}^T \left(\boldsymbol{\xi} \times (\mathbf{a}\,\mathrm{d}m - \mathrm{d}\mathbf{F}^{(e,K)} - \mathrm{d}\mathbf{F}^{(o,K|H\bar{H})})\right) \\ &+ \int_K \delta\boldsymbol{\varphi}^T \left(- \mathrm{d}\mathbf{M}^{(e,K)} - \mathrm{d}\mathbf{M}^{(o,K|H\bar{H})}\right).\end{aligned} \tag{2.16}$$

Note that (2.15) and (2.16) are already stated in terms of the independent virtual displacements $\delta\mathbf{s}$ and $\delta\boldsymbol{\varphi}$; that means $\delta\mathbf{r}$ has already been eliminated by (2.12). Further, $\delta\mathbf{s}$ and $\delta\boldsymbol{\varphi}$ are virtual displacements in the classical sense, that means small and independent of time. Compatibility of the virtual displacements with any constraints is *not yet* required; this restriction will occur later as a consequence on the force laws of classical perfect bilateral constraints.

Proposition 2.4.1. *If we demand that $\delta W^{(K)}$ is constant for any $\delta\mathbf{s}$, $\delta\boldsymbol{\varphi}$ and any choice of complementary subsystems $(H, \bar{H})$ of K, then*

(i) $\delta W^{(K)} = 0$

(ii) the Newton-Euler equations (2.7) apply for K,

$$\begin{aligned}&\int_K \mathbf{a}\,\mathrm{d}m - \mathrm{d}\mathbf{F}^{(e,K)} = 0 \quad \textit{and} \\ &\int_K \boldsymbol{\xi} \times (\mathbf{a}\,\mathrm{d}m - \mathrm{d}\mathbf{F}^{(e,K)}) - \mathrm{d}\mathbf{M}^{(e,K)} = 0\end{aligned} \tag{2.17}$$

(iii) the Newton-Euler equations (2.7) apply for the subsystem H of K,

$$\begin{aligned} &\int_H \mathbf{a}\,dm - \mathrm{d}\mathbf{F}^{(e,H)} = 0 \quad \textit{and} \\ &\int_H \boldsymbol{\xi} \times (\mathbf{a}\,dm - \mathrm{d}\mathbf{F}^{(e,H)}) - \mathrm{d}\mathbf{M}^{(e,H)} = 0 \end{aligned} \tag{2.18}$$

(iv) K is in the dynamic equilibrium.

Proof.

(i) Set $\delta\mathbf{s} = \delta\boldsymbol{\varphi} = 0$ on K in (2.16).

(ii) First, set $0 \neq \delta\mathbf{s}$ constant but arbitrary on K and $\delta\boldsymbol{\varphi} \equiv 0$ on K. Because of $\delta W^{(K)} = 0$ by (i), equation (2.16) becomes

$$0 = \delta\mathbf{s}^T \int_K \mathbf{a}\,dm - \mathrm{d}\mathbf{F}^{(e,K)} - \mathrm{d}\mathbf{F}^{(o,K|H\bar{H})}. \tag{2.19}$$

Since $\delta\mathbf{s}$ is arbitrary, the integral itself must vanish. In addition, the term $\int_K \mathrm{d}\mathbf{F}^{(o,K|H\bar{H})}$ drops out due to the law of interaction (2.14). Hence, we have obtained the principle of linear momentum for K, that is the first equation in (2.17). In order to validate the principle of angular momentum for K, we set now $0 \neq \delta\boldsymbol{\varphi}$ constant on K and $\delta\mathbf{s} \equiv 0$ on K in (2.16), and proceed in the same manner but now taking the second equation of the law of interaction (2.14).

(iii) There is only one thing to do: To rearrange the forces in (2.16) such that we may rediscover the external forces of H and of $\bar{H}$, respectively. We take the representations of the external forces $(\mathrm{d}\mathbf{F}^{(e,K)}, \mathrm{d}\mathbf{M}^{(e,K)})$ and the internal forces $(\mathrm{d}\mathbf{F}^{(o,K|H\bar{H})}, \mathrm{d}\mathbf{M}^{(o,K|H\bar{H})})$ from (2.5) and (2.13), respectively. By summing up both terms we obtain

$$\begin{aligned} &(\mathrm{d}\mathbf{F}^{(e,K)}, \mathrm{d}\mathbf{M}^{(e,K)}) + (\mathrm{d}\mathbf{F}^{(o,K|H\bar{H})}, \mathrm{d}\mathbf{M}^{(o,K|H\bar{H})}) \\ &= (\mathrm{d}\mathbf{F}^{(e,H)}, \mathrm{d}\mathbf{M}^{(e,H)}) + (\mathrm{d}\mathbf{F}^{(e,\bar{H})}, \mathrm{d}\mathbf{M}^{(e,\bar{H})}) \end{aligned} \tag{2.20}$$

according to the rule (2.6) on how the external forces of H (and of $\bar{H}$, respectively) have to be built up. If we choose now in (2.16) the virtual displacements $(\delta\mathbf{s}, \delta\boldsymbol{\varphi})$ such that $(\delta\mathbf{s}, \delta\boldsymbol{\varphi}) \equiv (0,0)$ on $\bar{H}$, then we obtain by (2.20)

$$\begin{aligned} \delta W^{(K)} &= \int_H \delta\mathbf{s}^T \left(\mathbf{a}\,dm - \mathrm{d}\mathbf{F}^{(e,H)}\right) \\ &\quad + \int_H \delta\boldsymbol{\varphi}^T \left(\boldsymbol{\xi} \times (\mathbf{a}\,dm - \mathrm{d}\mathbf{F}^{(e,H)}) - \mathrm{d}\mathbf{M}^{(e,H)}\right), \end{aligned} \tag{2.21}$$

and the problem is reduced to (ii).

(iv) We have proven in (ii) that the Newton-Euler equations hold for K, and in (iii) that they hold for some subsystem H of K. Since H was arbitrary by assumption, the definition of the dynamic equilibrium applies on K. □

Free moment terms are not considered in text books on classical mechanics. Thus, by setting $\mathrm{d}\mathbf{M} \equiv 0$ in (2.15), the virtual work reduces to

$$\delta W^{(K)} = \int\limits_K \left(\delta\mathbf{s} + \delta\boldsymbol{\varphi} \times \boldsymbol{\xi}\right)^T \left(\mathbf{a}\,\mathrm{d}m - \mathrm{d}\mathbf{F}^{(e,K)} - \mathrm{d}\mathbf{F}^{(o,K|H\bar{H})}\right), \tag{2.22}$$

and only one set of linear virtual displacements is needed as a consequence. By re-substituting $\delta\mathbf{r}$ from (2.12), equation (2.22) then changes to

$$\delta W^{(K)} = \int\limits_K \delta\mathbf{r}^T \left(\mathbf{a}\,\mathrm{d}m - \mathrm{d}\mathbf{F}^{(e,K)} - \mathrm{d}\mathbf{F}^{(o,K|H\bar{H})}\right). \tag{2.23}$$

To obtain this expression it was sufficient to take the free moments equal to zero. The symmetry of the stress tensor, *if* it exists, results then as a consequence of this assumption. Note in particular that the introduction of the stress tensor was not needed for this result, and that the virtual work has been stated by explicitly taking into account the free moment terms.

One should also realize that the procedure from (2.22) to (2.23) via (2.12) may be reversed, since arbitrary virtual displacements $\delta\mathbf{r}$ always contain as a subset the induced virtual displacements $\delta\boldsymbol{\varphi} \times \boldsymbol{\xi}$, and this for any $\delta\boldsymbol{\varphi}$. From (2.22) one then obtains the Newton-Euler equations (2.17), (2.18) with $\mathrm{d}\mathbf{M} = 0$; thus both, the principles of linear and angular momentum are contained in (2.23) in this particular case.

To the end, let us formally write $\mathrm{d}\mathbf{F}^{(K)} := \mathrm{d}\mathbf{F}^{(e,K)} + \mathrm{d}\mathbf{F}^{(o,K|H\bar{H})}$ in (2.23). This yields

$$\delta W^{(K)} = \int\limits_K \delta\mathbf{r}^T (\mathbf{a}\,\mathrm{d}m - \mathrm{d}\mathbf{F}^{(K)}) \tag{2.24}$$

which occurs in books on classical mechanics [52], [107], [34], [20], [145], [141] in context with d'Alembert/Lagrange's principle. Note, however, that (2.24) is *not yet* the principle of d'Alembert/Lagrange, but it is just the virtual work of the subsystem K. In particular, $\delta\mathbf{r}$ are still arbitrary displacements without any restriction to physical admissibility, and in $\mathrm{d}\mathbf{F}^{(K)}$ are still contained all the external forces of K together with the pairs of internal forces of arbitrary complementary subsystems of K. To proceed from (2.24) to the *classical* principle of d'Alembert/Lagrange, in rough outline the following steps should be performed:

1. Definition of the constraint forces (Bindungskräfte) in the sense of classical mechanics: Forces which result as a consequence on smooth bilateral geometric constraints are called constraint forces. The couple set up by the constraint and the constraint force should be regarded as the force element of the constraint, and the functional relationship between them should be called the force law of the constraint. Such force laws are set-valued.
2. Restriction to perfect bilateral holonomic constraints: The force laws of the constraints are not given in any axiomatic way, but are based

on physical observations. Depending on their particular structure, one distinguishes between perfect and non-perfect constraints. Perfect bilateral holonomic constraints are defined to be smooth submanifolds together with the force laws as the generalized gradient mappings of the indicators of the submanifolds. Non-perfect constraints, and hence non-perfect constraint forces, are excluded from classical mechanics, as well as any set-valued force law except for the force laws of the perfect constraints. Nonholonomic constraints have to be excluded when dealing with d'Alembert/Lagrange's principle, since they can only be taken into account when using the principles of Jourdain or Gauß. Every force allowed in classical dynamics which is not a perfect constraint force (Zwangskraft) is called an impressed force (eingeprägte Kraft). By the above definition, perfect constraint forces are potential forces of a very special force potential, the indicator of the constrained manifold.

3. The principle of d'Alembert/Lagrange: This is to realize that the force laws of the perfect bilateral holonomic constraints may be equivalently expressed by a variational equality with values equal to zero if the virtual displacements are chosen to be physically admissible, i.e. if they are taken from the tangent space of the constrained manifold. This variational equality with values equal to zero defines the virtual work done by the constraint forces.
4. Beneath the constraint forces, there are other forces for which the virtual work vanishes due to the constraints. The totality of all these forces is called the passive forces of the system. Constraint forces are thus always passive, impressed forces may be passive or may not, depending on the overall constraint configuration. Forces which are not passive are called active. These forces are always part of the impressed forces and must remain in the virtual work expression (2.24) when stated in context with d'Alembert/Lagrange's principle.

All these steps have been formulated very vaguely in this general context and should only indicate the direction in which the theory will further be developed. We will re-formulate the above statements more precisely in the course of the succeeding chapters, when we restrict ourselves to rigid bodies; we will extend them to unilateral constrained dynamics; and we will later drop the strict classification of forces (constraint — impressed and active — passive), since it does not make sense as soon as forces of intermediate type are discussed, which is one of the main topics in this book. We have introduced all the terms above, since we need them to proceed first to rigid bodies and, in a second step, to classical multibody theory, on which we finally impose set-valued force interactions.

2.5 Resultant Force and Inertia Terms

In this final section we introduce resultants of the force and inertia terms which will smooth the way to rigid bodies that are treated in the next chapter. For a distribution of forces and moments (2.1) on a mechanical subsystem K we denote by $\mathbf{F}^{(K)}$ the *resultant force* and by $\mathbf{M}_A^{(K)}$ the *resultant moment*, both obtained from integrating the expression (2.1),

$$\mathbf{F}^{(K)} := \int_K \mathrm{d}\mathbf{F} \quad \text{and} \quad \mathbf{M}_A^{(K)} := \int_K \boldsymbol{\xi} \times \mathrm{d}\mathbf{F} + \mathrm{d}\mathbf{M}. \tag{2.25}$$

The lower index A at the resultant moment denotes the reference point defined by the radius vector $\boldsymbol{\xi}$ as depicted in Figure 2.1. The resultant moment with respect to point B, for example, becomes $\mathbf{M}_B^{(K)} = \int_K \boldsymbol{\zeta} \times \mathrm{d}\mathbf{F} + \mathrm{d}\mathbf{M}$ which leads with $\boldsymbol{\zeta} = \mathbf{r}_{BA} + \boldsymbol{\xi}$ to $\mathbf{M}_B^{(K)} = \mathbf{r}_{BA} \times \int_K \mathrm{d}\mathbf{F} + \int_K \boldsymbol{\xi} \times \mathrm{d}\mathbf{F} + \mathrm{d}\mathbf{M}$. By (2.25), this expression may be written as

$$\mathbf{M}_B^{(K)} = \mathbf{M}_A^{(K)} + \mathbf{r}_{BA} \times \mathbf{F}^{(K)}, \tag{2.26}$$

which is the transformation formula for moments when the point of reference is changed.

In the course of the following calculations we will need three integrals defining the *mass* $m^{(K)}$ of the subsystem K, the displacement $\mathbf{r}_{AS}$ of the *center of mass* S relative to the point of reference A, and a term $\mathbf{N}_A^{(K)}$ related to the angular momentum of K. These integrals are

$$m^{(K)} := \int_K dm, \quad m^{(K)} \mathbf{r}_{AS} := \int_K \boldsymbol{\xi}\, dm, \quad \mathbf{N}_A^{(K)} := \int_K \boldsymbol{\xi} \times \mathbf{v}\, dm. \tag{2.27}$$

We will see that $\mathbf{N}_A^{(K)}$ becomes the angular momentum with respect to point A when A is chosen to be either the center of mass S or any point with velocity $\mathbf{v}_A$ equal to zero. Note also that $\int_K \boldsymbol{\sigma}\, dm = 0$ when $\boldsymbol{\sigma}$ denotes the displacement of $x \in K$ relative to S. This can be seen immediately from $\boldsymbol{\xi} = \mathbf{r}_{AS} + \boldsymbol{\sigma}$ and the first two integrals in (2.27), since

$$\begin{aligned} 0 &= \int_K \boldsymbol{\xi}\, dm - m^{(K)} \mathbf{r}_{AS} \\ &= \mathbf{r}_{AS} \int_K dm + \int_K \boldsymbol{\sigma}\, dm - m^{(K)} \mathbf{r}_{AS} \\ &= \int_K \boldsymbol{\sigma}\, dm. \end{aligned} \tag{2.28}$$

Let $\mathbf{v}$ denote the velocity of the point $x \in K$ as indicated in Figure 2.1. The *linear momentum* $\mathbf{p}^{(K)}$ and the *angular momentum* $\mathbf{L}_A^{(K)}$ of the subsystem K with respect to any moving or non-moving point A are defined as

$$\mathbf{p}^{(K)} := \int_K \mathbf{v}\, dm \quad \text{and} \quad \mathbf{L}_A^{(K)} := \int_K \boldsymbol{\xi} \times \dot{\boldsymbol{\xi}}\, dm. \tag{2.29}$$

Both terms may be rewritten by using the integrals defined in (2.27). With $\mathbf{v} = \mathbf{v}_S + \dot{\boldsymbol{\sigma}}$ and (2.28) the integral in the linear momentum becomes

$$\begin{aligned}\int_K \mathbf{v}\, dm &= \int_K \mathbf{v}_S\, dm + \int_K \dot{\boldsymbol{\sigma}}\, dm \\ &= \mathbf{v}_S \int_K dm + \Big(\int_K \boldsymbol{\sigma}\, dm\Big)^{\bullet} \\ &= m^{(K)}\, \mathbf{v}_S\end{aligned} \tag{2.30}$$

for systems with mass distribution not depending on time. For the integral in the angular momentum we obtain with $\dot{\boldsymbol{\xi}} = \mathbf{v} - \mathbf{v}_A$ and the second and third integral in (2.27)

$$\begin{aligned}\int_K \boldsymbol{\xi} \times \dot{\boldsymbol{\xi}}\, dm &= \int_K \boldsymbol{\xi} \times \mathbf{v}\, dm - \int_K \boldsymbol{\xi}\, dm \times \mathbf{v}_A \\ &= \mathbf{N}_A^{(K)} - m^{(K)}\, \mathbf{r}_{AS} \times \mathbf{v}_A.\end{aligned} \tag{2.31}$$

Hence, the linear and the angular momentum (2.29) become

$$\mathbf{p}^{(K)} = m^{(K)}\, \mathbf{v}_S \quad \text{and} \quad \mathbf{L}_A^{(K)} + m^{(K)}\, \mathbf{r}_{AS} \times \mathbf{v}_A = \mathbf{N}_A^{(K)}, \tag{2.32}$$

and we see that $\mathbf{L}_A^{(K)} = \mathbf{N}_A^{(K)}$ if, for instance, $\mathbf{v}_A = 0$ or $\mathbf{r}_{AS} = 0$ $(S \equiv A)$. As in (2.26) we may change the point of reference in the angular momentum. With $\boldsymbol{\zeta} = \mathbf{r}_{BA} + \boldsymbol{\xi}$ and (2.27), (2.29) we get

$$\begin{aligned}\mathbf{N}_B^{(K)} &= \int_K \boldsymbol{\zeta} \times \mathbf{v}\, dm \\ &= \mathbf{r}_{BA} \times \int_K \mathbf{v}\, dm + \int_K \boldsymbol{\xi} \times \mathbf{v}\, dm \\ &= \mathbf{r}_{BA} \times \mathbf{p}^{(K)} + \mathbf{N}_A^{(K)},\end{aligned} \tag{2.33}$$

which leads by the second equation in (2.32) to the transformation formula

$$\big(\mathbf{L}_B^{(K)} + m^{(K)}\, \mathbf{r}_{BS} \times \mathbf{v}_B\big) = \big(\mathbf{L}_A^{(K)} + m^{(K)}\, \mathbf{r}_{AS} \times \mathbf{v}_A\big) + \mathbf{r}_{BA} \times \mathbf{p}^{(K)}. \tag{2.34}$$

From this equation one immediately sees that $\mathbf{L}_A^{(K)} = \mathbf{L}_B^{(K)} = \mathbf{L}_S^{(K)}$ as soon as $\mathbf{v}_A = \mathbf{v}_B = \mathbf{v}_S$, since $m^{(K)}\,(\mathbf{r}_{BS} \times \mathbf{v}_S - \mathbf{r}_{AS} \times \mathbf{v}_S) = m^{(K)}\, \mathbf{r}_{BA} \times \mathbf{v}_S = \mathbf{r}_{BA} \times \mathbf{p}^{(K)}$.

We are now going to state the Newton-Euler equations (2.7) in terms of resultants. We start with the inertia term $\int_K \mathbf{a}\, dm$ in the principle of linear momentum. With $\mathbf{a} = \dot{\mathbf{v}}$ and the linear momentum $\mathbf{p}^{(K)}$ as defined in (2.29), this integral becomes

$$\begin{aligned}\int_K \mathbf{a}\,\mathrm{d}m &= \int_K \dot{\mathbf{v}}\,\mathrm{d}m \\ &= \Big(\int_K \mathbf{v}\,\mathrm{d}m\Big)^{\bullet} \\ &= \dot{\mathbf{p}}^{(K)}\end{aligned} \tag{2.35}$$

for systems with constant mass. For stating the inertia term $\int_K \boldsymbol{\xi} \times \mathbf{a}\,\mathrm{d}m$ in the principle of angular momentum we take into account that $\mathbf{a} = \mathbf{a}_A + \ddot{\boldsymbol{\xi}}$ with $\mathbf{a}_A$ the acceleration of point A, and $\boldsymbol{\xi} \times \ddot{\boldsymbol{\xi}} = (\boldsymbol{\xi} \times \dot{\boldsymbol{\xi}})^{\bullet}$, since $\dot{\boldsymbol{\xi}} \times \dot{\boldsymbol{\xi}} = 0$. We then obtain

$$\begin{aligned}\int_K \boldsymbol{\xi} \times \mathbf{a}\,\mathrm{d}m &= \int_K \boldsymbol{\xi}\,\mathrm{d}m \times \mathbf{a}_A + \Big(\int_K \boldsymbol{\xi} \times \dot{\boldsymbol{\xi}}\,\mathrm{d}m\Big)^{\bullet} \\ &= m^{(K)}\,\mathbf{r}_{AS} \times \mathbf{a}_A + \dot{\mathbf{L}}_A^{(K)},\end{aligned} \tag{2.36}$$

where, again, we have interchanged integration and differentiation with respect to time, and we have used the second integral in (2.27) and the definition of $\mathbf{L}_A^{(K)}$ from (2.29). Thus, by (2.35) and (2.36) the Newton-Euler equations (2.7) for a subsystem K in terms of resultants become

$$\dot{\mathbf{p}}^{(K)} - \mathbf{F}^{(e,K)} = 0 \quad \text{and} \quad \dot{\mathbf{L}}_A^{(K)} + m^{(K)}\,\mathbf{r}_{AS} \times \mathbf{a}_A - \mathbf{M}_A^{(e,K)} = 0 \tag{2.37}$$

with $\mathbf{F}^{(e,K)}$ and $\mathbf{M}_A^{(e,K)}$ being the resultants of the external forces and moments of the subsystem K that have to be built up from $\mathrm{d}\mathbf{F}^{(e,K)}$ and $\mathrm{d}\mathbf{M}^{(e,K)}$ according to equation (2.25).

We have by (2.26) and (2.34) the transformation formulas for the moments and the angular momenta. It is useful to state the latter also for the time derivatives of the angular momenta, since this gives together with (2.26) an explicit formula on how to change the point of reference in the Newton-Euler equations (2.37). Direct differentiation of (2.34) with respect to time yields

$$\big(\dot{\mathbf{L}}_B^{(K)} + m^{(K)}\,\mathbf{r}_{BS} \times \mathbf{a}_B\big) = \big(\dot{\mathbf{L}}_A^{(K)} + m^{(K)}\,\mathbf{r}_{AS} \times \mathbf{a}_A\big) + \mathbf{r}_{BA} \times \dot{\mathbf{p}}^{(K)} \tag{2.38}$$

with $\mathbf{a}_B = \dot{\mathbf{v}}_B$, $\mathbf{a}_A = \dot{\mathbf{v}}_A$, where we have taken into account that

$$m^{(K)}\,\dot{\mathbf{r}}_{BS} \times \mathbf{v}_B = m^{(K)}\,\dot{\mathbf{r}}_{AS} \times \mathbf{v}_A + \dot{\mathbf{r}}_{BA} \times \mathbf{p}^{(K)} \tag{2.39}$$

as direct calculation with $\dot{\mathbf{r}}_{CD} = \mathbf{v}_D - \mathbf{v}_C$ and $\mathbf{p}^{(K)} = m^{(K)}\,\mathbf{v}_S$ shows. Further, we have $\dot{\mathbf{L}}_A^{(K)} = \dot{\mathbf{L}}_B^{(K)} = \dot{\mathbf{L}}_S^{(K)}$ when $\mathbf{a}_A = \mathbf{a}_B = \mathbf{a}_S$, a similar result as stated after equation (2.34) for $\mathbf{L}_{(\cdot)}$.

3. Rigid Body Systems

The investigation of a constrained multibody system starts with an appropriate formulation of the equations of motion. We consider systems of rigid bodies under the influence of set-valued force laws including impressed forces, unilateral constraints as well as bilateral constraints. The bilateral constraints act permanently on the system which results in a time-independent reduction of the degrees of freedom. It is thus often convenient to formulate the equations of motion already in terms of coordinates $\mathbf{q}$ which implicitly take into account some of the classical constraints, in order to proceed with a reduced number of equations. In this chapter we review this classical reduction process by starting with the equations for a single rigid body, by composing them to a system of bodies, and by restricting them under the action of bilateral holonomic perfect constraints.

3.1 Preliminaries on the Vector Product

Let $\mathbf{x}$, $\mathbf{y}$, $\mathbf{a}$ be from a three-dimensional linear space. The vector product $\mathbf{x} \to \mathbf{a} \times \mathbf{x}$ is a linear map and will thus be written as

$$\tilde{\mathbf{a}}\mathbf{x} := \mathbf{a} \times \mathbf{x}. \tag{3.1}$$

The operator $\tilde{\mathbf{a}}$ is skew-symmetric, $\tilde{\mathbf{a}}^T = -\tilde{\mathbf{a}}$, which follows from comparing $\mathbf{y}^T\tilde{\mathbf{a}}\mathbf{x} = \det(\mathbf{y}, \mathbf{a}, \mathbf{x}) = -\det(\mathbf{x}, \mathbf{a}, \mathbf{y}) = -\mathbf{x}^T\tilde{\mathbf{a}}\mathbf{y}$ with the expression $\mathbf{y}^T\tilde{\mathbf{a}}\mathbf{x} = \mathbf{x}^T\tilde{\mathbf{a}}^T\mathbf{y}$ obtained from transposition.

We will also need $-\tilde{\mathbf{a}}^2$ which is symmetric and positive semidefinite. The first property follows from $\mathbf{x}^T\tilde{\mathbf{a}}^T\tilde{\mathbf{a}}\mathbf{y} = \mathbf{y}^T\tilde{\mathbf{a}}^T\tilde{\mathbf{a}}\mathbf{x}$ by transposition with $-\tilde{\mathbf{a}}^2 = \tilde{\mathbf{a}}^T\tilde{\mathbf{a}}$ due to skew-symmetry, and the second from $\mathbf{x}^T\tilde{\mathbf{a}}^T\tilde{\mathbf{a}}\mathbf{x} = \|\tilde{\mathbf{a}}\mathbf{x}\|^2 \geq 0$, $\forall\mathbf{x}$.

When the vector product is applied twice, the resulting vector $\mathbf{x} \times (\mathbf{y} \times \mathbf{z})$ is complanar to $\mathbf{y}$ and $\mathbf{z}$ and may be written as

$$\mathbf{x} \times (\mathbf{y} \times \mathbf{z}) = \mathbf{y}\,(\mathbf{x}^T\mathbf{z}) - \mathbf{z}\,(\mathbf{x}^T\mathbf{y}). \tag{3.2}$$

We use this formula to show that

$$\mathbf{a} \times (\mathbf{c} \times (\mathbf{c} \times \mathbf{a})) = -\mathbf{c} \times (\mathbf{a} \times (\mathbf{a} \times \mathbf{c})) \qquad \text{or} \qquad \tilde{\mathbf{a}}\tilde{\mathbf{c}}^2\mathbf{a} = -\tilde{\mathbf{c}}\tilde{\mathbf{a}}^2\mathbf{c}, \tag{3.3}$$

an expression that will later occur when we derive the principle of angular momentum for rigid bodies. By applying (3.2) on the two terms in (3.3) we obtain

$$\begin{aligned}\mathbf{a}\times(\mathbf{c}\times(\mathbf{c}\times\mathbf{a})) &= \mathbf{c}\,(\mathbf{a}^T(\mathbf{c}\times\mathbf{a})) - (\mathbf{c}\times\mathbf{a})(\mathbf{a}^T\mathbf{c})\\ \mathbf{c}\times(\mathbf{a}\times(\mathbf{a}\times\mathbf{c})) &= \mathbf{a}\,(\mathbf{c}^T(\mathbf{a}\times\mathbf{c})) - (\mathbf{a}\times\mathbf{c})(\mathbf{c}^T\mathbf{a}),\end{aligned} \tag{3.4}$$

which proves already (3.3), since $\mathbf{a}^T(\mathbf{c}\times\mathbf{a}) = \mathbf{c}^T(\mathbf{a}\times\mathbf{c}) = 0$ and $\mathbf{a}\times\mathbf{c} = -\mathbf{c}\times\mathbf{a}$.

3.2 Rigid Body Kinematics

A *rigid body* B is a mechanical subsystem of S ($B \in \mathcal{S}$) with the property that the Euclidean distance between any two points $x_1, x_2 \in B$ is constant. If we denote by $\mathbf{r}_i$ the displacement of x_i relative to some inertially fixed point O, this condition becomes $\|\mathbf{r}_1(t) - \mathbf{r}_2(t)\| = \text{const.}$ with t being time.

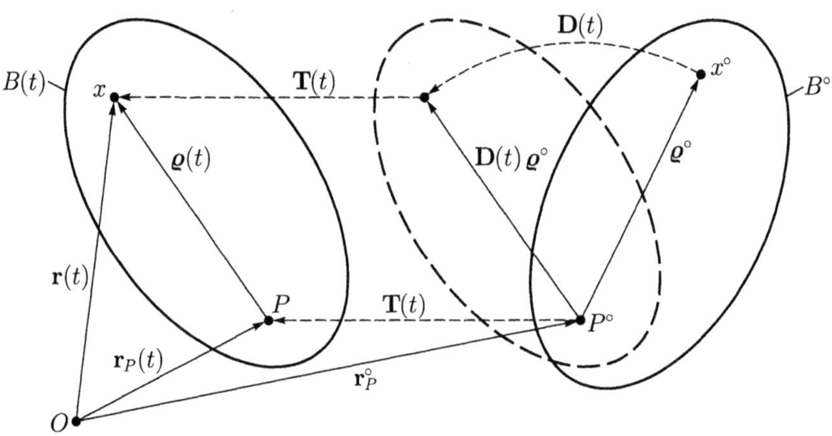

Figure 3.1. Rigid body motion

A *motion of a rigid body* relative to some reference configuration may always be composed of a rotation $\mathbf{D}(t)$ and a translation $\mathbf{T}(t)$. We choose the reference configuration of the body B to be inertially fixed, i.e. to be not changing with time, and we denote it by an upper index $\circ$. Further, let $P^\circ \in B^\circ$ and let $\boldsymbol{\varrho}^\circ$ denote the displacement of $x^\circ \in B^\circ$ relative to P° as indicated in Figure 3.1. The displacement of any point $x \in B$ relative to an inertially fixed point O may then be expressed as

$$\mathbf{r}(t) = \mathbf{r}_P^\circ + \mathbf{D}(t)\,\boldsymbol{\varrho}^\circ + \mathbf{T}(t), \tag{3.5}$$

where $\mathbf{r}_P^\circ$ and $\boldsymbol{\varrho}^\circ$, the displacements of P° relative to O and x° relative to P°, have constant values. Note that for $\mathbf{D} = \text{const.}$ (3.5) is a pure translation, whereas for $\mathbf{T} = \text{const.}$ it is a pure rotation with P as a fixed point. Further, (3.5) satisfies the rigid body property: For any two different points $x_1, x_2 \in B$ we obtain $\|\mathbf{r}_1(t) - \mathbf{r}_2(t)\| = \|\mathbf{D}(t)\,(\boldsymbol{\varrho}_1^\circ - \boldsymbol{\varrho}_2^\circ)\| = \|\boldsymbol{\varrho}_1^\circ - \boldsymbol{\varrho}_2^\circ\| = \text{const.}$, since rotations preserve distances.

Proposition 3.2.1. *As soon as three affinely independent points $P^\circ, x_1^\circ, x_2^\circ \in B^\circ$ with one of them being the point of reference for the rotation, say P°, have been chosen, the translation $\mathbf{T}$ and the rotation $\mathbf{D}$ in (3.5) are uniquely determined for any given rigid body motion $B^\circ \to B$.*

Proof. Suppose there are two different translations $\mathbf{T}_1$, $\mathbf{T}_2$ and rotations $\mathbf{D}_1$, $\mathbf{D}_2$ for the same motion $B^\circ \to B$, i.e. $\mathbf{r}(t) = \mathbf{r}_P^\circ + \mathbf{D}_1(t)\,\boldsymbol{\varrho}^\circ + \mathbf{T}_1(t) = \mathbf{r}_P^\circ + \mathbf{D}_2(t)\,\boldsymbol{\varrho}^\circ + \mathbf{T}_2(t)$ by (3.5) with $\boldsymbol{\varrho}^\circ \in \{0, \boldsymbol{\varrho}_1^\circ, \boldsymbol{\varrho}_2^\circ\}$ denoting the three displacements of $P^\circ, x_1^\circ, x_2^\circ$ relative to P°. In this equation, choose first $\boldsymbol{\varrho}^\circ = 0$ which gives $\mathbf{r}_P^\circ + \mathbf{T}_1(t) = \mathbf{r}_P^\circ + \mathbf{T}_2(t)$, hence $\mathbf{T}_1 = \mathbf{T}_2$. Now we have $\mathbf{D}_1(t)\,\boldsymbol{\varrho}^\circ = \mathbf{D}_2(t)\,\boldsymbol{\varrho}^\circ$ or, after setting $\mathbf{D}^\star(t) := \mathbf{D}_2^{-1}(t)\,\mathbf{D}_1(t)$, $(\mathbf{D}^\star(t) - \mathbf{E})\,\boldsymbol{\varrho}^\circ = 0$, where $\mathbf{E}$ denotes the identity and $\mathbf{D}^\star$ is another rotation. This equation holds if either $\mathbf{D}^\star$ is the identity or $\{\alpha\boldsymbol{\varrho}^\circ\}$ $(\alpha \in \mathbb{R})$ is the axis of rotation of $\mathbf{D}^\star$. Taking $\boldsymbol{\varrho}^\circ = \boldsymbol{\varrho}_1^\circ$ and $\boldsymbol{\varrho}^\circ = \boldsymbol{\varrho}_2^\circ$, the latter would require two different axes of rotation of $\mathbf{D}^\star$ because $\boldsymbol{\varrho}_1^\circ$ and $\boldsymbol{\varrho}_2^\circ$ are linearly independent by assumption. This is a contradiction, hence $\mathbf{D}^\star$ is the identity and $\mathbf{D}_1 = \mathbf{D}_2$. □

We are now going to state the velocity field of a rigid body which is done by differentiating (3.5) with respect to time,

$$\dot{\mathbf{r}}(t) = \dot{\mathbf{D}}(t)\,\boldsymbol{\varrho}^\circ + \dot{\mathbf{T}}(t). \tag{3.6}$$

The points O and P° are inertially fixed, hence $\dot{\mathbf{r}} =: \mathbf{v}$ is the velocity of point $x \in B$, and $\dot{\mathbf{T}} =: \mathbf{v}_P$ the velocity of point $P \in B$. Thus, (3.6) becomes $\mathbf{v}(t) = \mathbf{v}_P(t) + \dot{\mathbf{D}}(t)\,\boldsymbol{\varrho}^\circ$. We finally eliminate the terms $\boldsymbol{\varrho}^\circ$ referring to the (arbitrarily chosen) reference configuration B° to have an expression with all terms stated at time t: Since rotations belong to the group of orthogonal operators, the inverse rotation is obtained by transposition, $\mathbf{D}^{-1} = \mathbf{D}^T$, and $\boldsymbol{\varrho}(t) = \mathbf{D}(t)\,\boldsymbol{\varrho}^\circ$ from Figure 3.1 may hence be solved for $\boldsymbol{\varrho}^\circ$ to give $\boldsymbol{\varrho}^\circ = \mathbf{D}^T(t)\,\boldsymbol{\varrho}(t)$. By substituting this expression into (3.6) we obtain

$$\mathbf{v} = \mathbf{v}_P + \dot{\mathbf{D}}\mathbf{D}^T\boldsymbol{\varrho}, \tag{3.7}$$

where we have omitted to write down explicitly the dependence on time t, since it applies now for every term in (3.7). Let us denote $\dot{\mathbf{D}}\mathbf{D}^T$ by $\tilde{\boldsymbol{\Omega}}$. The operator $\tilde{\boldsymbol{\Omega}}$ is skew-symmetric, because by differentiation of the identity $\mathbf{D}\mathbf{D}^T = \mathbf{E}$ we obtain $\dot{\mathbf{D}}\mathbf{D}^T + \mathbf{D}\dot{\mathbf{D}}^T = 0$, and hence $\dot{\mathbf{D}}\mathbf{D}^T = -\mathbf{D}\dot{\mathbf{D}}^T = -(\dot{\mathbf{D}}\mathbf{D}^T)^T$. Further, it is known [7] that the action of the linear skew-symmetric operators on a three-dimensional oriented Euclidean space may be equivalently expressed by the vector product, i.e. for every skew-symmetric $\tilde{\boldsymbol{\Omega}}$ there is a unique vector $\boldsymbol{\Omega}$ such that

$$\dot{\mathbf{D}}\mathbf{D}^T \boldsymbol{\varrho} =: \tilde{\boldsymbol{\Omega}}\, \boldsymbol{\varrho} = \boldsymbol{\Omega} \times \boldsymbol{\varrho} \tag{3.8}$$

for all $\boldsymbol{\varrho}$. The $\boldsymbol{\Omega}$ defined in (3.8) is the angular velocity of the body B, and the velocity $\mathbf{v}$ of any point $x \in B$ may the be calculated from (3.7), (3.8),

$$\mathbf{v} = \mathbf{v}_P + \boldsymbol{\Omega} \times \boldsymbol{\varrho} \tag{3.9}$$

when $\mathbf{v}_P$ and $\boldsymbol{\Omega}$ are known.

To obtain the rigid body accelerations we differentiate (3.9) with respect to time,

$$\dot{\mathbf{v}} = \dot{\mathbf{v}}_P + \dot{\boldsymbol{\Omega}} \times \boldsymbol{\varrho} + \boldsymbol{\Omega} \times \dot{\boldsymbol{\varrho}}. \tag{3.10}$$

Here, $\dot{\mathbf{v}} =: \mathbf{a}$ is the acceleration of point $x \in B$, $\dot{\mathbf{v}}_P =: \mathbf{a}_P$ the acceleration of point $P \in B$, and $\dot{\boldsymbol{\Omega}} =: \boldsymbol{\Psi}$ the angular acceleration of the body B. Further, $\dot{\boldsymbol{\varrho}}$ is recognized from Figure 3.2 to be $\dot{\boldsymbol{\varrho}} = \mathbf{v} - \mathbf{v}_P$ which is by (3.9) $\dot{\boldsymbol{\varrho}} = \boldsymbol{\Omega} \times \boldsymbol{\varrho}$. Hence, (3.10) becomes

$$\mathbf{a} = \mathbf{a}_P + \boldsymbol{\Psi} \times \boldsymbol{\varrho} + \boldsymbol{\Omega} \times (\boldsymbol{\Omega} \times \boldsymbol{\varrho}), \tag{3.11}$$

which yields the acceleration $\mathbf{a}$ of any $x \in B$ when $\mathbf{a}_P$, $\boldsymbol{\Psi}$, $\boldsymbol{\Omega}$ are known. Figure 3.2 puts together the nomenclature that has been used for the kinematic terms of a rigid body.

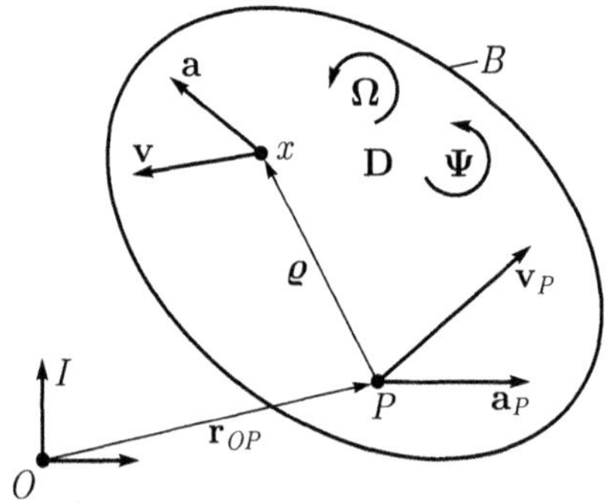

$\mathbf{D}$	rotation of B relative to I
$\tilde{\boldsymbol{\Omega}} = \dot{\mathbf{D}}\mathbf{D}^T$	angular velocity of B
$\boldsymbol{\Psi} = \dot{\boldsymbol{\Omega}}$	angular acceleration of B
$\mathbf{r}_{OP}$	displacement of P rel. to O
$\mathbf{v}_P = \dot{\mathbf{r}}_{OP}$	velocity of P
$\mathbf{a}_P = \dot{\mathbf{v}}_P$	acceleration of P
$\boldsymbol{\varrho}$	displacement of x relative to P
$\mathbf{v}$	velocity of x
$\mathbf{a}$	acceleration of x

Figure 3.2. Kinematic terms for rigid bodies

Let us finally introduce the rigid body virtual displacements, a restriction of the virtual displacements of Section 2.4 to physically admissible virtual rigid body motions respecting rigid body kinematics. Let $\mathbf{r}(t)$ be defined as in equation (3.5). In order to obtain the linear virtual displacement $\delta\mathbf{r}$ at point x (2.12) which is compatible with rigid body kinematics, we take for the same x° at the same time t a different rotation $\mathbf{D}^\star(t)$ and translation $\mathbf{T}^\star(t)$ to give

$$\mathbf{r}^\star(t) = \mathbf{r}_P^\circ + \mathbf{D}^\star(t)\, \boldsymbol{\varrho}^\circ + \mathbf{T}^\star(t). \tag{3.12}$$

Thus, $\mathbf{r}^\star$ differs from $\mathbf{r}$ in (3.5) by

$$\delta\mathbf{r} := \mathbf{r}^\star - \mathbf{r} = (\mathbf{D}^\star - \mathbf{D})\,\mathbf{D}^T\boldsymbol{\varrho} + (\mathbf{T}^\star - \mathbf{T}), \tag{3.13}$$

where we have expressed $\boldsymbol{\varrho}^\circ$ by $\mathbf{D}^T\boldsymbol{\varrho}$ as in (3.7). Equation (3.13) defines the *rigid body virtual displacements*, as soon as $\mathbf{D}^\star$ is chosen to be near $\mathbf{D}$: This turns $(\mathbf{D}^\star - \mathbf{D})\,\mathbf{D}^T$ into a skew-symmetric operator $\delta\tilde{\mathbf{\Phi}}$, an element of the tangent space $so(3)$ on the rotations $SO(3)$ at the identity $\mathbf{E}$. To see this we set

$$\mathbf{D}^\star := \mathrm{e}^{\delta\tilde{\mathbf{\Phi}}}\,\mathbf{D} \approx (\mathbf{E} + \delta\tilde{\mathbf{\Phi}})\,\mathbf{D} \tag{3.14}$$

for small rotations $\mathrm{e}^{\delta\tilde{\mathbf{\Phi}}}$ and obtain

$$(\mathbf{D}^\star - \mathbf{D})\,\mathbf{D}^T = ((\mathbf{E} + \delta\tilde{\mathbf{\Phi}})\,\mathbf{D} - \mathbf{D})\,\mathbf{D}^T = \delta\tilde{\mathbf{\Phi}}. \tag{3.15}$$

Finally, let us denote by $\delta\mathbf{r}_P := \mathbf{T}^\star - \mathbf{T}$ in (3.13) the virtual linear displacement at point P. Together with (3.14) we may then write (3.13) as

$$\delta\mathbf{r} = \delta\mathbf{r}_P + \delta\mathbf{\Phi} \times \boldsymbol{\varrho}, \tag{3.16}$$

which gives the physically admissible virtual displacements $\delta\mathbf{r}$ at point x in terms of the two sets of independent virtual displacements, the virtual translation $\delta\mathbf{r}_P$ and $\delta\mathbf{\Phi}$ from the virtual rotation $\mathbf{E}+\delta\tilde{\mathbf{\Phi}}$. By comparing (3.16) with (2.12) we see that the rigid body virtual displacements are obtained when setting $\delta\mathbf{s}$ and $\delta\boldsymbol{\varphi}$ constant on B with values $\delta\mathbf{s} = \delta\mathbf{r}_P$ and $\delta\boldsymbol{\varphi} = \delta\mathbf{\Phi}$.

3.3 Rigid Body Kinetics

In this section we evaluate the resultant inertia terms for the rigid body. In particular, we take the definitions of the linear and the angular momentum $\mathbf{p}^{(K)}$ and $\mathbf{L}_A^{(K)}$ of a mechanical subsystem K from Section 2.5 and apply them to a rigid body B; that means we combine them with the rigid body kinematics expressions from Section 3.2. We also state the Newton-Euler equations for rigid bodies. This requires in addition the evaluation of the time derivatives of $\mathbf{p}^{(K)}$ and $\mathbf{L}_A^{(K)}$, for which the rigid body accelerations are needed.

For the angular inertia terms as well as for the moments some point of reference has to be specified. It was taken in Section 2.5 to be any moving or non-moving point A. Since we know by equations (2.26), (2.34), (2.38) the transformation rules on how to change to another point, and since we know by Proposition 2.3.1 that the dynamic equilibrium is not influenced by a special choice of this point, we may choose it here to be the rigid body point P as defined in Figures 3.1 and 3.2.

We also omit the upper index (K) referring to the mechanical subsystem under consideration, since in this section it will be always the same, the rigid body $(B) = (K)$. As in (2.27) we denote by m the mass of the rigid body and by $\mathbf{r}_{PS}$ the displacement of the center of mass S with respect to the point of reference P. The two corresponding integrals are stated in equation (3.17), together with a third defining the *inertia operator* $\boldsymbol{\Theta}_P$ of the rigid body with respect to point P,

$$m = \int_B \mathrm{d}m, \quad m\mathbf{r}_{PS} = \int_B \boldsymbol{\varrho}\,\mathrm{d}m, \quad \boldsymbol{\Theta}_P := \int_B -\tilde{\boldsymbol{\varrho}}^2 \mathrm{d}m. \tag{3.17}$$

By the second paragraph of Section 3.1 the inertia operator $\boldsymbol{\Theta}_P$ is symmetric and at least positive semidefinite. It becomes positive definite as soon as the affine hull of the support of $\mathrm{d}m$ is two-dimensional.

We will now calculate the linear momentum $\mathbf{p}$ and the angular momentum $\mathbf{L}_P$ with respect to point P of the rigid body. With $\mathbf{p}$ as defined in (2.29) and the rigid body velocity field $\mathbf{v} = \mathbf{v}_P + \boldsymbol{\Omega} \times \boldsymbol{\varrho}$ from (3.9) we obtain

$$\mathbf{p} = \int_B \mathbf{v}\,\mathrm{d}m = \mathbf{v}_P \int_B \mathrm{d}m + \boldsymbol{\Omega} \times \int_B \boldsymbol{\varrho}\,\mathrm{d}m = m\mathbf{v}_P + m\boldsymbol{\Omega} \times \mathbf{r}_{PS}, \tag{3.18}$$

where we have taken into account the definition of the mass and of the center of mass in (3.17). In the same fashion we proceed with $\mathbf{L}_P$ in (2.29). With (3.9), (3.17) and $\dot{\boldsymbol{\varrho}} = \mathbf{v} - \mathbf{v}_P = \boldsymbol{\Omega} \times \boldsymbol{\varrho}$ we obtain

$$\mathbf{L}_P = \int_B \boldsymbol{\varrho} \times \dot{\boldsymbol{\varrho}}\,\mathrm{d}m = \int_B \boldsymbol{\varrho} \times (\boldsymbol{\Omega} \times \boldsymbol{\varrho})\,\mathrm{d}m = \int_B -\tilde{\boldsymbol{\varrho}}^2\,\mathrm{d}m\,\boldsymbol{\Omega} = \boldsymbol{\Theta}_P\boldsymbol{\Omega}. \tag{3.19}$$

Equations (3.18) and (3.19) are linear in $\mathbf{v}_P$ and $\boldsymbol{\Omega}$. Hence, we may write them as

$$\begin{pmatrix} \mathbf{p} \\ \mathbf{L}_P + m\mathbf{r}_{PS} \times \mathbf{v}_P \end{pmatrix} = \begin{pmatrix} m\mathbf{E} & m\tilde{\mathbf{r}}_{PS}^T \\ m\tilde{\mathbf{r}}_{PS} & \boldsymbol{\Theta}_P \end{pmatrix} \begin{pmatrix} \mathbf{v}_P \\ \boldsymbol{\Omega} \end{pmatrix} \tag{3.20}$$

which defines the *mass-inertia operator* of a rigid body with respect to point P. Apparently, this operator is symmetric and positive semidefinite, and it becomes positive definite under the same conditions as $\boldsymbol{\Theta}_P$.

In order to state the Newton-Euler equations (2.37) for a rigid body we first have to provide $\dot{\mathbf{p}}$ and $\dot{\mathbf{L}}_P + m\mathbf{r}_{PS} \times \mathbf{a}_P$ according to (2.35) and (2.36). Together with the rigid body accelerations $\mathbf{a}$ from (3.11) the first term becomes

$$\begin{aligned} \dot{\mathbf{p}} &= \int_B \mathbf{a}\,\mathrm{d}m \\ &= \int_B (\mathbf{a}_P + \boldsymbol{\Psi} \times \boldsymbol{\varrho} + \boldsymbol{\Omega} \times (\boldsymbol{\Omega} \times \boldsymbol{\varrho}))\,\mathrm{d}m \\ &= \mathbf{a}_P \int_B \mathrm{d}m + \boldsymbol{\Psi} \times \int_B \boldsymbol{\varrho}\,\mathrm{d}m + \boldsymbol{\Omega} \times (\boldsymbol{\Omega} \times \int_B \boldsymbol{\varrho}\,\mathrm{d}m) \\ &= m\mathbf{a}_P + m\boldsymbol{\Psi} \times \mathbf{r}_{PS} + m\boldsymbol{\Omega} \times (\boldsymbol{\Omega} \times \mathbf{r}_{PS}), \end{aligned} \tag{3.21}$$

where, again, we have used (3.17). The second term is evaluated in the same manner,

$$\begin{aligned}\dot{\mathbf{L}}_P + m\mathbf{r}_{PS} \times \mathbf{a}_P &= \int_B \boldsymbol{\varrho} \times \mathbf{a}\,\mathrm{d}m \\ &= \int_B \boldsymbol{\varrho} \times (\mathbf{a}_P + \boldsymbol{\Psi} \times \boldsymbol{\varrho} + \boldsymbol{\Omega} \times (\boldsymbol{\Omega} \times \boldsymbol{\varrho}))\,dm \\ &= \int_B \boldsymbol{\varrho}\,\mathrm{d}m \times \mathbf{a}_P + \int_B -\tilde{\boldsymbol{\varrho}}^2\,\mathrm{d}m\,\boldsymbol{\Psi} + \boldsymbol{\Omega} \times \int_B -\tilde{\boldsymbol{\varrho}}^2\,\mathrm{d}m\,\boldsymbol{\Omega} \\ &= m\mathbf{r}_{PS} \times \mathbf{a}_P + \boldsymbol{\Theta}_P\boldsymbol{\Psi} + \boldsymbol{\Omega} \times \boldsymbol{\Theta}_P\boldsymbol{\Omega} \qquad (3.22)\end{aligned}$$

where we have used formula (3.3) to proceed from the second to the third line. By substituting (3.21) and (3.22) into (2.37) we obtain the Newton-Euler equations for the rigid body

$$\begin{pmatrix}\dot{\mathbf{p}} \\ \dot{\mathbf{L}}_P + m\mathbf{r}_{PS} \times \mathbf{a}_P\end{pmatrix} = \begin{pmatrix} m\mathbf{E} & m\tilde{\mathbf{r}}_{PS}^T \\ m\tilde{\mathbf{r}}_{PS} & \boldsymbol{\Theta}_P\end{pmatrix}\begin{pmatrix}\mathbf{a}_P \\ \boldsymbol{\Psi}\end{pmatrix} + \begin{pmatrix} m\tilde{\boldsymbol{\Omega}}^2\mathbf{r}_{PS} \\ \tilde{\boldsymbol{\Omega}}\boldsymbol{\Theta}_P\boldsymbol{\Omega}\end{pmatrix} = \begin{pmatrix}\mathbf{F}^{(e)} \\ \mathbf{M}_P^{(e)}\end{pmatrix} \qquad (3.23)$$

with $\mathbf{F}^{(e)}$ the resultant of the external forces of the body, and $\mathbf{M}_P^{(e)}$ the resultant of the external free and induced moments.

We remark that $\dot{\mathbf{L}}_P$ in (3.22) is more directly obtained by differentiating $\mathbf{L}_P$ in (3.19) with respect to time,

$$\dot{\mathbf{L}}_P = \boldsymbol{\Theta}_P\boldsymbol{\Psi} + \dot{\boldsymbol{\Theta}}_P\boldsymbol{\Omega}. \qquad (3.24)$$

By comparing this expression with (3.22) we obtain the well-known relation

$$\dot{\boldsymbol{\Theta}}_P\boldsymbol{\Omega} = \boldsymbol{\Omega} \times \boldsymbol{\Theta}_P\boldsymbol{\Omega}. \qquad (3.25)$$

Finally, one might be interested in the transformation rule of the inertia operator $\boldsymbol{\Theta}_P$ when the point of reference is changed. For that we express $\boldsymbol{\varrho}$ as $\boldsymbol{\varrho} = \mathbf{r}_{PS} + \boldsymbol{\eta}$, where $\boldsymbol{\eta}$ denotes the displacement of $x \in B$ relative to the center of mass S. The inertia operator $\boldsymbol{\Theta}_P$ in (3.17) thus becomes $\boldsymbol{\Theta}_P = \int_B \tilde{\boldsymbol{\varrho}}^T\tilde{\boldsymbol{\varrho}}\,\mathrm{d}m = \tilde{\mathbf{r}}_{PS}^T\tilde{\mathbf{r}}_{PS}\int_B \mathrm{d}m + \tilde{\mathbf{r}}_{PS}^T\int_B \tilde{\boldsymbol{\eta}}\,\mathrm{d}m + \int_B \tilde{\boldsymbol{\eta}}^T\mathrm{d}m\,\tilde{\mathbf{r}}_{PS} + \int_B \tilde{\boldsymbol{\eta}}^T\tilde{\boldsymbol{\eta}}\,\mathrm{d}m$, where the last integral defines $\boldsymbol{\Theta}_S$. Hence, due to $\int_B \boldsymbol{\eta}\,\mathrm{d}m = 0$ as in (2.28) we have

$$\boldsymbol{\Theta}_P = \boldsymbol{\Theta}_S - m\tilde{\mathbf{r}}_{PS}^2 \qquad (3.26)$$

which is Steiner's formula for the three-dimensional case.

3.4 The Dynamic Equilibrium of a Rigid Body

Suppose that the Newton-Euler equations (3.23) hold for a rigid body B. Does this already mean that the rigid body is in the dynamic equilibrium?

Certainly not, because the dynamic equilibrium requires the Newton-Euler equations to hold in addition for any subsystem H of B. These subsystems, however, are again rigid bodies with the same kinematics as B which simplifies the above question considerably: The internal forces of B acting on H are responsible for the Newton-Euler equations to apply for H and, since H is arbitrary, for the dynamic equilibrium of B. In order to show how the internal forces have to be computed for a rigid body in the dynamic equilibrium we start with the Newton-Euler equations (3.23) for B,

$$\begin{pmatrix} m\mathbf{E} & m\tilde{\mathbf{r}}_{PS}^T \\ m\tilde{\mathbf{r}}_{PS} & \mathbf{\Theta}_P \end{pmatrix}^{(B)} \begin{pmatrix} \mathbf{a}_P \\ \mathbf{\Psi} \end{pmatrix}^{(B)} + \begin{pmatrix} m\tilde{\mathbf{\Omega}}^2\mathbf{r}_{PS} \\ \tilde{\mathbf{\Omega}}\mathbf{\Theta}_P\mathbf{\Omega} \end{pmatrix}^{(B)} = \begin{pmatrix} \mathbf{F} \\ \mathbf{M}_P \end{pmatrix}^{(e,B)} . \tag{3.27}$$

This equation determines the values of the accelerations $(\mathbf{a}_P^{(B)}, \mathbf{\Psi}^{(B)})$ for some point of reference $P \in B$, as soon as the mass distribution $\mathrm{d}m$ and the distribution of the external forces $(\mathrm{d}\mathbf{F}^{(e,B)}, \mathrm{d}\mathbf{M}^{(e,B)})$ is known on B. Let now H be a subsystem of B, this is another rigid body with the same P as an admissible point of reference and with the same kinematic magnitudes $(\mathbf{a}_P^{(B)}, \mathbf{\Psi}^{(B)}, \mathbf{\Omega}^{(B)})$. The Newton-Euler equations (3.23) for H thus read

$$\begin{pmatrix} m\mathbf{E} & m\tilde{\mathbf{r}}_{PS}^T \\ m\tilde{\mathbf{r}}_{PS} & \mathbf{\Theta}_P \end{pmatrix}^{(H)} \begin{pmatrix} \mathbf{a}_P \\ \mathbf{\Psi} \end{pmatrix}^{(B)} + \begin{pmatrix} m\tilde{\mathbf{\Omega}}^2\mathbf{r}_{PS} \\ \tilde{\mathbf{\Omega}}\mathbf{\Theta}_P\mathbf{\Omega} \end{pmatrix}^{(H)} - \begin{pmatrix} \mathbf{F} \\ \mathbf{M}_P \end{pmatrix}^{(e,B|H)} = \begin{pmatrix} \mathbf{F} \\ \mathbf{M}_P \end{pmatrix}^{(o,B|H)} \tag{3.28}$$

with the resultants of the external forces (e, H) of H having been split into the portion $(e, B|H)$ of the external forces of B acting on H and the internal forces $(o, B|H)$ of B acting from $\bar{H}$ on H according to (2.6). Further, in (3.28) $m^{(H)}$, $m^{(H)}\mathbf{r}_{PS}^{(H)}$, $\mathbf{\Theta}_P^{(H)}$ denote the three inertia terms obtained from (3.17) when integration is performed on H which yields the mass of the rigid body H, the displacement of the center of mass of H relative to point P, and the inertia operator of H with respect to P. The resultants of the internal forces $(o, B|H)$ may thus be computed from (3.28) when the mass distribution and the distribution of the external forces of B are given on H, and this also satisfies the Newton-Euler equations for H. Note, however, that for a rigid body only the resultants of the internal forces may be computed, but in general not the force distribution itself. By the above calculations we have proven the following:

Proposition 3.4.1. *A rigid body B satisfying the Newton-Euler equations is in the dynamic equilibrium, iff the internal forces acting on any subsystem H of B fulfill (3.28).*

We may turn this result to our advantage when dealing with the dynamic equilibrium of a rigid body, because the question on the internal forces may be postponed:

Corollary 3.4.1. *As soon as the Newton-Euler equations apply for a rigid body B one may consider it in the dynamic equilibrium, because for any subsystem H of B one may find internal forces such that the Newton-Euler equations are also satisfied for H.*

The next two assertions are straightforward consequences of Proposition 2.4.1 and rigid body kinematics. Although the proofs are more or less the same as in Proposition 2.4.1, we will repeat them because of their central importance for rigid body mechanics.

Proposition 3.4.2. *The virtual work done by the internal forces of a rigid body vanishes, as soon as the virtual displacements $(\delta\mathbf{s}, \delta\boldsymbol{\varphi})$ have been restricted to the rigid body virtual displacements $(\delta\mathbf{r}_P^{(B)}, \delta\boldsymbol{\Phi}^{(B)})$. This holds for any particular choice of $\delta\mathbf{r}_P^{(B)}, \delta\boldsymbol{\Phi}^{(B)}$.*

Proof. We need the definition of the virtual work, the rigid body virtual displacements and the law of interaction. By equation (2.16) the virtual work done by the internal forces is

$$\begin{aligned} -\delta W^{(o,B|H\bar{H})} &= \int_B \delta\mathbf{s}^T \mathrm{d}\mathbf{F}^{(o,B|H\bar{H})} \\ &+ \int_B \delta\boldsymbol{\varphi}^T (\boldsymbol{\varrho} \times \mathrm{d}\mathbf{F}^{(o,B|H\bar{H})} + \mathrm{d}\mathbf{M}^{(o,B|H\bar{H})}). \end{aligned} \tag{3.29}$$

When switching now to the rigid body virtual displacements $(\delta\mathbf{r}_P^{(B)}, \delta\boldsymbol{\Phi}^{(B)})$ we take into account that $\delta\mathbf{s} = \delta\mathbf{r}_P^{(B)}$ and $\delta\boldsymbol{\varphi} = \delta\boldsymbol{\Phi}^{(B)}$ with both virtual displacements being constant on B, see equation (3.16) and below. Thus, (3.29) can be written as

$$\begin{aligned} -\delta W^{(o,B|H\bar{H})} =& \delta\mathbf{r}_P^{(B)\,T} \int_B \mathrm{d}\mathbf{F}^{(o,B|H\bar{H})} \\ &+ \delta\boldsymbol{\Phi}^{(B)\,T} \int_B (\boldsymbol{\varrho} \times \mathrm{d}\mathbf{F}^{(o,B|H\bar{H})} + \mathrm{d}\mathbf{M}^{(o,B|H\bar{H})}). \end{aligned} \tag{3.30}$$

Both integrals in (3.30) are equal to zero due to the law of interaction (2.14), hence $\delta W^{(o,B|H\bar{H})} = 0$ independent of the choice of $\delta\mathbf{r}_P^{(B)}$ and $\delta\boldsymbol{\Phi}^{(B)}$. □

By this result the virtual work (2.16) of a rigid body becomes

$$\begin{aligned} \delta W^{(B)} =& \delta\mathbf{r}_P^{(B)\,T} \int_B (\mathbf{a}\,\mathrm{d}m - \mathrm{d}\mathbf{F}^{(e,B)}) \\ &+ \delta\boldsymbol{\Phi}^{(B)\,T} \int_B (\boldsymbol{\varrho} \times (\mathbf{a}\,\mathrm{d}m - \mathrm{d}\mathbf{F}^{(e,B)}) - \mathrm{d}\mathbf{M}^{(e,B)}) \end{aligned} \tag{3.31}$$

when the virtual displacements are restricted to the rigid body virtual displacements. Both integrals in (3.31) have already been evaluated in equations (3.21) and (3.22), thus (3.31) may be stated as

$$\delta W^{(B)} = \begin{pmatrix} \delta \mathbf{r}_P \\ \delta \mathbf{\Phi} \end{pmatrix}^{(B)\,T} \left[\begin{pmatrix} m\mathbf{E} & m\tilde{\mathbf{r}}_{PS}^T \\ m\tilde{\mathbf{r}}_{PS} & \mathbf{\Theta}_P \end{pmatrix}^{(B)} \begin{pmatrix} \mathbf{a}_P \\ \mathbf{\Psi} \end{pmatrix}^{(B)} + \begin{pmatrix} m\tilde{\mathbf{\Omega}}^2 \mathbf{r}_{PS} \\ \tilde{\mathbf{\Omega}} \mathbf{\Theta}_P \mathbf{\Omega} \end{pmatrix}^{(B)} - \begin{pmatrix} \mathbf{F} \\ \mathbf{M}_P \end{pmatrix}^{(e,B)} \right]. \tag{3.32}$$

Proposition 3.4.3. *If the Newton-Euler equations (3.27) hold for B, then $\delta W^{(B)} = 0$ in (3.32). Conversely, if $\delta W^{(B)} = 0$ for any rigid body virtual displacement $(\delta \mathbf{r}_P^T \, \delta \mathbf{\Phi}^T)^{(B)\,T}$, then the Newton-Euler equations are satisfied for B.*

Proof. Write (3.32) in the form $\delta W = \delta \mathbf{z}^T \mathbf{N}$. For $\mathbf{N} = 0$ we see that $\delta W = 0$. Suppose now that $\delta W = \delta \mathbf{z}^T \mathbf{N} = 0 \; \forall \delta \mathbf{z}$ and that $\mathbf{N} \neq 0$. This is a contradiction, because $\delta \mathbf{z}^T \mathbf{N} = 0$ would be violated for any $\delta \mathbf{z}$ not orthogonal to $\mathbf{N}$, hence $\mathbf{N} = 0$. □

We have seen by Proposition 3.4.2 that the internal forces of a rigid body do not contribute to the virtual work, as soon as the virtual displacements have been restricted to the rigid body virtual displacements. The internal forces of a rigid body are thus passive, but they may consist of more than just the constraint forces that hold together the particles of the body as indicated in Figure 3.3: For instance, a spring attached to a rigid body causes two forces acting on B for which the law of interaction holds. These forces belong to the internal forces of B and do therefore not contribute to the virtual work in (3.32), although forces resulting from springs are certainly not constraint forces. It is the overall constrained configuration that determines a force to be passive or not. A definition of passive forces and constraint forces of perfect constraints with respect to multibody systems is presented in the next sections.

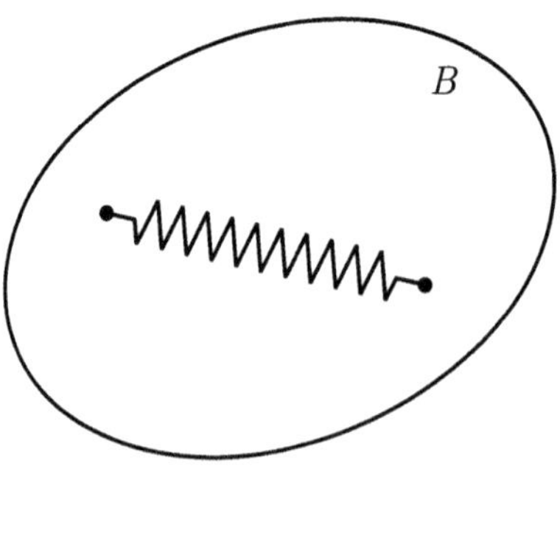

Figure 3.3. Internal forces of a rigid body

The rigid body virtual displacements are physically compatible with the virtual motion of an unconstrained rigid body, but not yet necessarily compatible with the virtual motion of a rigid body within a constrained multibody system. In the next section we will further restrict the rigid body virtual displacements which, however, does not affect Proposition 3.4.2 since the virtual work of the internal forces vanishes independent of the rigid body virtual displacements chosen. Note also that constraint forces acting between different rigid bodies are already contained in the Newton-Euler equations as members of the external forces of the body under consideration.

3.5 The Virtual Work of a Rigid Body System

We consider now a mechanical system S consisting of n rigid bodies B_i. By equations (2.16) and (3.32) the virtual work $\delta W^{(S)}$ of the *multibody system* S in terms of the n rigid body virtual displacements becomes

$$\delta W^{(S)} = \sum_{i=1}^{n} \left\{ \begin{pmatrix} \delta\mathbf{r}_P \\ \delta\boldsymbol{\Phi} \end{pmatrix}^T \left[\begin{pmatrix} m\mathbf{E} & m\tilde{\mathbf{r}}_{PS}^T \\ m\tilde{\mathbf{r}}_{PS} & \boldsymbol{\Theta}_P \end{pmatrix} \begin{pmatrix} \mathbf{a}_P \\ \boldsymbol{\Psi} \end{pmatrix} + \begin{pmatrix} m\tilde{\boldsymbol{\Omega}}^2\mathbf{r}_{PS} \\ \tilde{\boldsymbol{\Omega}}\boldsymbol{\Theta}_P\boldsymbol{\Omega} \end{pmatrix} - \begin{pmatrix} \mathbf{F} \\ \mathbf{M}_P \end{pmatrix}^{(e)} \right] \right\}^{(B_i)} . \tag{3.33}$$

Proposition 3.5.1. *If $\delta W^{(S)} = 0$ for every $(\delta\mathbf{r}_P^T\, \delta\boldsymbol{\Phi}^T)^{(B_i)T}$ $(i = 1, \dots, n)$, then the Newton-Euler equations are satisfied for each of the n rigid bodies B_i, and the system may be regarded as to be in the dynamic equilibrium.*

Proof. A consequence of Proposition 3.4.3 together with corollaries 3.4.1 and 2.3.1 (iii). □

We will now give a brief instruction on how to eliminate perfect bilateral holonomic constraints from the system and how to build up the equations of motion from the virtual work expression (3.33). A more detailed study of this process will be presented in the next two sections.

Let the degrees of freedom of the system S be restricted from $6n$ to $f := 6n - m$ due to m bilateral perfect holonomic rheonomic constraints. Let $\mathbf{q}$ be a set of generalized coordinates of S with minimal dimension, i.e. a set of *minimal coordinates* ($\mathbf{q} \in \mathbb{R}^f$) of S under the m constraints considered. By these coordinates, the positions and orientations of the n bodies B_i are uniquely determined with values $\mathbf{r}_P(\mathbf{q}, t)$ and $\mathbf{D}(\mathbf{q}, t)$, where explicit dependence on time is due to the rheonomic constraints and does not occur for scleronomic systems. From rigid body kinematics in Section 3.2 it is clear that the virtual displacements, the velocities and the accelerations take the form

$$\begin{aligned}
\begin{pmatrix} \delta\mathbf{r}_P \\ \delta\boldsymbol{\Phi} \end{pmatrix}^{(B_i)} &= \begin{pmatrix} \mathbf{J}_P \\ \mathbf{J}_R \end{pmatrix}^{(B_i)} \delta\mathbf{q} \\
\begin{pmatrix} \mathbf{v}_P \\ \boldsymbol{\Omega} \end{pmatrix}^{(B_i)} &= \begin{pmatrix} \mathbf{J}_P \\ \mathbf{J}_R \end{pmatrix}^{(B_i)} \dot{\mathbf{q}} + \begin{pmatrix} \tilde{\iota}_P \\ \tilde{\iota}_R \end{pmatrix}^{(B_i)} \\
\begin{pmatrix} \mathbf{a}_P \\ \boldsymbol{\Psi} \end{pmatrix}^{(B_i)} &= \begin{pmatrix} \mathbf{J}_P \\ \mathbf{J}_R \end{pmatrix}^{(B_i)} \ddot{\mathbf{q}} + \begin{pmatrix} \bar{\iota}_P \\ \bar{\iota}_R \end{pmatrix}^{(B_i)}
\end{aligned} \tag{3.34}$$

when $\mathbf{q}$ is considered as a function of time. For each body B_i, the operators $\mathbf{J}_P(\mathbf{q},t) \in \mathrm{I\!R}^{3,f}$ and $\mathbf{J}_R(\mathbf{q},t) \in \mathrm{I\!R}^{3,f}$ are called the *Jacobian of translation* at point P and the *Jacobian of rotation.* Further, $\tilde{\iota}_P(\mathbf{q},t)$ and $\tilde{\iota}_R(\mathbf{q},t)$ vanish for scleronomic systems, and $\bar{\iota}_P(\mathbf{q},\dot{\mathbf{q}},t)$ and $\bar{\iota}_R(\mathbf{q},\dot{\mathbf{q}},t)$ are determined by $\bar{\iota}_P = \dot{\mathbf{J}}_P\,\dot{\mathbf{q}} + \dot{\tilde{\iota}}_P$ and $\bar{\iota}_R = \dot{\mathbf{J}}_R\,\dot{\mathbf{q}} + \dot{\tilde{\iota}}_R$.

With the help of the first equation in (3.34) the rigid body virtual displacements in (3.33) may further be restricted to virtual displacements that are in addition compatible with the m holonomic rheonomic constraints. The set of all such physically admissible virtual displacements is obtained by varying $\delta\mathbf{q}$ over all of $\mathrm{I\!R}^f$, and the virtual work (3.33) becomes

$$\begin{aligned}\delta W_a^{(S)} = \delta\mathbf{q}^T \sum_{i=1}^{n} \Bigg\{ \begin{pmatrix}\mathbf{J}_P\\ \mathbf{J}_R\end{pmatrix}^T \Bigg[\begin{pmatrix} m\mathbf{E} & m\tilde{\mathbf{r}}_{PS}^T \\ m\tilde{\mathbf{r}}_{PS} & \mathbf{\Theta}_P \end{pmatrix} \begin{pmatrix}\mathbf{a}_P\\ \mathbf{\Psi}\end{pmatrix} \\ + \begin{pmatrix} m\tilde{\mathbf{\Omega}}^2\mathbf{r}_{PS} \\ \tilde{\mathbf{\Omega}}\mathbf{\Theta}_P\mathbf{\Omega}\end{pmatrix} - \begin{pmatrix}\mathbf{F}\\ \mathbf{M}_P\end{pmatrix}^{(e)} \Bigg] \Bigg\}^{(B_i)} .\end{aligned} \tag{3.35}$$

Let us now split for each body B_i the external forces (e) into the active forces (a) and the passive forces (p),

$$\begin{pmatrix}\mathbf{F}\\ \mathbf{M}_P\end{pmatrix}^{(e,B_i)} = \begin{pmatrix}\mathbf{F}\\ \mathbf{M}_P\end{pmatrix}^{(a,B_i)} + \begin{pmatrix}\mathbf{F}\\ \mathbf{M}_P\end{pmatrix}^{(p,B_i)} \qquad i = 1,\dots,n, \tag{3.36}$$

where the passive forces are understood as not to generate any virtual work

$$\forall \delta\mathbf{q} :\; 0 = \delta W_a^{(S,p)} = -\delta\mathbf{q}^T \sum_{i=1}^{n} \Bigg\{ \begin{pmatrix}\mathbf{J}_P\\ \mathbf{J}_R\end{pmatrix}^T \begin{pmatrix}\mathbf{F}\\ \mathbf{M}_P\end{pmatrix}^{(p)} \Bigg\}^{(B_i)} \tag{3.37}$$

or, as a consequence on the arbitrary choice of $\delta\mathbf{q}$, as the forces for which

$$0 = \sum_{i=1}^{n} \Bigg\{ \begin{pmatrix}\mathbf{J}_P\\ \mathbf{J}_R\end{pmatrix}^T \begin{pmatrix}\mathbf{F}\\ \mathbf{M}_P\end{pmatrix}^{(p)} \Bigg\}^{(B_i)} . \tag{3.38}$$

In order to satisfy the Newton-Euler equations of the multibody system on the unconstrained directions $\delta\mathbf{q}$ it is thus sufficient to take into account in (3.35) the active forces only, because the passive forces drop out by (3.37) when $\delta W_a^{(S)} = 0\ \forall \delta\mathbf{q}$ is demanded. By performing this step we therefore obtain

$$\begin{aligned} 0 = &\sum_{i=1}^{n} \Bigg\{ \begin{pmatrix}\mathbf{J}_P\\ \mathbf{J}_R\end{pmatrix}^T \begin{pmatrix} m\mathbf{E} & m\tilde{\mathbf{r}}_{PS}^T \\ m\tilde{\mathbf{r}}_{PS} & \mathbf{\Theta}_P \end{pmatrix} \begin{pmatrix}\mathbf{J}_P\\ \mathbf{J}_R\end{pmatrix} \Bigg\}^{(B_i)} \ddot{\mathbf{q}} \\ &+ \sum_{i=1}^{n} \Bigg\{ \begin{pmatrix}\mathbf{J}_P\\ \mathbf{J}_R\end{pmatrix}^T \Bigg[\begin{pmatrix} m\mathbf{E} & m\tilde{\mathbf{r}}_{PS}^T \\ m\tilde{\mathbf{r}}_{PS} & \mathbf{\Theta}_P \end{pmatrix} \begin{pmatrix}\bar{\iota}_P\\ \bar{\iota}_R\end{pmatrix} + \begin{pmatrix} m\tilde{\mathbf{\Omega}}^2\mathbf{r}_{PS} \\ \tilde{\mathbf{\Omega}}\mathbf{\Theta}_P\mathbf{\Omega}\end{pmatrix} \Bigg] \Bigg\}^{(B_i)} \\ &- \sum_{i=1}^{n} \Bigg\{ \begin{pmatrix}\mathbf{J}_P\\ \mathbf{J}_R\end{pmatrix}^T \begin{pmatrix}\mathbf{F}\\ \mathbf{M}_P\end{pmatrix}^{(a)} \Bigg\}^{(B_i)} , \end{aligned} \tag{3.39}$$

where the accelerations have been taken from the third line of (3.34). We will refer to (3.39) as the *projected Newton-Euler equations*, a term which becomes clear when re-substituting (3.23), i.e.

$$0 = \sum_{i=1}^{n} \left\{ \begin{pmatrix} \mathbf{J}_P \\ \mathbf{J}_R \end{pmatrix}^T \left[\begin{pmatrix} \dot{\mathbf{p}} \\ \dot{\mathbf{L}}_P + m\mathbf{r}_{PS} \times \mathbf{a}_P \end{pmatrix} - \begin{pmatrix} \mathbf{F} \\ \mathbf{M}_P \end{pmatrix}^{(a)} \right] \right\}^{(B_i)} . \tag{3.40}$$

Equation (3.39) will be written in the form

$$\mathbf{M}(\mathbf{q}, t)\, \ddot{\mathbf{q}} - \mathbf{h}(\mathbf{q}, \dot{\mathbf{q}}, t) - \mathbf{f}^{(a)} = 0, \tag{3.41}$$

where $\mathbf{M} \in \mathbb{R}^{f,f}$ denotes the symmetric and at least positive semidefinite mass matrix, $\mathbf{h} \in \mathbb{R}^f$ contains the gyroscopical accelerations, and $\mathbf{f}^{(a)} \in \mathbb{R}^f$ is the active portion of the external forces of the n rigid bodies. Further, we assume from now that the mass matrix is indeed positive definite; otherwise one could condense equation (3.41) to a dynamic subsystem of lower dimension such that this assumption holds, or to switch to another set of minimal coordinates if the parameterization becomes singular.

3.6 Classical Bilateral Constraints

In this section we consider, again, the system of n rigid bodies B_i stated in the form (3.33). At this stage the reduction process to the generalized coordinates $\mathbf{q}$ as sketched in the previous section has not yet been performed. The aim of this and the following section is to give a more detailed analysis of this process in order to end up with the same equations as already presented in (3.41). In this section we introduce classical holonomic perfect constraints, we define constraint forces and we state the classical principle of d'Alembert/Lagrange.

Let us start again with the virtual work (3.33) of the rigid body system. For each body B_i we choose a proper parameterization $\mathbf{z}_i \in \mathbb{R}^6$ of the linear and the angular displacements $\mathbf{r}_P^{(B_i)}(\mathbf{z}_i)$ and $\mathbf{D}^{(B_i)}(\mathbf{z}_i)$ in order to express the rigid body kinematics terms as

$$\begin{pmatrix} \delta\mathbf{r}_P \\ \delta\boldsymbol{\Phi} \end{pmatrix}^{(B_i)} = \mathbf{H}_i\, \delta\mathbf{z}_i, \quad \begin{pmatrix} \mathbf{v}_P \\ \boldsymbol{\Omega} \end{pmatrix}^{(B_i)} = \mathbf{H}_i\, \dot{\mathbf{z}}_i, \quad \begin{pmatrix} \mathbf{a}_P \\ \boldsymbol{\Psi} \end{pmatrix}^{(B_i)} = \mathbf{H}_i\, \ddot{\mathbf{z}}_i + \boldsymbol{\zeta}_i \tag{3.42}$$

with invertible $\mathbf{H}_i(\mathbf{z}_i)$ and $\boldsymbol{\zeta}_i(\mathbf{z}_i, \dot{\mathbf{z}}_i) = \dot{\mathbf{H}}_i(\mathbf{z}_i, \dot{\mathbf{z}}_i)\, \dot{\mathbf{z}}_i$. Further, we use the abbreviations

$$\begin{aligned} \mathbf{M}_i &:= \mathbf{H}_i^T \begin{pmatrix} m\mathbf{E} & m\tilde{\mathbf{r}}_{PS}^T \\ m\tilde{\mathbf{r}}_{PS} & \boldsymbol{\Theta}_P \end{pmatrix}^{(B_i)} \mathbf{H}_i, \quad \mathbf{f}_i^{(e)} := \mathbf{H}_i^T \begin{pmatrix} \mathbf{F} \\ \mathbf{M}_P \end{pmatrix}^{(e,B_i)} \\ \mathbf{h}_i &:= -\mathbf{H}_i^T \left[\begin{pmatrix} m\mathbf{E} & m\tilde{\mathbf{r}}_{PS}^T \\ m\tilde{\mathbf{r}}_{PS} & \boldsymbol{\Theta}_P \end{pmatrix}^{(B_i)} \boldsymbol{\zeta}_i + \begin{pmatrix} m\tilde{\boldsymbol{\Omega}}^2\mathbf{r}_{PS} \\ \tilde{\boldsymbol{\Omega}}\boldsymbol{\Theta}_P\boldsymbol{\Omega} \end{pmatrix}^{(B_i)} \right] \end{aligned} \tag{3.43}$$

and rewrite the virtual work (3.33) with the help of (3.42), (3.43) as

$$\delta W^{(S)} = \sum_{i=1}^{n} \delta \mathbf{z}_i^T (\mathbf{M}_i \ddot{\mathbf{z}}_i - \mathbf{h}_i - \mathbf{f}_i^{(e)}). \tag{3.44}$$

Let us finally introduce the terms

$$\bar{\mathbf{M}} := \begin{pmatrix} \mathbf{M}_1 & & \\ & \ddots & \\ & & \mathbf{M}_n \end{pmatrix} \quad \bar{\mathbf{h}} := \begin{pmatrix} \mathbf{h}_1 \\ \vdots \\ \mathbf{h}_n \end{pmatrix} \quad \mathbf{z} := \begin{pmatrix} \mathbf{z}_1 \\ \vdots \\ \mathbf{z}_n \end{pmatrix} \quad \bar{\mathbf{f}}^{(e)} := \begin{pmatrix} \mathbf{f}_1^{(e)} \\ \vdots \\ \mathbf{f}_n^{(e)} \end{pmatrix} \tag{3.45}$$

to state the virtual work (3.33) as

$$\delta W^{(S)} = \delta \mathbf{z}^T (\bar{\mathbf{M}}\, \ddot{\mathbf{z}} - \bar{\mathbf{h}} - \bar{\mathbf{f}}^{(e)}). \tag{3.46}$$

The components of $\mathbf{z} \in \mathbb{R}^{6n}$ are called the *system coordinates*, and $\bar{\mathbf{M}}(\mathbf{z})$ is the associated mass matrix. By taking $\delta W^{(S)} = 0\ \forall \delta \mathbf{z}$ in (3.46) we obtain

$$\bar{\mathbf{M}}\, \ddot{\mathbf{z}} - \bar{\mathbf{h}} - \bar{\mathbf{f}}^{(e)} = 0 \tag{3.47}$$

which are, as in Proposition 3.5.1, the Newton-Euler equations for the ensemble of n rigid bodies B_i.

Consider now m transversally intersecting constraints $g_i(\mathbf{z}, t) = 0$ which we write in vector notation as $\mathbf{g}(\mathbf{z}, t) = 0$ with $\mathbf{g} \in \mathbb{R}^m$. For every *fixed* time t the constraints restrict the motion of the system to a set

$$C := \{\mathbf{z} \,|\, \mathbf{g}_t(\mathbf{z}) = 0\} \tag{3.48}$$

which defines an f-dimensional submanifold of $\mathbb{R}^{6n}$ with $f = 6n - m$. For $\mathbf{z} \in C$ and under the condition that none of the m gradients $\frac{\partial g_{ti}}{\partial \mathbf{z}}$ vanishes, the tangent space $T_C(\mathbf{z})$ to C at $\mathbf{z}$, an f-dimensional subspace of $\mathbb{R}^{6n}$, is then represented by

$$T_C(\mathbf{z}) := \{\boldsymbol{\xi} \,|\, \mathbf{W}_t^T \boldsymbol{\xi} = 0\} \qquad \text{with } \mathbf{W}_t^T(\mathbf{z}) := \frac{\partial \mathbf{g}_t}{\partial \mathbf{z}}. \tag{3.49}$$

Further, we denote by $T_C^{\perp}(\mathbf{z})$ the m-dimensional subspace of $\mathbb{R}^{6n}$ which is the orthogonal complement of $T_C(\mathbf{z})$ with respect to the inner product $(\boldsymbol{\xi}, \boldsymbol{\eta})_{\bar{\mathbf{M}}} = \boldsymbol{\xi}^T\, \bar{\mathbf{M}}(\mathbf{z})\, \boldsymbol{\eta}$,

$$T_C^{\perp}(\mathbf{z}) := \{\boldsymbol{\eta} \,|\, \boldsymbol{\xi}^T \bar{\mathbf{M}} \boldsymbol{\eta} = 0,\ \forall \boldsymbol{\xi} \in T_C(\mathbf{z})\}, \tag{3.50}$$

and we introduce the duals of $T_C^{\perp}(\mathbf{z})$ and $T_C(\mathbf{z})$,

$$N_C(\mathbf{z}) := \bar{\mathbf{M}}\, T_C^{\perp}(\mathbf{z}), \qquad N_C^{\perp}(\mathbf{z}) := \bar{\mathbf{M}}\, T_C(\mathbf{z}), \tag{3.51}$$

two linear spaces with dimensions m and f, respectively. Note by (3.50) and (3.51) the orthogonality of the elements of $N_C(\mathbf{z})$ and $T_C(\mathbf{z})$,

$$\boldsymbol{\zeta}^T\boldsymbol{\xi} = 0, \quad \forall\boldsymbol{\zeta} \in N_C(\mathbf{z}), \quad \forall\boldsymbol{\xi} \in T_C(\mathbf{z}), \tag{3.52}$$

which immediately shows that the m linearly independent rows of $\mathbf{W}_t^T$ in (3.49) form a basis of $N_C(\mathbf{z})$. With the four linear spaces (3.49)–(3.51) we are now able to define constraint forces, passive forces and different types of virtual displacements.

Definition 3.6.1. *A bilateral holonomic rheonomic m-dimensional mechanical constraint is said to be* perfect, *if the forces $\mathbf{f}^{(c)}$ associated with the kinematical restrictions (3.48) satisfy the set-valued force law*

$$-\bar{\mathbf{f}}^{(c)} \in N_C(\mathbf{z}). \tag{3.53}$$

Classically, these forces $\bar{\mathbf{f}}^{(c)}$ are called the constraint forces (Zwangskräfte).

Remark: In German we distinguish between two different sorts of constraint forces: Generally, forces which occur due to constraints are called "Bindungskräfte." If these forces obey, in addition, the force law (3.53), they are called "Zwangskräfte." Permanently sliding bilateral constraints with Coulomb friction, for example, do not belong to perfect constraints, thus the associated forces are not called "Zwangskräfte."

A more general concept for perfect constraints is presented in Chapters 5–9. Perfect constraints are then regarded as kinematical restrictions to some set C together with the associated forces taken from the generalized gradient of a very special potential function, the indicator of C. Constraint forces are then potential forces and may thus be treated by potential theory.

Bilateral holonomic constraints are restrictions of the displacements to submanifolds. Constraint forces are physical reactions of the system against the constraints. There is a functional relation between constraints and constraint forces. This relation may or may not be of the form (3.53). In the first case we speak about perfect constraints, in the second case about constraints that are not perfect and, as a consequence, obey force laws different from (3.53), for example joints with Coulomb friction. Thus, there is nothing axiomatic behind the force law (3.53). It is just a definition of a certain class of interactions that are used when a multibody system is modeled: Based on physical observations one has to decide whether interconnections may or may not be approximated sufficiently precise by perfect constraints. In the latter case one has to choose other models. Further, it is always incomplete to regard constraints just as geometrical restrictions. Constraints, at least from the mechanical point of view, must always be understood as geometric restrictions *together* with additional forces in the Newton-Euler equations *and* a force law coupling together either of them.

With the constraint forces $\bar{\mathbf{f}}^{(c)}$ as part of the external forces $\bar{\mathbf{f}}^{(e)}$ of the bodies in the multibody system we may now write

$$\bar{\mathbf{f}}^{(e)} = \bar{\mathbf{f}}^{(c)} + \bar{\mathbf{f}}^{(r)}, \tag{3.54}$$

where $\bar{\mathbf{f}}^{(r)}$ contains all the remaining forces. In particular, these may be constraint forces of additional constraints which have not yet been considered in (3.48), (3.53), impressed forces which are classically understood as functions $(\mathbf{z}, \dot{\mathbf{z}}) \to \bar{\mathbf{f}}^{(i)}$ of the system's displacements and velocities, or forces of mixed type such as unilateral constraints or dry friction as to be discussed in the following chapters. We will see soon that forces contained in $N_C(\mathbf{z})$ drop out from the equations which determine the values of the accelerations $\ddot{\mathbf{q}}$ when C in (3.48) is parameterized by a minimal set of coordinates $\mathbf{q}$. Thus, not every member contained in $\bar{\mathbf{f}}^{(r)}$, even if it may be an impressed force, is really needed for this job, and we may therefore classify forces by the following concept:

By the four subspaces defined in (3.49)–(3.51) there exists a unique (orthogonal) decomposition of the external forces $\bar{\mathbf{f}}^{(e)}$ such that

$$\bar{\mathbf{f}}^{(e)} = \bar{\mathbf{f}}^{(a)} + \bar{\mathbf{f}}^{(p)}; \qquad -\bar{\mathbf{f}}^{(a)} \in N_C^{\perp}(\mathbf{z}), \quad -\bar{\mathbf{f}}^{(p)} \in N_C(\mathbf{z}) \tag{3.55}$$

and, in the same fashion, a unique (orthogonal) decomposition of any rigid body virtual displacement $\delta\mathbf{z}$ into

$$\delta\mathbf{z} = \delta\mathbf{z}_a + \delta\mathbf{z}_p; \qquad \delta\mathbf{z}_a \in T_C(\mathbf{z}), \quad \delta\mathbf{z}_p \in T_C^{\perp}(\mathbf{z}). \tag{3.56}$$

With respect to the constraints (3.48) we call $\bar{\mathbf{f}}^{(a)}$ the *active forces* and $\bar{\mathbf{f}}^{(p)}$ the *passive forces* of the system. Further, we note that constraint forces are always passive by (3.53), and that the kinematic restrictions (3.48) determine whether a force is passive or not. The concept of active and passive forces is thus not a classification related to the physical properties of force elements, but a classification with respect to the underlying geometry. In (3.56), we call $\delta\mathbf{z}_a$ the rigid body virtual displacements *compatible* with the constraints (3.48), and $\delta\mathbf{z}_p$ *incompatible*. To see that $\delta\mathbf{z}_a$ are indeed compatible or physically admissible, perform a Taylor series expansion of $\mathbf{g}_t(\mathbf{z}^\star)$ with $\mathbf{z}^\star \in C$ near $\mathbf{z} \in C$ and $\delta\mathbf{z}_a = \mathbf{z}^\star - \mathbf{z}$. This yields $\mathbf{g}_t(\mathbf{z}^\star) \approx \mathbf{g}_t(\mathbf{z}) + \mathbf{W}_t^T\,\delta\mathbf{z}_a$ and thus $\mathbf{W}_t^T\,\delta\mathbf{z}_a = 0$, because $\mathbf{g}_t(\mathbf{z}^\star) = \mathbf{g}_t(\mathbf{z}) = 0$ due to $\mathbf{z}, \mathbf{z}^\star \in C$. Hence, we have by (3.49) that $\delta\mathbf{z}_a \in T_C(\mathbf{z})$.

Proposition 3.6.1. *The virtual work done by the passive forces $\bar{\mathbf{f}}^{(p)}$ vanishes, as soon as the rigid body virtual displacements $\delta\mathbf{z}$ have been restricted to be compatible with the constraints, i.e.*

$$\delta W_a^{(S,p)} = -\delta\mathbf{z}_a^T\,\bar{\mathbf{f}}^{(p)} \equiv 0 \quad \forall \delta\mathbf{z}_a \in T_C(\mathbf{z}). \tag{3.57}$$

The principle of d'Alembert/Lagrange: *Constraint forces of perfect constraints do not produce any virtual work for compatible virtual displacements,*

$$\delta W_a^{(S,c)} = -\delta\mathbf{z}_a^T\,\bar{\mathbf{f}}^{(c)} \equiv 0 \quad \forall \delta\mathbf{z}_a \in T_C(\mathbf{z}). \tag{3.58}$$

Proof. Both, Proposition 3.6.1 and the principle of d'Alembert/Lagrange are immediate consequences on the orthogonality (3.52), because $-\bar{\mathbf{f}}^{(p)}, -\bar{\mathbf{f}}^{(c)} \in N_C(\mathbf{z})$ by (3.55), (3.53) and $\delta\mathbf{z}_a \in T_C(\mathbf{z})$ by (3.56). Note also that constraint forces of perfect constraints are always passive, but that the converse is *not* true in most situations. □

In classical mechanics equation (3.58) is used as the *definition* of the constraint forces of perfect constraints. This is, of course, equivalent to our definition in (3.53),

$$-\bar{\mathbf{f}}^{(c)} \in N_C(\mathbf{z}) \quad \Leftrightarrow \quad -\delta\mathbf{z}_a^T\,\bar{\mathbf{f}}^{(c)} = 0 \quad \forall \delta\mathbf{z}_a \in T_C(\mathbf{z}), \tag{3.59}$$

because $\delta\mathbf{z}$ span $T_C(\mathbf{z})$. In order to state now the constrained problem we need the Newton-Euler equations (3.47) together with the decomposition of the forces (3.54), the kinematical restrictions (3.48), and one of the two equivalent representations of the constraint forces in (3.59). By taking the first condition in (3.59) we obtain

$$\bar{\mathbf{M}}\,\ddot{\mathbf{z}} - \bar{\mathbf{h}} - \bar{\mathbf{f}}^{(r)} - \bar{\mathbf{f}}^{(c)} = 0, \quad \mathbf{z} \in C, \quad -\bar{\mathbf{f}}^{(c)} \in N_C(\mathbf{z}) \tag{3.60}$$

which yields after elimination of $\bar{\mathbf{f}}^{(c)}$ the differential inclusion

$$-(\bar{\mathbf{M}}\,\ddot{\mathbf{z}} - \bar{\mathbf{h}} - \bar{\mathbf{f}}^{(r)}) \in N_C(\mathbf{z}), \quad \mathbf{z} \in C. \tag{3.61}$$

Alternatively, by taking the second condition in (3.59) we get

$$\bar{\mathbf{M}}\,\ddot{\mathbf{z}} - \bar{\mathbf{h}} - \bar{\mathbf{f}}^{(r)} - \bar{\mathbf{f}}^{(c)} = 0, \quad \mathbf{z} \in C, \quad -\delta\mathbf{z}_a^T\,\bar{\mathbf{f}}^{(c)} = 0 \quad \forall \delta\mathbf{z}_a \in T_C(\mathbf{z}), \tag{3.62}$$

i.e. the classical variational equality

$$-\delta\mathbf{z}_a^T(\bar{\mathbf{M}}\,\ddot{\mathbf{z}} - \bar{\mathbf{h}} - \bar{\mathbf{f}}^{(r)}) = 0, \quad \mathbf{z} \in C, \quad \forall \delta\mathbf{z}_a \in T_C(\mathbf{z}). \tag{3.63}$$

Further, we may replace in (3.60)–(3.63) the forces $\bar{\mathbf{f}}^{(r)}$ by the active forces $\bar{\mathbf{f}}^{(a)}$ due to the *property* of the passive forces to lie in $N_C(\mathbf{z})$.

Note that (3.61) or (3.63) are *not* equivalent to the Newton-Euler equations (3.47): To obtain (3.47) from the virtual work $\delta W^{(S)}$ it was necessary in (3.46) to take $\delta W^{(S)} = 0$ for *all* $\delta\mathbf{z} \in \mathbb{R}^{6n}$. In (3.63), however, the virtual displacements $\delta\mathbf{z}_a$ are restricted to the f-dimensional subspace $T_C(\mathbf{z})$, thus the Newton-Euler equations will only hold on the corresponding f-dimensional space, that is $N_C^{\perp}(\mathbf{z})$. The remaining m dimensions necessary for the rigid body system to be in the dynamic equilibrium may then be taken into account in a second step by determining the constraint forces $-\bar{\mathbf{f}}^{(c)}$ on the m-dimensional $N_C(\mathbf{z})$. The *full* set of equations is obtained from (3.46) with the decompositions of $\delta\mathbf{z}$ and $\bar{\mathbf{f}}^{(e)}$ according to (3.56) and (3.55):

Proposition 3.6.2. *The Newton-Euler equations apply for each of the n bodies B_i of the constrained multibody system, iff*

$$\begin{aligned} \delta W_a^{(S)} &:= \delta \mathbf{z}_a^T \, (\bar{\mathbf{M}}\, \ddot{\mathbf{z}} - \bar{\mathbf{h}} - \bar{\mathbf{f}}^{(a)} - \bar{\mathbf{f}}^{(p)}) = 0 \quad \forall \delta \mathbf{z}_a \in T_C(\mathbf{z}), \\ \delta W_p^{(S)} &:= \delta \mathbf{z}_p^T \, (\bar{\mathbf{M}}\, \ddot{\mathbf{z}} - \bar{\mathbf{h}} - \bar{\mathbf{f}}^{(a)} - \bar{\mathbf{f}}^{(p)}) = 0 \quad \forall \delta \mathbf{z}_p \in T_C^{\perp}(\mathbf{z}). \end{aligned} \tag{3.64}$$

Proof. From the Newton-Euler equations (3.47) we see that (3.64) holds. To show the converse we note that $\delta \mathbf{z}_a$ and $\delta \mathbf{z}_p$ span T_C and $T_C^{\perp}$. By (3.64) the Newton-Euler equations hold on $N_C^{\perp}$ and N_C, hence on $\mathbb{R}^{6n} = N_C^{\perp} \oplus N_C$ which is (3.47). □

In the next section it will turn out that, indeed, only the first equation in (3.64) is needed to compute the generalized accelerations $\ddot{\mathbf{q}}$ for a parameterization of C in (3.48) via a minimal set of coordinates $\mathbf{q} \in \mathbb{R}^f$. For this, only the active forces are needed, because the passive drop out by $\delta \mathbf{z}_a^T \, \bar{\mathbf{f}}^{(p)} = 0$. The second equation in (3.64), independent of $\ddot{\mathbf{q}}$ determines the passive forces $\bar{\mathbf{f}}^{(p)}$ and via (3.54), (3.55) the constraint forces as solutions of $\bar{\mathbf{f}}^{(c)} = \bar{\mathbf{f}}^{(p)} + (\bar{\mathbf{f}}^{(a)} - \bar{\mathbf{f}}^{(r)})$ with $(\bar{\mathbf{f}}^{(a)} - \bar{\mathbf{f}}^{(r)}) \in N_C(\mathbf{z})$.

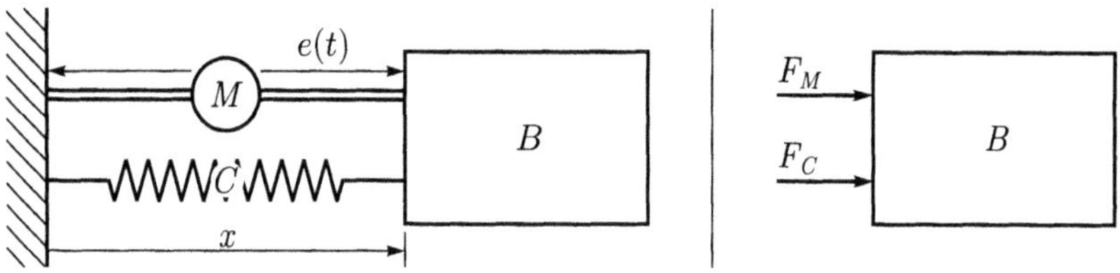

Figure 3.4. Constrained multibody system

Let us finally discuss the concept of active and passive forces by an example. The one-dimensional system depicted in Figure 3.4 consists of one rigid body B with the linear displacement x playing the role of the system coordinates. The body is connected by a spring C to the environment and is inertially driven by a powerful motor which provides, without any reactions, the kinematic motion $e(t)$. The displacement of the system is restricted by $m = 1$ holonomic rheonomic constraint $g(x,t) = x - e(t) = 0$. Thus, the system has $f = 0$ degrees of freedom, hence $\dim T_C = \dim N_C^{\perp} = 0$ and $\dim T_C^{\perp} = \dim N_C = 1$. In particular, the set of compatible virtual displacements consists of $\{0\}$ only, thus $\delta x_a \equiv 0$. The external forces of the body B are F_M and F_C, the forces acting from the motor and from the spring on B. F_M is the constraint force, F_C is an impressed force, both of them constitute the passive force which does not produce any virtual work, $(F_M + F_C)\, \delta x_a \equiv 0$ due to $\delta x_a \equiv 0$. The value of the passive force is $(F_M + F_C) = m\ddot{e}$ where m denotes the mass of the body, the impressed force becomes $F_C = -ce$ with c the stiffness of the spring, and the constraint force finally is $F_M = m\ddot{e} + ce$. Note that $F_M(t) \equiv 0$, if the motor is driven such that $e(t) = A \sin \omega t$ with $\omega = \sqrt{c/m}$. Although the virtual work of the passive force is equal to zero, the actual work clearly is not. Virtual work and actual work are two completely different concepts; the virtual work δW, for example, is always based on linear spaces, whereas for increments dW of the actual work the spaces may be affine.

3.7 Generalized Coordinates

The aim of this section is to state the Newton-Euler equations in terms of a set of generalized coordinates $\mathbf{q} \in \mathbb{R}^f$ which is minimal with respect to the system constrained by (3.48). Let therefore $\mathbf{q} \in \mathbb{R}^f$ be a set of coordinates such that any position $\mathbf{z}$ compatible with the constraints (3.48) is uniquely determined by the values of $\mathbf{q}$. By such coordinates the constraints are implicitly fulfilled, i.e. for $\mathbf{z}(\mathbf{q}, t)$ one has

$$\mathbf{g}(\,\mathbf{z}(\mathbf{q}, t),\, t) \equiv 0 \quad \forall \mathbf{q}, \forall t. \tag{3.65}$$

Further, at least locally we may assume the parameterization such that the Jacobian of $\mathbf{z}$ with respect to $\mathbf{q}$ has full rank f. For t being fixed we write (3.65), similar to (3.48), as

$$\mathbf{g}_t(\mathbf{z}_t(\mathbf{q})) \equiv 0 \quad \forall \mathbf{q}. \tag{3.66}$$

This equation holds for every $\mathbf{q}$. In particular, as soon as one $\mathbf{q}$ has been specified, it holds for every $\mathbf{q}^\star$ near $\mathbf{q}$. Expanding thus (3.66) in a Taylor series

$$\mathbf{g}_t(\mathbf{z}_t(\mathbf{q}^\star)) \approx \mathbf{g}_t(\mathbf{z}_t(\mathbf{q})) + \frac{\partial \mathbf{g}_t}{\partial \mathbf{z}_t}\frac{\partial \mathbf{z}_t}{\partial \mathbf{q}}\,(\mathbf{q}^\star - \mathbf{q}) \tag{3.67}$$

and taking into account that $\mathbf{g}_t(\mathbf{z}_t(\mathbf{q})) = \mathbf{g}_t(\mathbf{z}_t(\mathbf{q}^\star)) = 0$ by (3.66), we obtain from (3.67)

$$0 = \mathbf{W}^T\,\mathbf{Q}; \qquad \mathbf{W}^T(\mathbf{q}, t) = \frac{\partial \mathbf{g}}{\partial \mathbf{z}}, \quad \mathbf{Q}(\mathbf{q}, t) := \frac{\partial \mathbf{z}}{\partial \mathbf{q}} \tag{3.68}$$

with $\mathbf{z} = \mathbf{z}(\mathbf{q}, t)$, because (3.67) holds for every direction $\delta\mathbf{q} = \mathbf{q}^\star - \mathbf{q}$ in $\mathbb{R}^f$. We recall that the m columns of the $6n \times m$-matrix $\mathbf{W}$ are a basis of $N_C(\mathbf{z})$, see equations (3.49)–(3.52). Thus, it follows from the orthogonality (3.52) that the f columns of the $6n \times f$-matrix $\mathbf{Q}$ are a basis of $T_C(\mathbf{z})$. Bases for $N_C^\perp$ and $T_C^\perp$ are then available through (3.51). Together we have:

basis of T_C :	f columns of $\mathbf{Q}$
basis of $T_C^\perp$:	m columns of $\bar{\mathbf{M}}^{-1}\mathbf{W}$
basis of $N_C^\perp$:	f columns of $\bar{\mathbf{M}}\mathbf{Q}$
basis of N_C :	m columns of $\mathbf{W}$

Elements of these spaces may now be written as linear combinations of the associated basis vectors. The virtual displacements $\delta\mathbf{z}_a$ from T_C and $\delta\mathbf{z}_p$ from $T_C^\perp$ in (3.64), for example, become

$$\delta \mathbf{z}_a = \mathbf{Q}\,\delta \mathbf{q}, \qquad \delta \mathbf{z}_p = \bar{\mathbf{M}}^{-1}\mathbf{W}\,\delta \mathbf{p} \tag{3.69}$$

with $\delta \mathbf{q} = \mathbf{q}^\star - \mathbf{q} \in \mathbb{R}^f$ from (3.67) and $\delta \mathbf{p} \in \mathbb{R}^m$. Properties to hold $\forall \delta \mathbf{z}_a \in T_C(\mathbf{z})$ and $\forall \delta \mathbf{z}_p \in T_C^\perp(\mathbf{z})$ are now transformed by (3.69) to hold $\forall \delta \mathbf{q}$ and $\forall \delta \mathbf{p}$, respectively, such as $\delta W_a^{(S)} = 0$ and $\delta W_p^{(S)} = 0$ in (3.64) The active and passive forces as members of $N_C^\perp$ and N_C, see equation (3.55), may be written as

$$\bar{\mathbf{f}}^{(a)} = \bar{\mathbf{M}}\mathbf{Q}\,\boldsymbol{\nu}, \qquad \bar{\mathbf{f}}^{(p)} = \mathbf{W}\,\boldsymbol{\mu} \tag{3.70}$$

with $\boldsymbol{\nu} \in \mathbb{R}^f$ and $\boldsymbol{\mu} \in \mathbb{R}^m$. Finally, the velocities $\dot{\mathbf{z}}$ and the accelerations $\ddot{\mathbf{z}}$ become by differentiation of $\mathbf{z}(\mathbf{q}, t)$

$$\dot{\mathbf{z}} = \mathbf{Q}\,\dot{\mathbf{q}} + \tilde{\boldsymbol{\kappa}}, \qquad \ddot{\mathbf{z}} = \mathbf{Q}\,\ddot{\mathbf{q}} + \bar{\boldsymbol{\kappa}}, \tag{3.71}$$

where $\mathbf{Q}(\mathbf{q}, t) = \frac{\partial \mathbf{z}}{\partial \mathbf{q}}$, $\tilde{\boldsymbol{\kappa}}(\mathbf{q}, t) = \frac{\partial \mathbf{z}}{\partial t}$ and $\bar{\boldsymbol{\kappa}}(\mathbf{q}, \dot{\mathbf{q}}, t) = \dot{\mathbf{Q}}\dot{\mathbf{q}} + \dot{\tilde{\boldsymbol{\kappa}}}$.

Let us now return to equations (3.64). First, we take into account that $\delta \mathbf{z}_a^T\, \bar{\mathbf{f}}^{(p)} = \delta \mathbf{z}_p^T\, \bar{\mathbf{f}}^{(a)} = 0$. Then we substitute $\delta \mathbf{z}_a$ and $\delta \mathbf{z}_p$ from (3.69), $\bar{\mathbf{f}}^{(p)}$ from (3.70) and $\ddot{\mathbf{z}}$ from (3.71) to obtain

$$\begin{aligned} \delta W_a^{(S)} &= \delta \mathbf{q}^T\, \mathbf{Q}^T (\bar{\mathbf{M}}\mathbf{Q}\,\ddot{\mathbf{q}} + \bar{\mathbf{M}}\bar{\boldsymbol{\kappa}} - \bar{\mathbf{h}} - \bar{\mathbf{f}}^{(a)}) = 0 \quad \forall \delta \mathbf{q}, \\ \delta W_p^{(S)} &= \delta \mathbf{p}^T\, \mathbf{W}^T \bar{\mathbf{M}}^{-1} (\bar{\mathbf{M}}\mathbf{Q}\,\ddot{\mathbf{q}} + \bar{\mathbf{M}}\bar{\boldsymbol{\kappa}} - \bar{\mathbf{h}} - \mathbf{W}\boldsymbol{\mu}) = 0 \quad \forall \delta \mathbf{p}. \end{aligned} \tag{3.72}$$

Since these equations are required to hold for any $\delta \mathbf{q}$ and any $\delta \mathbf{p}$, it follows that

$$\begin{aligned} &\mathbf{Q}^T \bar{\mathbf{M}}\mathbf{Q}\,\ddot{\mathbf{q}} - \mathbf{Q}^T(\bar{\mathbf{h}} - \bar{\mathbf{M}}\bar{\boldsymbol{\kappa}}) - \mathbf{Q}^T \bar{\mathbf{f}}^{(a)} = 0 \\ &\mathbf{W}^T \bar{\mathbf{M}}^{-1}\mathbf{W}\boldsymbol{\mu} + \mathbf{W}^T(\bar{\mathbf{M}}^{-1}\bar{\mathbf{h}} - \bar{\boldsymbol{\kappa}}) = 0 \end{aligned} \tag{3.73}$$

for the Newton-Euler equations to apply for each of the n bodies B_i. For a given kinematic state $(\mathbf{q}, \dot{\mathbf{q}})$ at time t and known active forces $\bar{\mathbf{f}}^{(a)}$ the generalized accelerations $\ddot{\mathbf{q}}$ are determined by only the first equation in (3.73). Note in particular that the constraint forces as part of the passive forces are not required to calculate $\ddot{\mathbf{q}}$. The second equation in (3.73) determines the values of $\boldsymbol{\mu}$, hence the values of the passive forces $\bar{\mathbf{f}}^{(p)} = \mathbf{W}\boldsymbol{\mu}$. Obviously, they can be calculated by only knowing the kinematic state $(\mathbf{q}, \dot{\mathbf{q}})$ at time t. The active forces as well as the generalized accelerations $\ddot{\mathbf{q}}$ are not needed, because they dropped out by the orthogonality (3.68). The constraint forces $\bar{\mathbf{f}}^{(c)} =: \mathbf{W}\boldsymbol{\lambda}$, if of interest, may be calculated in a postponed step from (3.54), (3.55), i.e. $\mathbf{W}\boldsymbol{\lambda} = \mathbf{W}\boldsymbol{\mu} + (\bar{\mathbf{f}}^{(a)} - \bar{\mathbf{f}}^{(r)})$ with $\boldsymbol{\lambda}$ known as the classical Lagrange multipliers.

In the following chapters only the first equation in (3.73) will further be discussed, because it contains sufficient information to determine the generalized accelerations $\ddot{\mathbf{q}}$. As soon as this equation is satisfied, we may regard the system as to be in the dynamic equilibrium, since we know that there is a second equation independent of $\ddot{\mathbf{q}}$ which determines the values of the passive forces and which ensures that the Newton-Euler equations can be satisfied for each body. To shorten the notation we set

$$\mathbf{M} := \mathbf{Q}^T \bar{\mathbf{M}} \mathbf{Q}, \qquad \mathbf{h} := \mathbf{Q}^T (\bar{\mathbf{h}} - \bar{\mathbf{M}} \bar{\kappa}), \qquad \mathbf{f}^{(a)} := \mathbf{Q}^T \bar{\mathbf{f}}^{(a)} \tag{3.74}$$

in the first equation of (3.73) This yields the projected Newton-Euler equations

$$\mathbf{M}(\mathbf{q}, t)\, \ddot{\mathbf{q}} - \mathbf{h}(\mathbf{q}, \dot{\mathbf{q}}, t) - \mathbf{f}^{(a)} = 0 \tag{3.75}$$

already known from (3.41) in Section 3.5. Finally, we drop the index (a) from the external active forces and denote by $\mathbf{u} = \dot{\mathbf{q}}$ the generalized velocities,

$$\mathbf{M}(\mathbf{q}, t)\, \dot{\mathbf{u}} - \mathbf{h}(\mathbf{q}, \mathbf{u}, t) = \mathbf{f} \quad \text{with } \mathbf{u} = \dot{\mathbf{q}}. \tag{3.76}$$

Different symbols for the displacements $\mathbf{q}$ and the velocities $\mathbf{u}$ will be useful for studying the dynamics of non-smooth systems, because as functions of time they will have substantial different properties.

Remark: For application problems a strict split of $\bar{\mathbf{f}}^{(e)}$ into $\bar{\mathbf{f}}^{(a)}$ and $\bar{\mathbf{f}}^{(p)}$ according to equation (3.55) may be cumbersome and ineffective, and may be relaxed for computing the generalized accelerations $\ddot{\mathbf{q}}$ in (3.75): Instead of the forces $\bar{\mathbf{f}}^{(a)}$ that *are* active, one takes into account all forces $\bar{\mathbf{f}}^{(\tilde{a})}$ which *may* become active, a bigger class (properly) containing $\bar{\mathbf{f}}^{(a)}$ and physically easier accessible, because the passive portions contained in $\bar{\mathbf{f}}^{(\tilde{a})}$ drop out in any case when $\mathbf{Q}^T \bar{\mathbf{f}}^{(\tilde{a})}$ is performed.

4. Motion and Discontinuity Events

In this chapter we discuss discontinuity events that might occur in non-smooth mechanics. In particular, these are jumps in the velocities, usually associated with collisions and treated within impact theory, but also jumps in the accelerations as they appear, for example, at the transitions from sliding to stiction when dry friction is prevalent. Within the framework of integration we discuss the meaning of displacements, velocities and accelerations, and also their interconnections as stated in [91]. Special attention is paid to one-sided accelerations, as they will be used in all of the remaining chapters to state the directional Newton-Euler equations.

4.1 Preliminaries on Integration of Functions

In order to motivate the use of absolutely continuous functions as the solution space of second order ordinary (measure-) differential inclusions in rigid body dynamics, some basic definitions and properties concerning absolute continuity, bounded variations, and differential measures are put together. This section has to be understood just as a scratch of these topics, but it should help the reader to accept the kinematical framework of non-smooth dynamics presented in [91]. Most of the material has been taken from [90] and [31], but also [127] and [84] have been used. For an introduction to integration theory we refer to [127], an advanced exposition may be found in [31]. Functions of bounded variations defined on $\mathbb{R}$ with values in $\mathbb{R}^f$ are treated in [90], and this is what one needs for finite-dimensional evolution problems. In continuum mechanics one requires even BV-functions defined on some n-dimensional sets. For these cases we refer to [143].

One-sided limits and derivatives: Let $I \subset \mathbb{R}$ and $\mathbf{u} : I \to \mathbb{R}^f$. We call $\mathbf{u}^+(t)$ and $\mathbf{u}^-(t)$ the *right limit* and the *left limit* of $\mathbf{u}$ at t if

$$\mathbf{u}^+(t) = \lim_{\tau \downarrow 0} \mathbf{u}(t+\tau) \quad \text{and} \quad \mathbf{u}^-(t) = \lim_{\tau \uparrow 0} \mathbf{u}(t+\tau) \tag{4.1}$$

exist in $\mathbb{R}^f$. The *right derivative* $\dot{\mathbf{u}}^+$ and the *left derivative* $\dot{\mathbf{u}}^-$ of $\mathbf{u}$ at t are defined by

$$\dot{\mathbf{u}}^+(t) = \lim_{\tau \downarrow 0} \frac{\mathbf{u}(t+\tau) - \mathbf{u}(t)}{\tau} \quad \text{and} \quad \dot{\mathbf{u}}^-(t) = \lim_{\tau \uparrow 0} \frac{\mathbf{u}(t+\tau) - \mathbf{u}(t)}{\tau} \tag{4.2}$$

if these limits exist with values in $\mathbb{R}^f$. For the left end of I, if it belongs to I, only the right limit and the right derivative is defined, and, analogously, for the right end of I only the left limits. For $\mathbf{u}^+(t) = \mathbf{u}^-(t)$, $\mathbf{u}$ is continuous at t, and, for $\dot{\mathbf{u}}^+(t) = \dot{\mathbf{u}}^-(t)$, $\mathbf{u}$ is differentiable at t.

Almost everywhere: Let $(X, \mathcal{A}, \mu)$ be a measure space, i.e. a σ-algebra $\mathcal{A}$ in a set X together with a measure $\mu : \mathcal{A} \to \bar{\mathbb{R}}$ defined on $\mathcal{A}$. We recall that a property is said to hold μ-*almost everywhere* (μ-a.e.) on X or for μ-almost every $x \in X$, if there exist a μ-negligible set $N \in \mathcal{A}$ (a set N for which $\mu(N) = 0$) such that the property applies for every $x \in X \backslash N$. For the one-dimensional Lebesgue measure λ on $\mathbb{R}$, for example, every singleton $\{a\}$ ($a \in \mathbb{R}$) is λ-negligible and also, as a consequence on the σ-additivity of measures, any countable collection of singletons. There are also uncountable sets with Lebesgue measure zero such as the Cantor set described below. With respect to integration we recall that the μ-integral of a function $f(x)$ is not affected when the values of f are changed on a μ-negligible set: If f and g are two measurable and integrable functions, then for every $A \in \mathcal{A}$

$$\int_A f \, \mathrm{d}\mu = \int_A g \, \mathrm{d}\mu \quad \text{when } f = g \ \mu\text{-a.e. on } X. \tag{4.3}$$

This holds in particular for the Lebesgue measure λ on $\mathbb{R}$ if f and g are different on an at most countable number of points.

The Cantor set: In order to construct the Cantor set we start with the closed interval $C_0 = [0, 1]$ and remove from it the open middle third $(\frac{1}{3}, \frac{2}{3})$. The set C_1 obtained in this way consists of the two disjoint closed intervals $[0, \frac{1}{3}]$ and $[\frac{2}{3}, 1]$, see Figure 4.1. From each of these intervals we remove, again, the open middle thirds in order to obtain $C_2 = [0, \frac{1}{9}] \cup [\frac{2}{9}, \frac{1}{3}] \cup [\frac{2}{3}, \frac{7}{9}] \cup [\frac{8}{9}, 1]$. Proceeding in this manner, the set $C = \lim_{n \to \infty} C_n$ is called the *Cantor set* and has the following properties [31]: It is a nowhere dense, closed, bounded subset of $[0, 1]$ with Hausdorff dimension $\delta(C) = \log 2 / \log 3 \approx 0.6309$. Moreover, it is not countable since it still consists of (a countable number of) "very, very short intervals," and its complement with respect to $[0, 1]$ has Lebesgue measure $\lambda([0, 1] \backslash C) = 1$ giving the overall length of the segments which has been removed from $[0, 1]$. The Cantor set is thus a not countable but Lebesgue negligible subset of $[0, 1]$.

The Cantor function: The graph of this function $x(t) = \lim_{n \to \infty} x_n(t)$ is constructed in the following way: We set $x_0(t) = 0$ for $t < 0$ and $x_0(t) = 1$ for $t > 1$, and connect the points $(0, 0)$ and $(1, 1)$ by a straight line. This makes a monotonic continuous function which is piecewise constant on $\mathbb{R} \backslash C_0$ and linearly increasing on C_0 with $x_0(C_0) = [0, 1]$. To construct $x_n(t)$ for

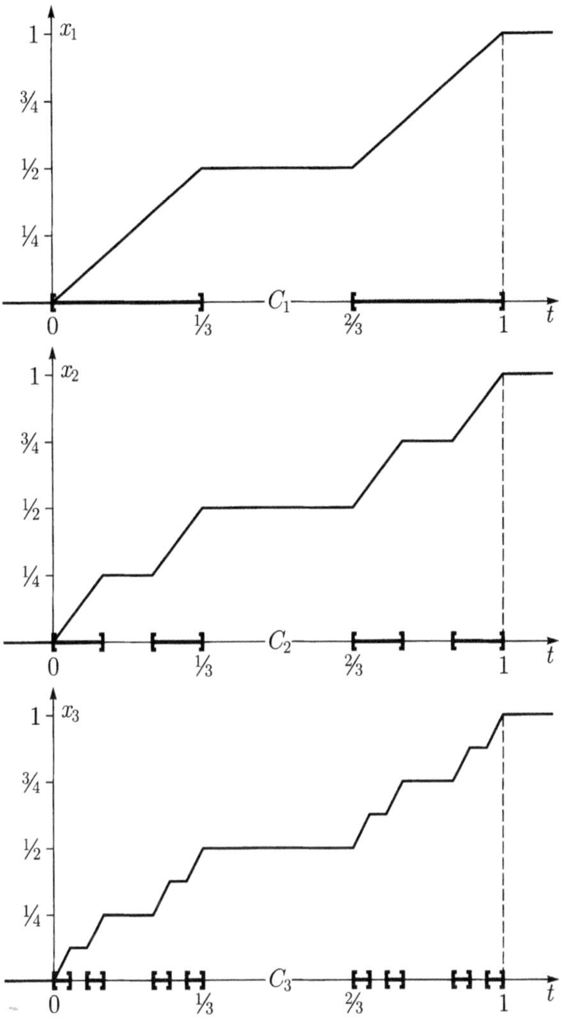

Figure 4.1. On the construction of the Cantor set $C \subset [0,1]$ and Cantor function $x(t)$

$n = 1, 2, \ldots$ from $x_{n-1}(t)$ we proceed as depicted in Figure 4.1: We take $x_n(t)$ to be piecewise constant on $\mathbb{R}\backslash C_n$, i.e. constant on each of the open intervals obtained in this way, with values $x_n(t) = x_{n-1}(t)$ for $t \in \mathbb{R}\backslash C_{n-1}$ and $x_n(t) = \frac{1}{2}(x_{n-1}(t_l) + x_{n-1}(t_r))$ for $t \in C_{n-1}\backslash C_n$, where t_l and t_r is any member of the next neighboring interval in $\mathbb{R}\backslash C_{n-1}$ on the left and the right of t, respectively. So far, $x_n(t)$ is determined up to the values in C_n. To complete the graph of $x_n(t)$ we join the endpoints of the just specified horizontal line segments by straight lines. This gives a monotonic continuous function on $\mathbb{R}$, piecewise constant on $\mathbb{R}\backslash C_n$, and linearly increasing on C_n with $x_n(C_n) = [0,1]$. For $n \to \infty$ the sequence $\{x_n(t)\}$ converges uniformly to a function $x(t)$, called the *Cantor function* [31], with the following properties: The Cantor function is monotonic continuous with $x(0) = 0$, $x(1) = 1$ and $x(C) = [0,1]$. It is piecewise constant on the open set $\mathbb{R}\backslash C$, hence differentiable with $\dot{x}(t) = 0$ for all $t \notin C$. Since the Cantor set is Lebesgue negligible, we have even $\dot{x}(t) = 0$ for almost every t. As an important consequence, the Cantor function may not be gained back by just integrating its (almost everywhere defined) derivative, since

$$1 = x(1) - x(0) \neq \int_0^1 \dot{x}(\tau)\,\mathrm{d}\tau = \int_{[0,1]\backslash C} \dot{x}(\tau)\,\mathrm{d}\tau = 0. \tag{4.4}$$

The key which makes integration (4.4) to work as usual is absolute continuity, as a necessary and a sufficient condition.

Absolutely continuous functions (AC): A real function $q(t)$ on $I = [t_A, t_E]$ is said to be ***absolutely continuous*** on I if for every $\varepsilon > 0$ there is a $\delta > 0$ such that

$$\sum_{i=1}^{n} |\, q(\beta_i) - q(\alpha_i)\,| < \varepsilon$$

for any n and any disjoint collection of intervals $(\alpha_1, \beta_1), \dots, (\alpha_n, \beta_n)$ in I satisfying

$$\sum_{i=1}^{n} (\beta_i - \alpha_i) < \delta.$$

Absolute continuity is thus stronger than ordinary continuity, since the latter is obtained when setting $n = 1$. As an example, the Cantor function is continuous on $[0, 1]$ but not absolutely continuous, since all its growth is gained on the Cantor set, a set of Lebesgue measure zero. The above definition may be extended to mappings from I into Hilbert spaces, Banach spaces, or even metric spaces as done in [90]. For f-dimensional real Banach spaces with norm $\|\cdot\|$, for example, one has just to replace the expression $|\, q(\beta_i) - q(\alpha_i)\,|$ in the definition by $\|\, \mathbf{q}(\beta_i) - \mathbf{q}(\alpha_i)\,\|$ when a mapping $\mathbf{q} : I \to \mathbb{R}^f$ is considered. For absolute continuous functions, integration always "works":

Theorem 4.1.1. *If $\mathbf{q}(t)$ is **absolutely continuous** on $I = [t_A, t_E]$, then $\mathbf{q}(t)$ is differentiable at almost every $t \in I$, $\dot{\mathbf{q}}(t) \in L^1$, and for every interval $[t_k, t_l] \subset I$ one has*

$$\mathbf{q}(t_l) - \mathbf{q}(t_k) = \int_{t_k}^{t_l} \dot{\mathbf{q}}(\tau)\,\mathrm{d}\tau. \tag{4.5}$$

Functions of bounded variation (BV): Let $u : I = [t_A, t_E] \to \mathbb{R}$ be a real function. For $t \in I$, define $U(t)$ to be

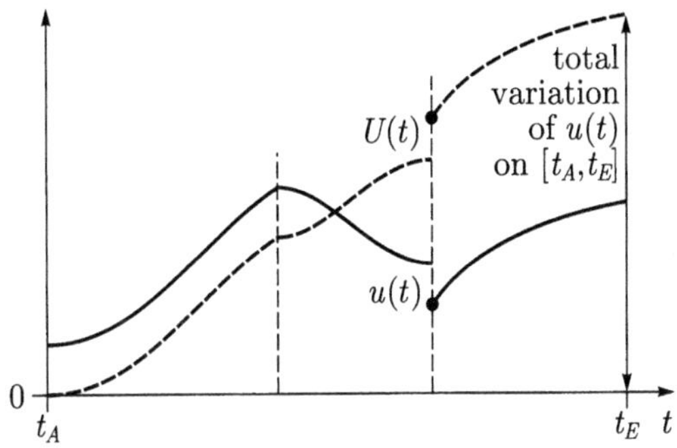

Figure 4.2. The function $u(t)$ and associated variation function $U(t)$

$$U(t) = \sup \sum_{i=1}^{n} |\, u(t_i) - u(t_{i-1}) \,|,$$

where the supremum is taken over all n and over all choices of $\{t_i\}$ such that

$$t_A = t_0 < t_1 < \ldots < t_n = t_E.$$

This $U(t)$ is called the *total variation function* of $u(t)$. For $U(t_E) < \infty$, u is said to have *bounded variations* on I, and the value $U(t_E)$ is called the *total variation* of u on I, see Figure 4.2. In the case of an $\mathbb{R}^f$-valued map $\mathbf{u}$ the variation function $U(t)$ is defined when replacing $|\, u(t_i) - u(t_{i-1}) \,|$ by $\|\, \mathbf{u}(t_i) - \mathbf{u}(t_{i-1}) \,\|$, see e.g. [90] and [84]. For functions of bounded variations the following properties apply:

(a) Every absolutely continuous function is of bounded variations.

(b) A function $\mathbf{u}(t) \in BV([t_A, t_E])$ has an at most countable number of discontinuities on $[t_A, t_E]$. At every point $t \in (t_A, t_E]$ the left limit $\mathbf{u}^-(t)$ exists, as well as the right limit $\mathbf{u}^+(t)$ for every $t \in [t_A, t_E)$.

(c) Every $\mathbf{u}(t) \in BV$ may be decomposed into $\mathbf{u} = \mathbf{u}_S + \mathbf{u}_A$, where $\mathbf{u}_S \in BV$ is a continuous function (*not* necessarily AC!), and $\mathbf{u}_A \in BV$ is a step function (a piecewise constant function with a countable number of discontinuities, and $\dot{\mathbf{u}}_A = 0$ for dt-almost every t as a consequence, see e.g. [90]). This decomposition is unique up to an additive constant.

(d) The continuous function $\mathbf{u}_S$ from (c) may be further decomposed into $\mathbf{u}_S = \mathbf{u}_L + \mathbf{u}_C$, where $\mathbf{u}_L \in BV$ is absolutely continuous (hence $\dot{\mathbf{u}}_L$ exists almost everywhere and $\dot{\mathbf{u}}_L \in L^1$), and $\mathbf{u}_C \in BV$ is a singular function (i.e. a continuous function with $\dot{\mathbf{u}}_C = 0$ almost everywhere such as the Cantor function). Again, this decomposition is unique up to an additive constant.

(e) A function $\mathbf{u}$ of bounded variations is differentiable almost everywhere. This means that the limit $\lim_{\tau \to 0} \frac{1}{\tau}(\mathbf{u}(t+\tau) - \mathbf{u}(t))$ with values in $\mathbb{R}^f$ exists for almost every t.

When integrating the derivative $\dot{\mathbf{u}}$ of a function of bounded variations by the formula (4.5), the same problem as already in (4.4) turns up. We know from (e) that $\dot{\mathbf{u}}$ exists almost everywhere, and that the Lebesgue integral is not affected by changing the values of the function to be integrated on an at most countable set of points. We may thus choose any values at the points at which $\dot{\mathbf{u}}$ does not exist in order to make it well-defined on the interval for which integration should be performed. In general, one then obtains

$$\int_{[t_k,t_l]} \dot{\mathbf{u}}\,\mathrm{d}\tau = \mathbf{u}_L(t_l) - \mathbf{u}_L(t_k) \neq \mathbf{u}^+(t_l) - \mathbf{u}^-(t_k), \tag{4.6}$$

which gives only the portion related to the absolutely continuous part, because the singular and the step function do not contribute, since $\dot{\mathbf{u}}_C = 0$ and $\dot{\mathbf{u}}_A = 0$ almost everywhere by (c) and (d). Thus, one seeks for an expression, the integration of which yields back the overall bounded variation function. This job is done by the differential measure.

The differential measure: The following definition is taken from [90], see also [84]: Let $\mathcal{S}$ be the set of finite subsets of I. Every member $S \in \mathcal{S}$ consists of a finite number n_S of points t_i^S which may be enumerated as a strictly increasing finite sequence, say $S : t_0^S < t_1^S < \ldots < t_{n_S}^S$. The elements of $\mathcal{S}$ themselves may be ordered by inclusion: If $S, P \in \mathcal{S}$ there exists $Q \in \mathcal{S}$ such that $Q \supset S$ and $Q \supset P$, namely $Q = S \cup P$, and one writes $Q \geq S$ iff $Q \supset S$. This turns $\mathcal{S}$ into a directed set, providing the concept of convergence of mappings: One may take limits of families indexed on $\mathcal{S}$, that is limits of sequences H_S with $\mathcal{S} \ni S \to H_S$, called generalized sequences. An *intercalator* ϑ^S on $S \in \mathcal{S}$ is an assignment which associates with every two neighbors t_{i-1}^S, t_i^S $(1 \leq i \leq n_S)$ in S an element $\vartheta_i^S \in [t_{i-1}^S, t_i^S]$. Let now $\mathbf{u} : I \to \mathrm{I\!R}^f$ be a function of bounded variation and φ a continuous real function on I with compact support relative to I, denoted by $\varphi \in \mathcal{K}(I)$. Further, take any $S \in \mathcal{S}$ and any intercalator ϑ^S on S. With these ingredients one defines the (Riemann-Stieltjes) sum

$$\mathbf{H}(S, \vartheta^S, \varphi; \mathbf{u}) = \sum_{i=1}^{n_S} \varphi(\vartheta_i^S)\,(\mathbf{u}(t_i) - \mathbf{u}(t_{i-1})). \tag{4.7}$$

In [90] it is then proven that, for every $\varphi \in \mathcal{K}(I)$, the mapping $S \to \mathbf{H}(S, \vartheta^S, \varphi; \mathbf{u})$ converges to a limit independent of the choice of ϑ^S, if $\mathbf{u}$ is of bounded variation. This limit is denoted by

$$\lim_{S \in \mathcal{S}} \mathbf{H}(S, \vartheta^S, \varphi; \mathbf{u}) =: \int \varphi\,\mathrm{d}\mathbf{u} \tag{4.8}$$

and is called the integral of $\varphi \in \mathcal{K}(I)$ with respect to the *differential measure* $\mathrm{d}\mathbf{u}$ of $\mathbf{u}$. With that measure one has, for every compact subinterval $[t_k, t_l]$ of I,

$$\int_{[t_k,t_l]} \mathrm{d}\mathbf{u} = \mathbf{u}^+(t_l) - \mathbf{u}^-(t_k) \tag{4.9}$$

and, in particular, if $[t_k, t_l]$ reduces to a singleton $\{t_k\}$,

$$\int_{\{t_k\}} \mathrm{d}\mathbf{u} = \mathbf{u}^+(t_k) - \mathbf{u}^-(t_k). \tag{4.10}$$

The differential measure $\mathrm{d}\mathbf{u}$ is thus capable of regarding all the three portions contained in $\mathbf{u} \in BV$, these are the absolutely continuous function, the step function, and the singular function, and integration works in the usual way. Equation (4.9) (and (4.10) as a special case of (4.9), respectively) is obtained when (4.8) is evaluated for a sequence $\varphi_p \in \mathcal{K}(I)$ converging pointwise to the characteristic function $\chi_{[t_k,t_l]}$ of $[t_k, t_l]$, and the dominated convergence theorem is applied, i.e. $\lim_{p\to\infty} \int \varphi_p \, \mathrm{d}\mathbf{u} = \mathbf{u}^+(t_l) - \mathbf{u}^-(t_k)$ on the one hand, and $\lim_{p\to\infty} \int \varphi_p \, \mathrm{d}\mathbf{u} = \int \lim_{p\to\infty} \varphi_p \, \mathrm{d}\mathbf{u} = \int \chi_{[t_k,t_l]} \, \mathrm{d}\mathbf{u} = \int_{[t_k,t_l]} \mathrm{d}\mathbf{u}$ on the other hand, see [90].

4.2 Displacements, Velocities, and Accelerations

We are now going to choose appropriate function spaces for the displacements, velocities and accelerations as functions of time of a rigid multibody system. Suppose there is a velocity jump at time t_i, for instance caused by a collision. Intuitively, the velocities just before and just after the collision, $\mathbf{u}^-(t_i)$ and $\mathbf{u}^+(t_i)$, make sense from the physical point of view, whereas the velocity *at* the collision time, $\mathbf{u}(t_i)$, is meaningless. On the other hand, this value is not needed by (4.3) when the velocities are integrated to obtain the displacements $\mathbf{q}$ of the system. At least the displacements, as functions of time, should be continuous. There are only a few examples for which a model allowing displacement jumps is reasonable. These facts lead quite naturally to the following kinematical setting:

In order to study the motion of a holonomic multibody system we denote, as in the previous chapter, by $\mathbf{q} \in \mathbb{R}^f$ the f generalized coordinates of the bodies. Motion is defined by making $\mathbf{q}$ dependent on time t. We are interested in a motion of the system on a compact time interval $I = [t_A, t_E]$. Following [91] we do not suppose the function $\mathbf{q} : I \to \mathbb{R}^f$ to be differentiable everywhere. Instead, we assume the velocities to be functions of bounded variations (BV) on I, admitting a countable number of finite jumps. This means, we introduce a *generalized velocity function* $\mathbf{u} : I \to \mathbb{R}^f$, $\mathbf{u} \in BV(I, \mathbb{R}^f)$ such that

$$\forall t \in I : \quad \mathbf{q}(t) = \mathbf{q}(t_A) + \int_{t_A}^{t} \mathbf{u}(\tau) \, \mathrm{d}\tau, \tag{4.11}$$

leading to displacements $\mathbf{q}(t)$ that are absolutely continuous on I (see e.g. [127], [31]).

For a function $\mathbf{u}$ of bounded variations the left limit $\mathbf{u}^-(t)$ exists at every point in $(t_A, t_E]$, the right limit $\mathbf{u}^+(t)$ exists at every point in $[t_A, t_E)$, and the set of points at which $\mathbf{u}$ is discontinuous is at most countable and will be denoted by $\{t_i\}$. It is known [90] that $\mathbf{u}^-(t)$ is a left-continuous function which we call the *left velocity* and $\mathbf{u}^+(t)$ is a right-continuous function which we call the *right velocity.* As in [91] we understand $\mathbf{u}(t_A)$ to be equal to the left limit $\mathbf{u}^-(t_A)$ in order to allow a first velocity jump to occur already at t_A, because there is no reason why motion should not start with an impact. Symmetrically, we set $\mathbf{u}^+(t_E) = \mathbf{u}(t_E)$. Via (4.11) we have $\dot{\mathbf{q}} = \mathbf{u}$ for Lebesgue-almost every $t \in [t_A, t_E]$, hence $\mathrm{d}\mathbf{q} = \mathbf{u}\,\mathrm{d}t$. Moreover, $\dot{\mathbf{q}}^+ = \mathbf{u}^+$ and $\dot{\mathbf{q}}^- = \mathbf{u}^-$ holds for any $t \in (t_A, t_E)$ with $\dot{\mathbf{q}}^+$ and $\dot{\mathbf{q}}^-$ being the right and the left derivative of $\mathbf{q}$, respectively. Due to (4.3) the integration (4.11) may be uniquely performed by choosing any finite values for $\mathbf{u}$ at the points t_i where $\mathbf{u}$ is not defined, for example by taking $\mathbf{u}^+(t)$ or $\mathbf{u}^-(t)$ instead of $\mathbf{u}$, because the value of the integral is not affected by changes of Lebesgue-measure zero.

Since $\mathbf{u}$ is of bounded variations its derivative $\dot{\mathbf{u}}$, which we call the *generalized acceleration*, exists for $[\mathrm{d}t]$-a.e. t, and $\dot{\mathbf{u}} \in L^1(I, \mathbb{R}^f)$. However, due to the discontinuities of $\mathbf{u}$ which are taken into account by the step function $\mathbf{u}_A$, and due to the singular portion $\mathbf{u}_C$ contained in $\mathbf{u}$, it is not possible to obtain the velocities by just integrating $\dot{\mathbf{u}}$ as it has been done for the displacements in (4.11), since then one would obtain only the absolutely continuous part $\mathbf{u}_L$ of $\mathbf{u}$ as already indicated in (4.6). Instead, one has to use the differential measure $\mathrm{d}\mathbf{u}$ of $\mathbf{u}$ as introduced in (4.8), leading to the integration formulas (4.9) and (4.10). Since $[t_k, t_l] = \{t_k\} \cup (t_k, t_l) \cup \{t_l\}$, this admits a representation of the velocities in the form

$$\mathbf{u}^+(t_k) = \mathbf{u}^-(t_k) + \int\limits_{\{t_k\}} \mathrm{d}\mathbf{u}, \qquad \mathbf{u}^-(t_l) = \mathbf{u}^+(t_k) + \int\limits_{(t_k, t_l)} \mathrm{d}\mathbf{u}. \tag{4.12}$$

Note especially that the term $\int_{\{t_k\}} \mathrm{d}\mathbf{u}$ in the first equation vanishes almost everywhere, except of the discontinuity points t_i of $\mathbf{u}$. An extensive theoretical treatise on functions of bounded variations and associated measures may be found in [90].

It is known that, for every function $\mathbf{u} \in BV(I, \mathbb{R}^f)$, there exist a decomposition into functions of bounded variations and a corresponding decomposition of the differential measure $\mathrm{d}\mathbf{u}$ into

$$\mathbf{u} = \mathbf{u}_L + \mathbf{u}_A + \mathbf{u}_C \quad \text{and} \quad \mathrm{d}\mathbf{u} = \mathrm{d}\mathbf{u}_L + \mathrm{d}\mathbf{u}_A + \mathrm{d}\mathbf{u}_C \tag{4.13}$$

with the following properties:

The function $\mathbf{u}_L$ is absolutely continuous, and so is the measure $\mathrm{d}\mathbf{u}_L$ with respect to the 1-dimensional Lebesgue measure $\mathrm{d}t$, admitting as density the function $\dot{\mathbf{u}}$, i.e.

$$\mathrm{d}\mathbf{u}_L = \dot{\mathbf{u}}\,\mathrm{d}t. \tag{4.14}$$

Integration of $\mathrm{d}\mathbf{u}_L$ over a one-dimensional cell I_{kl} yields

$$\int_{I_{kl}} \mathrm{d}\mathbf{u}_L = \int_{t_k}^{t_l} \dot{\mathbf{u}}\,\mathrm{d}t = \mathbf{u}_L(t_l) - \mathbf{u}_L(t_k), \tag{4.15}$$

where a one-dimensional cell I_{kl} is understood to be an interval of any sort with endpoints t_k and t_l, i.e. any open or closed or half-open interval (t_k, t_l), $[t_k, t_l]$, $(t_k, t_l]$, $[t_k, t_l)$.

The function $\mathbf{u}_A$ is a step function. It is constant on $I\backslash\{t_i\}$ and takes into account the discontinuities of $\mathbf{u}$ by finite jumps at points t_i. Hence, $\dot{\mathbf{u}}_A = 0$ for almost every $t \in I$. The corresponding measure $\mathrm{d}\mathbf{u}_A$ is therefore purely atomic and 0-dimensional, and it may be represented by

$$\mathrm{d}\mathbf{u}_A = (\mathbf{u}^+ - \mathbf{u}^-)\,\mathrm{d}\eta, \tag{4.16}$$

where $\mathrm{d}\eta$ is concentrated on the set of discontinuities $\{t_i\}$ of $\mathbf{u}$. It turns out that $\mathrm{d}\eta$ is the sum of the Dirac point measures $\mathrm{d}\delta_i$,

$$\mathrm{d}\eta = \sum_i \mathrm{d}\delta_i \qquad \text{with} \int_{I_{kl}} \mathrm{d}\delta_i = \begin{cases} 1 & \text{if } t_i \in I_{kl} \\ 0 & \text{if } t_i \notin I_{kl} \end{cases} \tag{4.17}$$

where I_{kl} is any one-dimensional cell in I. The evaluation of the atomic measure thus yields

$$\int_{[t_k, t_l]} \mathrm{d}\mathbf{u}_A = \sum_{\tau \in [t_k, t_l] \cap \{t_i\}} (\mathbf{u}^+(\tau) - \mathbf{u}^-(\tau)) = \mathbf{u}_A^+(t_l) - \mathbf{u}_A^-(t_k) \tag{4.18}$$

when $[t_k, t_l] \subset I$ and $\{t_i\}$ denotes the discontinuity points of $\mathbf{u}$ on I as above.

The third term in (4.13), $\mathbf{u}_C$, is called the singular function. It is continuous and not constant, but $\dot{\mathbf{u}}_C = 0$ for almost every t. The corresponding measure $\mathrm{d}\mathbf{u}_C$ is singular with respect to $\mathrm{d}t$, and it may have support on sets with Hausdorff dimensions between 0 and 1, see [4]. The only thing which can be said about its integral is

$$\int_{I_{kl}} \mathrm{d}\mathbf{u}_C = \mathbf{u}_C(t_l) - \mathbf{u}_C(t_k). \tag{4.19}$$

We assume $\mathrm{d}\mathbf{u}_C = 0$ because we are basically interested in evolution problems that are composed of intervals of classical smooth motion and impacts. Forces of fractal type resulting from set-valued as well as from single-valued force laws, leading to additional fine oscillations of the accelerations, are excluded with that assumption.

As mentioned above the derivative $\dot{\mathbf{u}}$ of $\mathbf{u}$ in (4.14) exists almost everywhere in I, except for a countable set of points that we will denote by $\{t_j\}$, probably different from $\{t_i\}$. This set may be further reduced by considering, instead of $\dot{\mathbf{u}}$, only directional derivatives. Applying the first equation in (4.2) on the function $\mathbf{u}^+(t)$ we define the *right acceleration* $\dot{\mathbf{u}}^+$ to be the right derivative of the right-continuous function $\mathbf{u}^+(t)$ and, in the same fashion, we introduce the *left acceleration* $\dot{\mathbf{u}}^-$ to be the left derivative of the left-continuous velocity $\mathbf{u}^-(t)$. Obviously, the sets $\{t_{j,\text{right}}\}$ and $\{t_{j,\text{left}}\}$ where the right and the left accelerations do not exist are both contained in $\{t_j\}$. Choosing arbitrary values at these points and summarizing the assumptions and results obtained so far we may now express the three measures in (4.13) to be

$$\begin{aligned} \mathrm{d}\mathbf{u}_L &= \dot{\mathbf{u}}\,\mathrm{d}t = \dot{\mathbf{u}}^+\,\mathrm{d}t = \dot{\mathbf{u}}^-\,\mathrm{d}t \\ \mathrm{d}\mathbf{u}_A &= (\mathbf{u}^+ - \mathbf{u}^-)\,\mathrm{d}\eta \\ \mathrm{d}\mathbf{u}_C &= 0. \end{aligned} \tag{4.20}$$

Once again, let us discuss how the discontinuity points $\{t_i\}$ of the velocities $\mathbf{u}$ and $\{t_j\}$ of the accelerations $\dot{\mathbf{u}}$ are connected together. We suppose that $\{t_{j,\text{right}}\} = \{t_{j,\text{left}}\} = \emptyset$, i.e. $\dot{\mathbf{u}}^+$ and $\dot{\mathbf{u}}^-$ exist everywhere in I. For almost every t we have $\mathbf{u}^+ = \mathbf{u}^-$ and $\dot{\mathbf{u}}^+ = \dot{\mathbf{u}}^-$, that means continuity. At the discontinuity points $\{t_i\}$ of the velocities we have $\mathbf{u}^+ \neq \mathbf{u}^-$. These *velocity jumps* may be accompanied by continuous accelerations $\dot{\mathbf{u}}^+ = \dot{\mathbf{u}}^-$ or even discontinuous accelerations $\dot{\mathbf{u}}^+ \neq \dot{\mathbf{u}}^-$ as it can already be seen by the most primitive impact system:

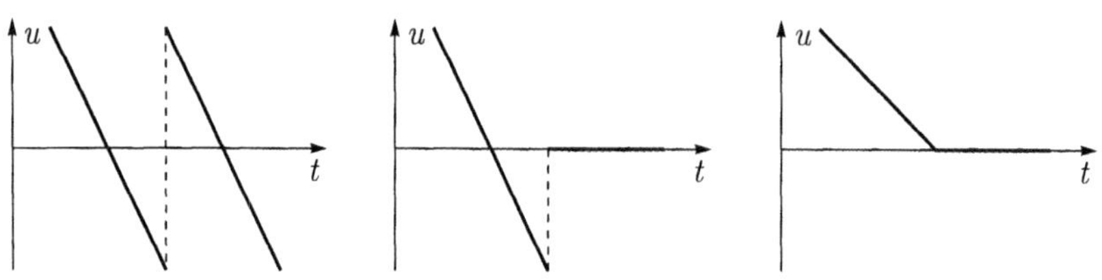

Figure 4.3. Discontinuities in the velocities u and accelerations $\dot{u}$

We consider the *bouncing ball problem* with completely elastic impact (Figure 4.3, left diagram). At the impact time the velocity $u^- < 0$ is reversed, $u^+ = -u^-$. The acceleration at the impact time is continuous because it is obtained by the equation of motion for the free-flight state $\dot{u} = \dot{u}^+ = \dot{u}^- = -g$ where g denotes gravity. Now we consider the same system but under the influence of a completely inelastic impact (Figure 4.3, middle diagram). The acceleration before the impact is $\dot{u} = -g$. At the impact the velocity $u^- < 0$ is changed to $u^+ = 0$, i.e. after the impact the ball remains at the impact surface and hence $\dot{u}^+ = 0$.

Suppose now that the velocities are continuous ($\mathbf{u} = \mathbf{u}^+ = \mathbf{u}^-$, i.e. $t \notin \{t_i\}$). Even in this case there might be an *acceleration jump* $\dot{\mathbf{u}}^+ \neq \dot{\mathbf{u}}^-$ as it can be seen by the following example (Figure 4.3, right diagram): We consider a mass m sliding on a plane with a velocity $u > 0$ under the influence of dry friction (here: Coulomb friction with a given normal load). The equations of motion during sliding are $\dot{u} = -\mu g$ where μ is the coefficient of friction and g denotes gravity as above. The mass is decelerated until it comes to a rest. After it has stopped moving we have $\dot{u} = 0$, and hence for the transition point $\dot{u}^- = -\mu g \neq 0 = \dot{u}^+$ without any velocity jump.

4.3 Restriction to Finite Numbers of Discontinuities

The main purpose of all of the following is the determination of the left and the right accelerations if they exist. Knowing these magnitudes and assuming, for example, an impact at time t_k followed by an impact-free interval of motion (t_k, t_l), equation (4.12) together with (4.20) may now be rewritten as

$$\mathbf{u}^+(t_k) - \mathbf{u}^-(t_k) = \int\limits_{\{t_k\}} (\mathbf{u}^+ - \mathbf{u}^-)\,\mathrm{d}\eta \tag{4.21}$$

and

$$\mathbf{u}^-(t_l) - \mathbf{u}^+(t_k) = \int\limits_{(t_k,t_l)} \dot{\mathbf{u}}\,\mathrm{d}t = \int\limits_{(t_k,t_l)} \dot{\mathbf{u}}^+\mathrm{d}t = \int\limits_{(t_k,t_l)} \dot{\mathbf{u}}^-\mathrm{d}t. \tag{4.22}$$

With that decomposition of a motion into a finite number of impact events and intervals of impact-free motion, called a *motion of finite sort*, one is able to use standard integration routines for solving the differential inclusions that will be presented in the following chapters. In the case of a collision, one has only to determine the post-impact velocities $\mathbf{u}^+$ by the use of certain impact laws in order to continue with ordinary integration until the next impact occurs. Acceleration jumps are taken into account by computing the values of the right accelerations $\dot{\mathbf{u}}^+$ at every time point during numerical integration.

This approach, however, is restricted to the treatment of only a finite number of discontinuity events in a time interval. More or less serious problems are caused when a dynamic behavior is observed such as in the systems depicted in Figure 4.4. For the bouncing ball problem (left diagrams in Figure 4.4), for example, one recognizes that the impact times constitute a geometric sequence with limit $t^\star$ if the impact law has been chosen to be $u^+ = -\varepsilon u^-$, $0 < \varepsilon < 1$, with $u = \dot{y}$. Although the right acceleration is well defined ($\dot{u}^+ = -g$ for $t < t^\star$, $\dot{u}^+ = 0$ for $t \geq t^\star$), there is no way to pass $t^\star$ by the numerical approach discussed above since the time intervals between two succeeding impacts tend to zero. One might "solve" this difficulty, for example, by making the coefficient of restitution ε dependent on u^- such that a completely inelastic impact should take place if $|u|$ becomes less than a given number δ, i.e. $\varepsilon(u^-) = 0$ if $|u^-| < \delta$, in order to stop the impact

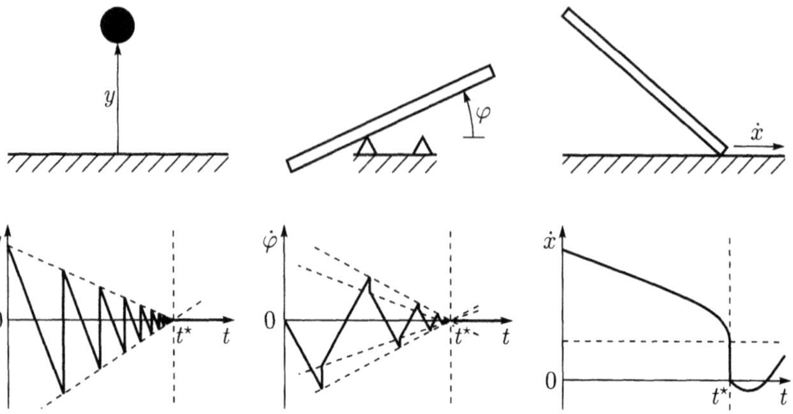

Figure 4.4. Some types of velocity jumps

sequence after a finite number of collisions. The outcome of the last impact would then be like in the middle diagram of Figure 4.3, and one would be able to continue with integration since $t^\star$ has been passed.

A very similar distribution of the impact times can be observed at the *rocking rod*, depicted in the middle diagrams of Figure 4.4. Rocking is caused by a periodic switching between two supporting obstacles, where the event of changing from one closed contact to the next is the outcome of a completely inelastic impact ($\varepsilon = 0$), see [37], [116]. As in the bouncing ball example the rod stops rocking after an infinite number of impacts that occur in a finite time interval (lower diagram; $|\varphi| \ll 1$ assumed). Contrary to the bouncing ball, this impact sequence can not be stopped by setting $\varepsilon = 0$ since the restitution coefficients are already equal to zero. In order to stop the movement of the rod when $|\dot{\varphi}| < \delta$ one might apply a "correction" that sets $\dot{\varphi} = 0$ and that is usually done by projections. In dynamics, however, an instantaneous change in velocities, it might be called correction or anything else, is always the outcome of some impact impulsions. In our example, this additional impulsion would act as a tensional magnitude in order to prevent separation at one of the contacts. This is an unsatisfactory fact, at least from the mechanical point of view, since we are used to understand impulsions occurring from unilateral constraints as compressive magnitudes.

Things become worst for the third example, the *sliding rod* (right diagrams in Figure 4.4). It is known that, under the influence of Coulomb friction with friction coefficients big enough, the accelerations of the system may tend to infinity, leading to an *impact without collision.* This phenomenon has been called in [91] a *frictional catastrophe*, see also [21], [37], [116], [36]. The horizontal velocity $\dot{x} = u$ of the rod's contact point is depicted in the lower right diagram of Figure 4.4, [37]. Note that the left acceleration at the singularity $t^\star$ exists in $\bar{\mathbb{R}}$, i.e. $\dot{u}^- = -\infty$, but not in $\mathbb{R}$. This means that the evolution of the system may be computed by the aforementioned approach up to $t^\star$, but it is neither possible to reach $t^\star$ nor to pass it.

All the difficulties encountered in these examples may be overcome when using integration algorithms that are based on the velocity level as introduced in [91] and also presented in [70], [136], because they do not use the accelerations explicitly. Instead, the differential measure $d\mathbf{u}$ in (4.9) is directly evaluated for a given integration step Δt, leading finally to an equilibrium of all the accelerations and forces, impulsive and non-impulsive. From the mechanical point of view this procedure may be interpreted in the sense that, also for smooth motion, the velocity increments obtained after discretization may be generally regarded as the outcome of impacts. It is then easy to handle impulsions resulting from collisions within the same framework. On the other hand, we think that the left and the right accelerations are crucial for the understanding of non-smooth mechanics because a lot of concepts used in classical analytical dynamics may be carried over in this way.

5. Displacement and Velocity Potentials

The projected Newton-Euler equations (3.76) from Chapter 3 do not yet provide a complete description of the dynamics of the multibody system, because some force laws connecting the external forces $\mathbf{f}$ with the system's state $(\mathbf{q}, \mathbf{u})$ are still missing. The topic of this chapter is thus to introduce a certain class of force laws, broad enough to cover most of the standard interaction models used in multibody dynamics. In particular, the theory should feature classical impressed forces, bilaterally and unilaterally constrained motion as well as force laws of mixed type at which the system passively switches from impressed forces to constrained motion. A concept general enough to handle these demands is provided by the generalized gradient mapping; we will thus treat force laws as set-valued maps within extended potential theory.

5.1 Directional Newton-Euler Equations

In this section we state the Newton-Euler equations for future and past events as well as the impact equations of the system. As it is obvious from the examples presented in Chapter 4, discontinuity events in the velocities and the accelerations of the system have to be allowed. The projected Newton-Euler equations (3.76) require both, the existence of the velocities $\mathbf{u}$ and the accelerations $\dot{\mathbf{u}}$. Hence, on the time interval I the projected Newton-Euler equations are only defined up to the discontinuity points $\{t_i\}$ of $\mathbf{u}$ and the points $\{t_j\}$ where $\dot{\mathbf{u}}$ does not exist, and have therefore, more precisely, be stated as

$$\mathbf{M}(\mathbf{q}, t)\,\dot{\mathbf{u}} - \mathbf{h}(\mathbf{q}, \mathbf{u}, t) = \mathbf{f} \quad \text{for } [dt]\text{-a.e. } t \in I. \tag{5.1}$$

We recall that $\mathbf{M}$ is the symmetric and positive definite generalized mass matrix depending smoothly on $(\mathbf{q}, t)$, and $\mathbf{h}$ is a smooth function of $(\mathbf{q}, \mathbf{u}, t)$ containing the gyroscopical accelerations of the multibody system.

Equation (5.1) is not suitable for further investigations because exactly the points of interest, i.e. the discontinuity points of $\mathbf{u}$ and of its derivative $\dot{\mathbf{u}}$ are excluded. Thus we try to get rid of the restriction "almost everywhere" and replace equation (5.1) as in [91] by the corresponding *equality of measures*

$$\mathbf{M}(\mathbf{q}, t)\,\mathrm{d}\mathbf{u} - \mathbf{h}(\mathbf{q}, \mathbf{u}, t)\,\mathrm{d}t = \mathrm{d}\mathbf{R} \tag{5.2}$$

which holds for every $t \in I$. From the mechanical point of view this equality should be understood as an equilibrium of momenta and impulsions at the impact, and as a balance of changes in momenta and of forces for impact free motion. For the force measure $\mathrm{d}\mathbf{R}$ in (5.2) we take naturally the same decomposition as for the differential measure $\mathrm{d}\mathbf{u}$ in (4.13),

$$\mathrm{d}\mathbf{R} = \mathbf{f}\,\mathrm{d}t + \mathbf{F}\,\mathrm{d}\eta + \mathrm{d}\mathbf{R}_C, \tag{5.3}$$

i.e. it may consist of Lebesgue-measurable forces $\mathbf{f}$, purely atomic impact impulsions $\mathbf{F}$, and any singular force measure $\mathrm{d}\mathbf{R}_C$ which we also assume to vanish. Upon substitution of (5.3) into (5.2) and with $\mathrm{d}\mathbf{u} = \dot{\mathbf{u}}\,\mathrm{d}t + (\mathbf{u}^+ - \mathbf{u}^-)\,\mathrm{d}\eta$ from (4.20) and (4.13), equation (5.2) becomes

$$\mathbf{M}(\mathbf{q},t)\,\dot{\mathbf{u}}\,\mathrm{d}t + \mathbf{M}(\mathbf{q},t)\,(\mathbf{u}^+ - \mathbf{u}^-)\,\mathrm{d}\eta - \mathbf{h}(\mathbf{q},\mathbf{u},t)\,\mathrm{d}t = \mathbf{f}\,\mathrm{d}t + \mathbf{F}\,\mathrm{d}\eta \tag{5.4}$$

and can be split into the atomic and the Lebesgue part, i.e.

$$\begin{aligned} \mathbf{M}(\mathbf{q},t)\,(\mathbf{u}^+ - \mathbf{u}^-)\,\mathrm{d}\eta &= \mathbf{F}\,\mathrm{d}\eta, \\ \mathbf{M}(\mathbf{q},t)\,\dot{\mathbf{u}}\,\mathrm{d}t - \mathbf{h}(\mathbf{q},\mathbf{u},t)\,\mathrm{d}t &= \mathbf{f}\,\mathrm{d}t. \end{aligned} \tag{5.5}$$

From the atomic part we obtain after evaluation of $\mathrm{d}\eta$ with respect to (4.17) the *impact equations* of the system

$$\mathbf{M}(\mathbf{q}_i,t_i)\,(\mathbf{u}_i^+ - \mathbf{u}_i^-) = \mathbf{F}_i, \tag{5.6}$$

where $\mathbf{q}_i = \mathbf{q}(t_i)$, $\mathbf{u}_i^\pm = \mathbf{u}^\pm(t_i)$, $\mathbf{F}_i = \mathbf{F}(t_i)$, and t_i is any of the discontinuity points of the velocity $\mathbf{u}$. Since the right equation in (5.5) is not affected by any changes of Lebesgue measure zero we may rewrite it with the help of (4.20) in order to obtain the two equivalent expressions

$$\begin{aligned} \mathbf{M}(\mathbf{q},t)\,\dot{\mathbf{u}}^+\mathrm{d}t - \mathbf{h}(\mathbf{q},\mathbf{u}^+,t)\,\mathrm{d}t &= \mathbf{f}^+\mathrm{d}t, \\ \mathbf{M}(\mathbf{q},t)\,\dot{\mathbf{u}}^-\mathrm{d}t - \mathbf{h}(\mathbf{q},\mathbf{u}^-,t)\,\mathrm{d}t &= \mathbf{f}^-\mathrm{d}t. \end{aligned} \tag{5.7}$$

Note that the points where $\mathbf{u}^+ \neq \mathbf{u}^-$ and $\dot{\mathbf{u}}^+ \neq \dot{\mathbf{u}}^-$ are immaterial and hence Lebesgue-negligible. In this setting it is quite natural to assume similar properties for the forces $\mathbf{f}$ in the right-hand sides of eqs. (5.5) and (5.7). Considered as functions of time we propose $\mathbf{f}^+ = \mathbf{f}^- = \mathbf{f}$ for $[dt]$-a.e. t, where $\mathbf{f}^+$ and $\mathbf{f}^-$ denotes the right and left limit of $\mathbf{f}$ with respect to time, respectively. Now we "divide" both equations in (5.7) by $\mathrm{d}t$ in order to obtain the *directional Newton-Euler equations*

$$\begin{aligned} \mathbf{M}(\mathbf{q},t)\,\dot{\mathbf{u}}^+ - \mathbf{h}(\mathbf{q},\mathbf{u}^+,t) &= \mathbf{f}^+, \\ \mathbf{M}(\mathbf{q},t)\,\dot{\mathbf{u}}^- - \mathbf{h}(\mathbf{q},\mathbf{u}^-,t) &= \mathbf{f}^- \end{aligned} \tag{5.8}$$

with $\dot{\mathbf{q}}^+ = \mathbf{u}^+$, $\dot{\mathbf{q}}^- = \mathbf{u}^-$, where the first equation in (5.8) is defined on $I\backslash\{t_{j,\text{right}}\}$ and the second on $I\backslash\{t_{j,\text{left}}\}$. For $\{t_{j,\text{right}}\} = \{t_{j,\text{left}}\} = \emptyset$ we therefore have the following result: Every motion $\mathbf{q}(t)$ which fulfills the impact equation (5.6) and one of the equations in (5.8), is also a solution of the measure equality (5.2).

In the following we are only interested in the values of the right and left accelerations. Thus the impact equation (5.6) will no longer be considered. The left equation in (5.8) represents the evolution of the multibody system with respect to the future, because $\dot{\mathbf{q}}^+ = \mathbf{u}^+$ and $\dot{\mathbf{u}}^+$ are the right limits of the velocities and the right accelerations, respectively. Suppose, for example, that $\mathbf{q}$, $\mathbf{u}^+$, and $\mathbf{f}^+$ are given at a certain time point. In this case we may directly compute $\dot{\mathbf{u}}^+$ from the left equation in (5.8) which describes the behavior of the solution for succeeding times. Analogously, the right equation in (5.8) contains the evolution of the system pointing into the past, i.e. the evolution after reversal of the time arrow.

5.2 Set-Valued Force Laws

Up to now we have obtained a decomposition of the measure equality (5.2) into the Newton-Euler equations for future and past events (5.8) and the impact equations (5.6) of the system. The system's dynamics, however, is not yet completely determined by equations (5.6) and (5.8), since some force laws have still to be specified in order to express the forces $(\mathbf{f}^+, \mathbf{f}^-)$ and the impulsions $\mathbf{F}_i$ in terms of the system's kinematic state $(\mathbf{q}, \mathbf{u})$ and time t. These force laws are also the only source for the discontinuities in $\mathbf{u}(t)$ and $\dot{\mathbf{u}}(t)$, since all the terms introduced so far in (5.6), (5.8) depend smoothly on $(\mathbf{q}, \mathbf{u}, t)$. From now we are only interested in the impact-free motion of the system, described by equations (5.8). Impacts and impact laws are beyond the scope of the book, hence equation (5.6) is not further discussed.

By the common understanding of mechanics, the accelerations of a system are caused by the impressed forces but not vice versa. Impressed forces may therefore depend on displacements, on velocities, and on time, leading to certain values of the accelerations, but they never depend on the acceleration itself. This concept may be slightly modified to apply for *all* forces, including the classical constraint forces and even forces of intermediate type such as dry friction or the like: *Every* force in a dynamic system may be expressed by a force law depending exclusively on $(\mathbf{q}, \mathbf{u}, t)$. The most general force laws without hysteresis are therefore set-valued maps $D : \mathbb{R}^f \times \mathbb{R}^f \times \mathbb{R} \to \bar{\mathbb{R}}^f$ of the form

$$-\mathbf{f} \in D(\mathbf{q}, \mathbf{u}, t), \tag{5.9}$$

where we assume upper hemicontinuity of D as an essential property of force laws in multibody dynamics. Indeed, it is hardly imaginable for passive force elements that the set of admissible forces $-D(\mathbf{q}_0, \mathbf{u}_0, t_0)$ at a given point

$(\mathbf{q}_0, \mathbf{u}_0, t_0)$ does not contain the limits of the sets $-D(\mathbf{q}, \mathbf{u}, t)$ of neighboring points approaching $(\mathbf{q}_0, \mathbf{u}_0, t_0)$. In context with (5.8) the force law (5.9) becomes

$$\begin{aligned} -\mathbf{f}^+ &\in D(\mathbf{q}, \mathbf{u}^+, t), \\ -\mathbf{f}^- &\in D(\mathbf{q}, \mathbf{u}^-, t) \end{aligned} \tag{5.10}$$

where $D(\mathbf{q}, \mathbf{u}^+, t)$ and $D(\mathbf{q}, \mathbf{u}^-, t)$ are obviously different when $\mathbf{u}^+ \neq \mathbf{u}^-$. For continuous velocities ($\mathbf{u}^+ = \mathbf{u}^-$) both, $-\mathbf{f}^-$ and $-\mathbf{f}^+$, belong to the same set $D(\mathbf{q}, \mathbf{u}, t)$, but even in this case they might be different. One might say that the (probably different) values of $\mathbf{f}^+$ and $\mathbf{f}^-$ are "chosen by the dynamics of the system," as we will see. Equality $\mathbf{f}^+ = \mathbf{f}^-$ holds, for example, if the map D is additionally single-valued.

The inclusion (5.9) is too general for applications. We therefore restrict ourselves to the investigation of set-valued force laws that are derived by generalized differentiation from non-smooth, non-convex potential functions

$$D(\mathbf{q}, \mathbf{u}, t) = \bar{\partial}_q V(\mathbf{q}, t) + \bar{\partial}_u \Phi(\mathbf{q}, \mathbf{u}, t), \tag{5.11}$$

providing convex sets where the forces are from. Hereby, $\bar{\partial}$ denotes the generalized gradient [123] which is an upper hemicontinuous map [9], V is the *displacement potential*, and Φ is the non-integrable *velocity potential*, often also called the dissipation function. Although we have introduced potential functions in (5.11) we are not yet interested in solving any minimization problem as it is done at this point, for example, in statics. We have primarily stated D in terms of generalized gradients since the latter provide a convenient mathematical tool to express set-valuedness of a certain kind of force laws that appear frequently in applications. We additionally assume that V is lower semicontinuous in $\mathbf{q}$ and a smooth function of t and, similarly, that Φ is lower semicontinuous in $\mathbf{u}$ and it depends smoothly on $(\mathbf{q}, t)$.

By equation (5.11) we have excluded any classical force laws with twisted vector fields such as they appear in rotor dynamics, oil bearings, and the like, as well as Coulomb friction which shares similar properties and which also can not be represented by potential theory. Dry friction (i.e. Coulomb friction with given normal force), however, is still included. Even the assumption of convex force domains is a restriction with respect to application problems. Non-convex sets occur, for example, in a very reasonable manner when prestressed bilateral constraints with Coulomb friction are modeled. We refer to [126] for comments on the formulation and evaluation. Note also that maximal monotony does not apply to (5.11) since up to now neither convexity nor even subdifferentially regularity is assumed for the potentials V and Φ to hold. Due to (5.11) the forces $\mathbf{f}$ in (5.9) may now be split into

$$\mathbf{f} = \mathbf{f}_V + \mathbf{f}_\Phi \quad \text{with} \quad \begin{cases} -\mathbf{f}_V \in \bar{\partial}_q V(\mathbf{q}, t), \\ -\mathbf{f}_\Phi \in \bar{\partial}_u \Phi(\mathbf{q}, \mathbf{u}, t). \end{cases} \tag{5.12}$$

which yields, after substitution of (5.9)–(5.12) into (5.8), the differential inclusions of the impact-free motion for future and past events,

$$\begin{aligned} \mathbf{M}(\mathbf{q},t)\,\dot{\mathbf{u}}^+ - \mathbf{h}(\mathbf{q},\mathbf{u}^+,t) &\in -\bar{\partial}_q V(\mathbf{q},t) - \bar{\partial}_{u^+}\Phi(\mathbf{q},\mathbf{u}^+,t), \\ \mathbf{M}(\mathbf{q},t)\,\dot{\mathbf{u}}^- - \mathbf{h}(\mathbf{q},\mathbf{u}^-,t) &\in -\bar{\partial}_q V(\mathbf{q},t) - \bar{\partial}_{u^-}\Phi(\mathbf{q},\mathbf{u}^-,t). \end{aligned} \tag{5.13}$$

5.3 Scalar Potential Functions

In application problems one frequently is confronted with the class of one-dimensional force laws. They might be treated in a convenient and rather general manner by potential theory because rotations do not exist in one-dimensional spaces. We therefore discuss potentials of the form

$$\begin{aligned} V(\mathbf{q},t) &= \sum_{k=1}^{l} v_k(g_k(\mathbf{q},t)), \\ \Phi(\mathbf{q},\mathbf{u},t) &= \sum_{k=l+1}^{n} \varphi_k(\gamma_k(\mathbf{q},\mathbf{u},t)), \end{aligned} \tag{5.14}$$

which are also the main subject of these notes. Hereby, $v_k(g_k)$ and $\varphi_k(\gamma_k)$ denote any of l scalar displacement and $(n-l)$ scalar velocity potentials, and g_k and γ_k are the relative displacements and the relative velocities within the corresponding force elements, respectively. Generally, g_k depends smoothly on $\mathbf{q}$ and t and is strictly differentiable, whereas γ_k in the velocity potentials φ_k can be assumed to be an affine function on $\mathbf{u}$. Indeed, up to now there is not a single counter-example of an (uncontrolled) multibody system where this property does not apply, i.e. where a nonlinear dependence of γ_k on $\mathbf{u}$ has been observed. This discussion has been extensively performed in classical mechanics for the existence of nonholonomic nonlinear bilateral constraints. We can therefore write the relative velocities γ_k for both, the displacement and the velocity potentials, together with their time derivatives being some kind of relative accelerations [116], as

$$\begin{aligned} \gamma_k &= \mathbf{w}_k^T \mathbf{u} + \tilde{w}_k, \\ \dot{\gamma}_k &= \mathbf{w}_k^T \dot{\mathbf{u}} + \bar{w}_k. \end{aligned} \tag{5.15}$$

Here, $\mathbf{w}_k(\mathbf{q},t)$ and $\tilde{w}_k(\mathbf{q},t)$ result from differentiation of $g_k(\mathbf{q},t)$ for $k = 1,\dots,l$, i.e. $\mathbf{w}_k = \partial g_k/\partial \mathbf{q}$ and $\tilde{w}_k = \partial g_k/\partial t$ with $\gamma_k = \dot{g}_k$ almost everywhere, and they are given as smooth functions of $\mathbf{q}$ and t for $k = l+1,\dots,n$. For $\bar{w}_k$ in the right equation of (5.15) we obtain $\bar{w}_k = \mathbf{u}^T\dot{\mathbf{w}}_k + \dot{\tilde{w}}_k$ with $\dot{\mathbf{w}}_k = (\partial \mathbf{w}_k/\partial \mathbf{q})\,\mathbf{u} + (\partial \mathbf{w}_k/\partial t)$ and $\dot{\tilde{w}}_k = (\partial \tilde{w}_k/\partial \mathbf{q})\,\mathbf{u} + (\partial \tilde{w}_k/\partial t)$. Note that all the properties of $\mathbf{q}(t)$, $\mathbf{u}(t)$, $\dot{\mathbf{u}}(t)$ from Chapter 4 do also apply to g_k, γ_k, $\dot{\gamma}_k$ when considered as functions of time. In particular, the relative displacements $g_k(\mathbf{q}(t),t)$ are absolutely continuous, the relative velocities $\gamma_k(\mathbf{q}(t),\mathbf{u}(t),t)$

are of bounded variations with $\gamma_k = \dot{g}_k$ a.e., and $\dot{\gamma}_k(\mathbf{q}(t), \mathbf{u}(t), \dot{\mathbf{u}}(t), t)$ exists whenever $\mathbf{u}(t)$ and $\dot{\mathbf{u}}(t)$ do.

Applying the chain rule [24] on (5.14) we are now able to state the generalized gradients of V and Φ as

$$\begin{aligned} \bar{\partial}_q V(\mathbf{q}, t) &\subset \sum_{k=1}^{l} \mathbf{w}_k \, \bar{\partial} v_k(g_k), \\ \bar{\partial}_u \Phi(\mathbf{q}, \mathbf{u}, t) &\subset \sum_{k=l+1}^{n} \mathbf{w}_k \, \bar{\partial} \varphi_k(\gamma_k), \end{aligned} \tag{5.16}$$

where equality holds if, for non-vanishing $\mathbf{w}_k$, the potentials $v_k(g_k)$ and $\varphi_k(\gamma_k)$ are in addition subdifferentially regular, i.e. if the contingent and the tangent cone to the epigraphs of v_k and φ_k coincide at every point $(g_k, v_k(g_k))$ and $(\gamma_k, \varphi_k(\gamma_k))$, respectively. Under this assumption the functions $\mathbf{q} \to V(\mathbf{q}, t)$ and $\mathbf{u} \to \Phi(\mathbf{q}, \mathbf{u}, t)$ become also regular in this sense, and one has excluded, for example, the so-called "reintrant corner points" in the domain of $\mathbf{q} \to V(\mathbf{q}, t)$ at which d'Alembert/Lagrange's principle fails [80], [39]. Setting

$$\begin{aligned} \mathbf{f}_V &= \sum_{k=1}^{l} \mathbf{w}_k \, \lambda_k, \\ \mathbf{f}_\Phi &= \sum_{k=l+1}^{n} \mathbf{w}_k \, \lambda_k \end{aligned} \tag{5.17}$$

the overall force $\mathbf{f} = \mathbf{f}_V + \mathbf{f}_\Phi$ then equals the sum of n single forces $\mathbf{w}_k \lambda_k$ with $\mathbf{w}_k$ being a given force direction in $\mathrm{I\!R}^f$, called a *generalized force direction*, and λ_k being the associated unknown scalar value of the force. Upon substitution of (5.12), (5.17) into (5.16) and by comparing the left and the right hand members of the resulting inclusion, we obtain for regular potentials the n scalar set-valued force laws

$$\begin{aligned} -\lambda_k &\in \bar{\partial} v_k(g_k) \qquad && \text{for } k = 1, \ldots, l, \\ -\lambda_k &\in \bar{\partial} \varphi_k(\gamma_k) \qquad && \text{for } k = l+1, \ldots, n. \end{aligned} \tag{5.18}$$

Finally, we state the complete set of equations that describe the impact-free motion of a system being subjected to scalar force potentials. After having expressed $\mathbf{f}^{\pm}$ in (5.8) by (5.12) and (5.17) we obtain for future and past events

$$\begin{aligned} & \mathbf{M}\dot{\mathbf{u}}^+ - \mathbf{h}^+ - \sum_{k=1}^{n} \mathbf{w}_k \lambda_k^+ = 0 \\ & \quad -\lambda_k^+ \in \bar{\partial} v_k(g_k), \quad g_k = g_k(\mathbf{q}, t), \quad k = 1, \ldots, l \\ & \quad -\lambda_k^+ \in \bar{\partial} \varphi_k(\gamma_k^+), \quad \gamma_k^+ = \mathbf{w}_k^T \mathbf{u}^+ + \tilde{w}_k, \quad k = l+1, \ldots, n \\ & \mathbf{M}\dot{\mathbf{u}}^- - \mathbf{h}^- - \sum_{k=1}^{n} \mathbf{w}_k \lambda_k^- = 0 \\ & \quad -\lambda_k^- \in \bar{\partial} v_k(g_k), \quad g_k = g_k(\mathbf{q}, t), \quad k = 1, \ldots, l \\ & \quad -\lambda_k^- \in \bar{\partial} \varphi_k(\gamma_k^-), \quad \gamma_k^- = \mathbf{w}_k^T \mathbf{u}^- + \tilde{w}_k, \quad k = l+1, \ldots, n \end{aligned} \tag{5.19}$$

where the scalar contact laws have been taken from (5.18), and the relative kinematic terms g_k and $\gamma_k^{\pm}$ obey (5.15), see e.g. [37]. Note that all the terms involved in the force laws and in the relative kinematics expressions are well-defined physical magnitudes. We will show by example (Figure 5.1) how these terms have to be derived and how they may be interpreted from the physical point of view.

5.4 On the Modeling of Force Laws

The introduction of forces and force laws into a multibody system is an abstract process that chooses reasonable mathematical descriptions of the physical interactions between bodies in a real device. This process belongs to the so-called "modeling" of the mechanical system, and is only based on the physical understanding of the real device and on the special problem that has to be investigated. There are different models for the same physical device, depending on what has to be examined, and there is no general rule on which model should be chosen and how it should look like. However, as soon as one class of models has been judged to be suitable, for example a rigid body model, there are several restrictions on the continuing process of modeling in order not to leave the chosen framework, and to ensure consistency as far as any possible.

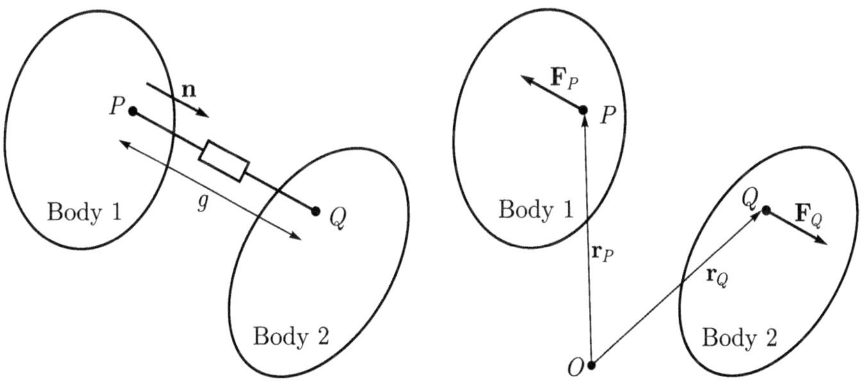

Figure 5.1. Scalar force element in rigid body dynamics

For one-dimensional force elements connecting two rigid bodies as in Figure 5.1, for example, the following steps have to be done: First of all, since we are dealing with forces but not with moments, the points P, Q at which the force element acts have to be specified which is done by defining the corresponding displacements $\mathbf{r}_P(\mathbf{q}, t)$ and $\mathbf{r}_Q(\mathbf{q}, t)$ relative to some inertially fixed point O. Secondly, the force direction and the direction in which the relative displacement g (or velocity γ) of the force element is measured have to be defined. These directions may be different in general, as for example in the case of a sliding contact with Coulomb friction, but coincide for one-dimensional potential functions. We define this direction by the unit vector $\mathbf{n}(\mathbf{q}, t) \in \mathbb{R}^3$. With $(\mathbf{r}_P, \mathbf{r}_Q, \mathbf{n})$ given, the relative displacement g of the force element becomes

$$g = \mathbf{n}^T (\mathbf{r}_Q - \mathbf{r}_P), \tag{5.20}$$

and the forces $\mathbf{F}_P = -\mathbf{F}_Q$ may now be expressed as

$$\mathbf{F}_P = -\mathbf{n}\lambda, \qquad \mathbf{F}_Q = +\mathbf{n}\lambda. \tag{5.21}$$

By this formulation, λ is not only proportional to the scalar value of $\mathbf{F}_Q$ and $-\mathbf{F}_P$ as often treated in context with Lagrange's equations of first kind, but it is the scalar value itself. Note that for Coulomb friction problems one must use the force value itself, because otherwise the equations of motion become wrong. Finally, one has to choose a force law which is, in our case, of the form

$$-\lambda \in \bar{\partial} v(g). \tag{5.22}$$

With equations (5.20)–(5.22) the force interaction between the two bodies in Figure 5.1 is fully determined. We still want to show how the generalized force direction $\mathbf{w}$ occurring in the Newton-Euler equations as well as in the relative velocity γ is obtained from (5.20)–(5.22). This may be done conveniently by using the invariant virtual work δW of the forces which reads in the configuration space

$$\delta W = \mathbf{f}^T \delta \mathbf{q} \quad \text{with } \mathbf{f} = \mathbf{w}\lambda \tag{5.23}$$

and, on the other hand, in $\mathbb{R}^3$

$$\delta W = \mathbf{F}_P^T \delta \mathbf{r}_P + \mathbf{F}_Q^T \delta \mathbf{r}_Q. \tag{5.24}$$

By introducing the Jacobians of translation at points P and Q, i.e. $\mathbf{J}_P = \partial \mathbf{r}_P / \partial \mathbf{q}$ and $\mathbf{J}_Q = \partial \mathbf{r}_Q / \partial \mathbf{q}$, we are able to express the virtual displacements in (5.24) as $\delta \mathbf{r}_P = \mathbf{J}_P \, \delta \mathbf{q}$ and $\delta \mathbf{r}_Q = \mathbf{J}_Q \, \delta \mathbf{q}$. With (5.21) we therefore obtain

$$\delta W = \lambda \mathbf{n}^T (\mathbf{J}_Q - \mathbf{J}_P) \, \delta \mathbf{q} \tag{5.25}$$

and we may identify $\mathbf{w}$ from (5.23) as

$$\mathbf{w} = (\mathbf{J}_Q - \mathbf{J}_P)^T \mathbf{n}. \tag{5.26}$$

Finally, we show that $\delta g = \mathbf{w}^T \delta \mathbf{q}$. We obtain from (5.20) the virtual relative displacements δg as

$$\delta g = \mathbf{n}^T (\delta \mathbf{r}_Q - \delta \mathbf{r}_P) + (\mathbf{r}_Q - \mathbf{r}_P)^T \delta \mathbf{n}, \tag{5.27}$$

where the second term vanishes since $\|\mathbf{n}\| = 1$, hence $\delta\mathbf{n} \perp \mathbf{n} \parallel (\mathbf{r}_Q - \mathbf{r}_P)$. Expressing $\delta\mathbf{r}_P = \mathbf{J}_P\,\delta\mathbf{q}$, $\delta\mathbf{r}_Q = \mathbf{J}_Q\,\delta\mathbf{q}$ as above we obtain

$$\delta g = \mathbf{n}^T (\mathbf{J}_Q - \mathbf{J}_P)\,\delta\mathbf{q}, \tag{5.28}$$

which is the desired solution, i.e. $\delta g = \mathbf{w}^T\,\delta\mathbf{q}$ with $\mathbf{w}$ from (5.26).

Note that P and Q in Figure 5.1 are not necessarily body-fixed points. They may also move relatively to the bodies, for example at their contours. To every fixed time t, every solution $(\mathbf{r}_P, \mathbf{r}_Q)$ of the minimization problem

$$g = \inf_{\substack{P\in\text{Body 1}\\ Q\in\text{Body 2}}} \|\mathbf{r}_Q - \mathbf{r}_P\| \tag{5.29}$$

with $\text{int(Body 1)} \cap \text{int(Body 2)} = \emptyset$ then defines a pair of possible contact points with g being the distance of the bodies and $\mathbf{n}$ being, for example, the outward normal of body 1 at P. Depending on the physical requirements of the contact model one might then choose a compliance element, a unilateral constraint, or anything else between the so defined points P and Q. For more comments on the evaluation of the kinematic contact problem we refer to [37], [116] and especially to Chapter 10 where the spatial contact problem is treated.

This approach is also not restricted to forces only. Moments as they appear, for example, if drilling friction is modeled, or in frictional clutches and one-way clutches, may be handled in exactly the same manner. The term $\mathbf{w}$ then denotes the interaction direction of the moments in $\mathbb{R}^f$, λ becomes the scalar value of the moment, and $(\mathbf{J}_P, \mathbf{J}_Q)$ in (5.26) have to be substituted by the corresponding Jacobians of rotation, see e.g. [48]. Screw-type force elements with a kinematic coupling of translational and rotational motion can be also treated, as well as force elements that depend on the velocities γ, instead of the relative displacements g, as already mentioned. Equation (5.19) covers one-dimensional force elements only. Note that the most general interaction laws may be f-dimensional, agreeing with the dimension of the configuration space. Force laws between two single rigid bodies have at most six dimensions, and spatial Coulomb friction, for example, has to be regarded as three-dimensional, since up to now no decomposition into independent force laws of lower dimensions is known.

We want to point out again that the use of physical Lagrange multipliers λ, i.e. the scalar values of the forces or moments, should be preferred in context with bilateral or unilateral constraints. By doing so the generalized force direction $\mathbf{w}(\mathbf{q}, t)$ stated in (5.26) is a well-defined geometrical magnitude that usually does not vanish at any point $\mathbf{q}$, a necessary condition on the decomposition (5.16) to hold as an equality. Moreover, ill-posedness of the system (5.19), probably occurring due to the values of λ, is then physically based but not introduced artificially by improper but correct mathematical formulation.

6. Representation of Scalar Force Laws

In this chapter we are mainly interested in a further discussion of the force laws from the last two equations in (5.19). We will present a reasonable decomposition of the potentials v_k and φ_k with respect to application problems in multibody dynamics, and we will put together some equivalent representations of the force laws as it has been done, for example, in [101] by using concepts from non-smooth and (non-)convex analysis [9], [123], [24], [121]. We will see that some basic set-valued force element, called a unilateral primitive, plays a fundamental role in this decomposition procedure, a pleasant property that allows engineers to arrange and to design new force characteristics in a conventional manner. As a result, the connection of classical force elements and unilateral primitives parallel to each other and in series will lead to well-known displacement and velocity force elements that appear frequently in practical applications.

6.1 Decomposition into Unilateral Primitives

This section treats the decomposition of one-dimensional force laws into parallel and serial connections of *unilateral primitives*, i.e. set-valued force laws which can be expressed by one single *complementarity condition* $x \geq 0$, $y \geq 0$, $xy = 0$. Mechanical examples are later shown in Figures 6.6 and 6.7. The force laws from the last two equations in (5.19) are of the form

$$-\lambda \in \bar{\partial}\pi(s), \tag{6.1}$$

where the indices + and k have been omitted for brevity. In all what follows one may set $(\pi, s) = (v, g)$ for displacement potentials or $(\pi, s) = (\varphi, \gamma)$ if velocity potentials are of interest. We assume a decomposition of the potential π into three parts: A differentiable function π_D which covers the classical force laws that are expressed by functions; an indicator function π_S that restricts the relative displacement (or velocity) s of the force element to an interval $S = [s_A, s_E] \subset \mathrm{I\!R}$ with $-\infty \leq s_A \leq s_E \leq +\infty$; and a convex potential π_P with polyhedral epigraph. We therefore obtain

$$\pi(s) = \pi_D(s) + \pi_S(s) + \pi_P(s) \tag{6.2}$$

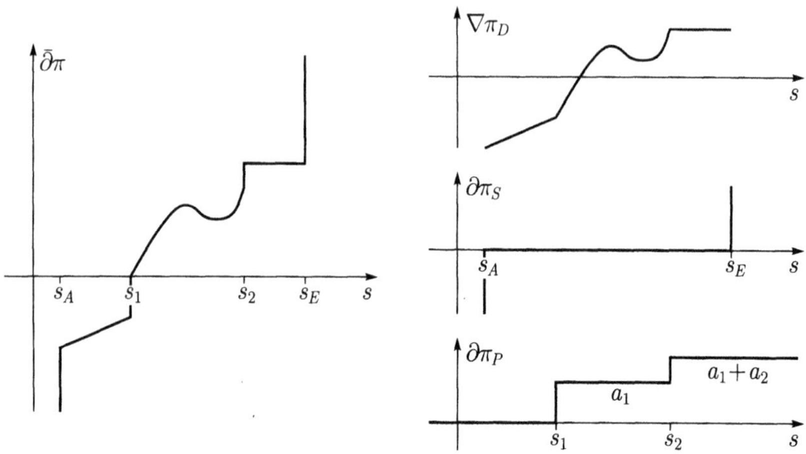

Figure 6.1. Decomposition of $\bar{\partial}\pi$ into $\nabla\pi_D$, $\partial\pi_S$, and $\sum\partial\pi_{Pi}$

with the indicator π_S being defined by

$$\pi_S(s) = \begin{cases} 0 & \text{if } s_A \leq s \leq s_E, \\ +\infty & \text{else.} \end{cases} \tag{6.3}$$

Without loss of generality with respect to the class of force laws obtainable by generalized differentiation of (6.2), we may assume the convex potential π_P with polyhedral epigraph to be of the form $\pi_P = \sum_{i=1}^m \pi_{Pi}$ with

$$\pi_{Pi}(s) = \begin{cases} 0 & \text{if } s \leq s_i \\ a_i(s - s_i) & \text{if } s \geq s_i \end{cases} \tag{6.4}$$

where $s_A \leq s_1 \leq \ldots \leq s_m \leq s_E$ and $a_i \geq 0$. The potentials $\pi_{Pi}(s - s_i)$ are therefore the support functions of the sets $P_i = [0, a_i]$ that will later express the set-valued property of the force law (6.1) at the points s_i. Assigning forces to each of the members in (6.2),

$$\lambda = \lambda_D + \lambda_S + \textstyle\sum_{i=1}^m \lambda_{Pi}, \tag{6.5}$$

and performing differentiation we arrive with the $m + 2$ force laws

$$\begin{aligned} -\lambda_D &= \nabla\pi_D(s), \\ -\lambda_S \in \partial\pi_S(s) &= \begin{cases} 0 & \text{if } s_A < s < s_E, \\ [-\infty, 0] & \text{if } s_A = s < s_E, \\ [0, +\infty] & \text{if } s_A < s = s_E, \\ [-\infty, +\infty] & \text{if } s_A = s = s_E, \end{cases} \\ -\lambda_{Pi} \in \partial\pi_{Pi}(s) &= \begin{cases} 0 & \text{if } s < s_i, \\ [0, a_i] & \text{if } s = s_i, \\ a_i & \text{if } s > s_i. \end{cases} \end{aligned} \tag{6.6}$$

Hereby, the generalized gradient $\bar{\partial}$ in (6.1) has been replaced by the gradient ∇ and the subdifferential ∂ since π_D is differentiable and π_S, π_{Pi} are convex.

The resulting decomposition of the force law (6.1) is depicted in Figure 6.1. On the right hand side one can see the three different portions of the force law corresponding to the three equations in (6.6), where the monotone set-valued step function in the lower figure is itself composed of the m single steps of height a_i, represented by $\partial\pi_{Pi}$. From the mechanical point of view this decomposition has to be understood as an arrangement of $m + 2$ force elements with characteristics (6.6) parallel to each other, since the same relative displacement s is used but the forces λ_D, λ_S, λ_{Pi} are different. Note that we have allowed only non-negative steps ($a_i \geq 0$) resulting in a maximal monotone operator $\partial\pi_P$ which will finally secure uniqueness of the solution. Negative steps might also occur in applications, for example if the breaking of fibers in cables is modeled [6]. The following decomposition into unilateral primitives, however, is not affected by negative steps; above all, we have them excluded in order not to violate the assumed subdifferentially regularity of the potential functions, and because of their restricted importance for multibody systems. For an extensive overview of set-valued and non-monotone force laws we especially refer to [6].

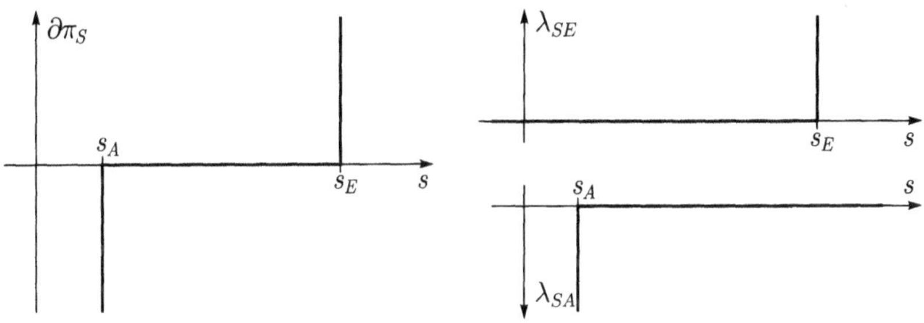

Figure 6.2. Boundary forces as a parallel connection of two unilateral primitives

We will now decompose the indicator force law $\partial\pi_S$ into two unilateral primitives, each of them being represented by one complementarity condition. As depicted in Figure 6.2, this can be done by a *parallel connection* of the unilateral primitives, since different forces λ_{SA} and $-\lambda_{SE}$ have to be *added* for the *same* relative displacement s. The resulting decomposition then writes

$$-\lambda_S \in \partial\pi_S(s) \quad \Leftrightarrow \quad \lambda_S = \lambda_{SA} - \lambda_{SE} \text{ with:} \tag{6.7}$$

$$\lambda_{SA} \geq 0,\ (s - s_A) \geq 0,\ \lambda_{SA}\,(s - s_A) = 0,$$
$$\lambda_{SE} \geq 0,\ (s_E - s) \geq 0,\ \lambda_{SE}\,(s_E - s) = 0,$$

and corresponds, with respect to the admissible displacements s, to the *intersection* of the two sets $S_A := \{s \mid s \leq s_E\}$ and $S_E := \{s \mid s \geq s_A\}$ stated in equation (6.7), i.e. $s \in S_A \cap S_E$.

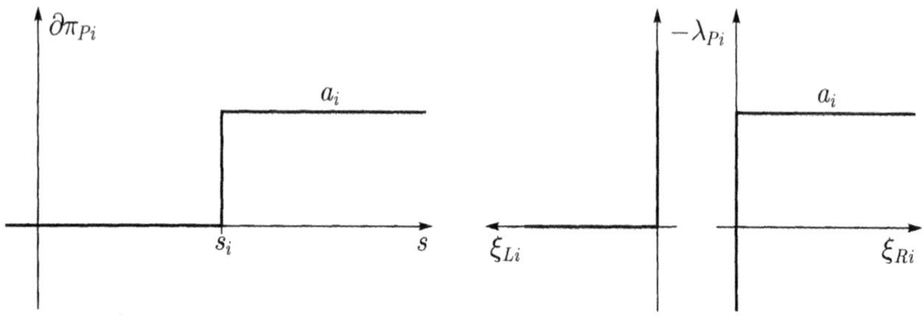

Figure 6.3. Representation of a single step by two unilateral primitives in series

In contrast, the decomposition of a single step $\partial\pi_{Pi}$ works different, see Figure 6.3. The unilateral primitives have now to be arranged in *series*, i.e. the force λ_{Pi} in both characteristics is the *same*, accompanied by *different* displacements ξ_{Li} and ξ_{Ri},

$$-\lambda_{Pi} \in \partial\pi_{Pi}(s) \quad \Leftrightarrow \quad s - s_i = \xi_{Ri} - \xi_{Li} \text{ with:} \qquad (6.8)$$
$$\xi_{Ri} \geq 0,\ (\lambda_{Pi} + a_i) \geq 0,\ \xi_{Ri}\,(\lambda_{Pi} + a_i) = 0,$$
$$\xi_{Li} \geq 0,\ -\lambda_{Pi} \geq 0,\ \xi_{Li}\,\lambda_{Pi} = 0.$$

Note that this decomposition requires an additional kinematic variable, say ξ_{Li}, similar as for classical force elements where inner variables have to be used when connected in series. As a result, the admissible displacements $s \in \mathbb{R}$ are now obtained by the *union* of the sets $\Xi_{Ri} := \{s \,|\, s \geq s_i\}$ and $\Xi_{Li} := \{s \,|\, s \leq s_i\}$, i.e. they are *not* restricted to any subset of $\mathbb{R} = \Xi_{Ri} \cup \Xi_{Li} \ni s$.

6.2 Variational Formulations and Upper Subderivatives

In this section we show by the example of the force law (6.1), (6.6) how to switch between the generalized gradient and an equivalent variational representation that uses upper subderivatives. By the definition of the generalized gradient [9], equation (6.1) can be stated as a hemivariational inequality

$$-\lambda \in \bar{\partial}\pi(s) \quad \Leftrightarrow \quad \pi^{\uparrow}(s, s^\star - s) \geq -\lambda\,(s^\star - s), \quad s \in S, \quad \forall s^\star. \qquad (6.9)$$

We recall that $\pi : \mathbb{R} \to \mathbb{R} \cup \{+\infty\}$ has been assumed to be a proper lower semicontinuous (lsc.) and subdifferentially regular function with effective domain $S = [s_A, s_E] \subset \mathbb{R}$, and that $\bar{\partial}$ is the generalized gradient, a closed convex set. We denote by $\pi^{\uparrow}(s, \delta s)$ the upper subderivative [123] of π at s in the direction of the virtual displacements $\delta s := s^\star - s$, which is identical to the (more general) epiderivative [9] when applied to lsc. functions. Note that the virtual displacements δs are not restricted to any subset of $\mathbb{R}$, since (6.9) has to hold for every $s^\star$. The function $\delta s \to \pi^{\uparrow}(s, \delta s)$ is positively homogeneous, lsc., and convex, since $\pi^{\uparrow}(s, \delta s) > -\infty\ \forall \delta s \in \mathbb{R}$ in our case, and its epigraph is the tangent cone to epi π at $(s, \pi(s))$. Since π is subdifferentially regular, we could also use, instead of $\pi^{\uparrow}(s, \delta s)$, the epicontingent derivative $\pi^{\downarrow}(s, \delta s)$ which gives the same value in this case. Generally, $\delta s \to \pi^{\downarrow}(s, \delta s)$ is also positively homogeneous and lsc. but not convex, since its epigraph is the contingent cone to epi π at $(s, \pi(s))$. From (6.9) it is obvious that the epiderivative of π may be also written as

$$\pi^{\uparrow}(s, \delta s) = \sup\{-\lambda\, \delta s \mid -\lambda \in \bar{\partial}\pi(s)\} \tag{6.10}$$

if the generalized gradient $\bar{\partial}\pi(s)$ is known. Vice versa, the generalized gradient may be expressed by

$$\bar{\partial}\pi(s) = \{-\lambda \mid \pi^{\uparrow}(s, \delta s) \geq -\lambda\, \delta s \quad \forall \delta s\} \tag{6.11}$$

for $s \in S$ and given $\pi^{\uparrow}(s, \delta s)$.

Let us now show by example how $\bar{\partial}\pi(s)$ in (6.9) is obtained from $\pi^{\uparrow}(s, \delta s)$. The epiderivatives of $\pi_D(s)$, $\pi_S(s)$, $\pi_P(s)$ may be directly written down by using (6.3) and (6.4), compare also Figure 6.1, or, equivalently, by applying (6.10) on the gradients (6.6). We obtain

$$\begin{aligned}
\pi_D^{\uparrow}(s, \delta s) &= \nabla \pi_D(s)\, \delta s \\
\pi_S^{\uparrow}(s, \delta s) &= \begin{cases} +\infty & \text{if } s = s_A,\ \delta s < 0 \qquad \text{or } s = s_E,\ \delta s > 0 \\ 0 & \text{if } s_A \leq s < s_E,\ \delta s \geq 0 \quad \text{or } s_A < s \leq s_E,\ \delta s \leq 0 \end{cases} \\
\pi_{Pi}^{\uparrow}(s, \delta s) &= \begin{cases} a_i\, \delta s & \text{if } s > s_i,\ \forall \delta s \quad \text{or } s = s_i,\ \delta s \geq 0 \\ 0 & \text{if } s < s_i,\ \forall \delta s \quad \text{or } s = s_i,\ \delta s \leq 0 \end{cases}
\end{aligned} \tag{6.12}$$

and, since additivity of the epiderivatives holds in our case,

$$\pi^{\uparrow}(s, \delta s) = \begin{cases} (\nabla \pi_D(s) + \sum_{j=0}^{i} a_j)\, \delta s & \text{if } s_i \leq s < s_{i+1},\ \delta s \geq 0 \\ & \text{or } s_i < s \leq s_{i+1},\ \delta s \leq 0 \\ +\infty & \text{if } s = s_{m+1},\ \delta s > 0 \\ & \text{or } s = s_0,\ \delta s < 0, \end{cases} \tag{6.13}$$

where i runs from 0 to m, and (s_0, s_{m+1}, a_0) are defined to be $(s_A, s_E, 0)$. Equation (6.13) may now be used to derive $\bar{\partial}\pi(s)$ from (6.11). For $s \neq s_i$ the inequality $\pi^{\uparrow}(s, \delta s) \geq -\lambda\, \delta s$ in (6.11) becomes

$$\nabla \pi_D(s) + \sum_{j=0}^{i} a_j \leq -\lambda \leq \nabla \pi_D(s) + \sum_{j=0}^{i} a_j, \quad s_i < s < s_{i+1} \tag{6.14}$$

which is actually an equality that covers the single-valued case of the force law. For $s = s_i$, $i = 1, \ldots, m$ the evaluation of (6.11) with respect to (6.13) yields

$$\nabla \pi_D(s_i) + \sum_{j=0}^{i-1} a_j \leq -\lambda \leq \nabla \pi_D(s_i) + \sum_{j=0}^{i} a_j, \quad i = 1, \ldots, m, \tag{6.15}$$

an interval of length a_i where the forces $-\lambda$ are taken from. The last two cases take into account the force law at $s_0 = s_A$ and $s_{m+1} = s_E$,

$$-\infty \leq -\lambda \leq \nabla\pi_D(s_0) \quad \text{and} \quad \nabla\pi_D(s_{m+1}) + \sum_{j=0}^{m} a_j \leq -\lambda \leq +\infty, \tag{6.16}$$

providing unbounded sets due to the unilateral constraints that are expressed by the indicator potential function π_S. As a result, (6.14)–(6.16) is again the generalized gradient $\bar{\partial}\pi(s) = \nabla\pi_D(s) + \partial\pi_S(s) + \partial\pi_P(s)$ as in (6.6).

6.3 The Convex Case: Conjugate Potentials and Duality

We turn now to convex potentials. In this case, the generalized gradient agrees with the subdifferential as proven in [122] and reviewed in (6.17)–(6.20). For a closed proper convex potential $\pi(s)$ there are, indeed, at least five equivalent ways to state the force law (6.1). These five expressions are presented in (6.24) and require the conjugate of the potential function. The main purpose of this section is thus to provide the conjugate of our force potential, which will be done in the paragraphs after equation (6.24).

Suppose now that $\pi(s)$ is a convex function. This affects only π_D since π_S and π_P are already convex. We additionally assume strict convexity and co-finiteness of π_D in order to express later the conjugate of π_D by the Legendre transformation, see [121]. We denote by $\pi'(s, \delta s)$ the directional derivative of π at s in the direction δs (with values $\pm\infty$ as acceptable possibilities). With [123]

$$\pi^{\uparrow}(s, \delta s) = \liminf_{\delta s' \to \delta s} \pi'(s, \delta s') \tag{6.17}$$

one always has $\pi^{\uparrow}(s, \delta s) \leq \pi'(s, \delta s)$ and, due to the convexity of π, $\pi'(s, s^\star - s) \leq \pi(s^\star) - \pi(s)$. The elements $-\lambda$ of the generalized gradient $\bar{\partial}\pi(s)$ are therefore related to the values of π by

$$-\lambda\,(s^\star - s) \leq \pi^{\uparrow}(s, s^\star - s) \leq \pi'(s, s^\star - s) \leq \pi(s^\star) - \pi(s) \tag{6.18}$$

which holds for any $s^\star$, and which can be seen immediately from the right equation in (6.9). On the other hand, the subdifferential $\partial\pi(s)$ for convex functions is defined by

$$-\lambda \in \partial\pi(s) \quad \Leftrightarrow \quad \pi(s^\star) - \pi(s) \geq -\lambda\,(s^\star - s), \quad s \in S, \quad \forall s^\star, \tag{6.19}$$

see e.g. [121], or, equivalently, in terms of the directional derivative by

$$-\lambda \in \partial\pi(s) \quad \Leftrightarrow \quad -\lambda\,(s^\star - s) \leq \pi'(s, s^\star - s), \quad s \in S, \quad \forall s^\star. \tag{6.20}$$

From (6.9), (6.18), and (6.20) one immediately obtains $\bar{\partial}\pi(s) \subset \partial\pi(s)$. Moreover, $\bar{\partial}\pi(s) = \partial\pi(s)$ since (6.9) and (6.20) are required to hold for every $s^\star$, leading by (6.17) to an equality $\pi^{\dagger}(s, s_1^\star - s) = \pi'(s, s_2^\star - s)$ as the most restrictive case in (6.18).

For convex functions π the force law may therefore be expressed as

$$-\lambda \in \partial\pi(s) \quad \text{or} \quad \pi(s^\star) - \pi(s) \geq -\lambda\,(s^\star - s), \quad s \in S, \quad \forall s^\star \tag{6.21}$$

with $\bar{\partial}\pi(s) \equiv \partial\pi(s)$. It is also possible to solve the force law $-\lambda \in \partial\pi(s)$ for s in order to obtain the *inverse* inclusion $s \in (\partial\pi)^{-1}(-\lambda)$. For π being closed, proper, convex, the set $(\partial\pi)^{-1}$ equals the subdifferential $\partial\pi^\star$ of a closed, proper, convex function $\pi^\star$ that is defined by

$$\pi^\star(-\lambda) = \sup_s \{-s\lambda - \pi(s)\} \tag{6.22}$$

and that is called the conjugate of π, see [121]. From this definition one directly obtains Fenchel's inequality

$$-s\lambda \leq \pi(s) + \pi^\star(-\lambda) \quad \forall s,\ \forall \lambda \tag{6.23}$$

that is proven to hold as an equality if and only if $-\lambda \in \partial\pi(s)$. By these results, the force law (6.21) may now be stated by any of the five equivalent expressions

$$\begin{array}{ll} \text{for } s \in S,\ -\lambda \in P: & -s\lambda = \pi(s) + \pi^\star(-\lambda), \\ \text{for } s \in S: & -\lambda \in \partial\pi(s), \\ \text{for } s \in S,\ \forall s^\star: & -\lambda\,(s^\star - s) \leq \pi(s^\star) - \pi(s), \\ \text{for } -\lambda \in P: & s \in \partial\pi^\star(-\lambda), \\ \text{for } -\lambda \in P,\ \forall\lambda^\star: & s\,(-\lambda^\star + \lambda) \leq \pi^\star(-\lambda^\star) - \pi^\star(-\lambda), \end{array} \tag{6.24}$$

where P denotes the effective domain of $\pi^\star$. The conjugate $\pi^\star$, closely related to the Legendre transformation in the smooth case, may be also interpreted as follows: As a consequence of the convexity of π, the graph $\Gamma := \{(s, -\lambda)\,|\, -\lambda \in \partial\pi(s)\}$ of the subdifferential mapping $s \to \partial\pi(s)$ is a complete non-decreasing curve in $\mathbb{R}^2$, and so is its inverse $\Gamma^\star = \{(-\lambda, s)\,|\,(s, -\lambda) \in \Gamma\}$, see Figure 6.4. This curve $\Gamma^\star$ may then be regarded as the subdifferential mapping $-\lambda \to \partial\pi^\star(-\lambda)$ of some convex function $\pi^\star$, uniquely determined up to an additive constant c, that equals the conjugate of π for a proper choice of c.

We now compute the conjugate $\pi^\star(s)$ of the force potential $\pi(s)$ from (6.2). This will be done by first applying (6.22) on each of the members π_D, π_S, π_P of π in order to obtain the conjugates $\pi_D^\star$, $\pi_S^\star$, $\pi_P^\star$, and then by performing infimal convolution on $\pi_D^\star$, $\pi_S^\star$, $\pi_P^\star$ that finally gives $\pi^\star$.

Let us start with the smooth potential π_D. Since $\pi_D(s)$ is differentiable, strictly convex and co-finite by assumption, conjugacy turns out to be the same as the classical Legendre transformation. The second line in (6.24) then simply is $-\lambda_D = \nabla\pi_D(s)$, and $\partial\pi_D^\star(-\lambda_D) = (\partial\pi_D)^{-1}(-\lambda_D)$ from the fourth

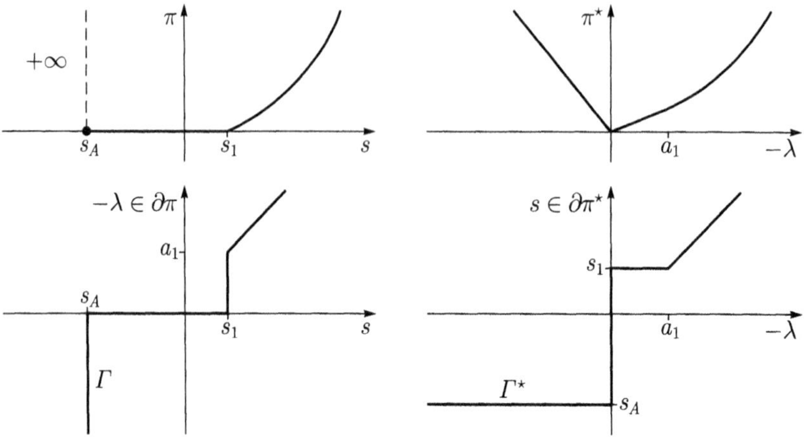

Figure 6.4. Convex force potential, conjugate function, and subdifferentials

line may therefore be rewritten as $(\nabla\pi_D)^{-1}(-\lambda_D) = s$. From the first equation in (6.24) we then obtain, after elimination of s, the Legendre conjugate of π_D,

$$\pi_D^\star(-\lambda_D) = -(\nabla\pi_D)^{-1}(-\lambda_D)\,\lambda_D - \pi_D((\nabla\pi_D)^{-1}(-\lambda_D)), \tag{6.25}$$

a likewise differentiable, strictly convex and co-finite function with π as Legendre conjugate in turn.

By applying (6.22) on the indicator function π_S (6.3) of the closed interval $S = [s_A, s_E]$ with $s_A \leq s_E$ we calculate the conjugate $\pi_S^\star(-\lambda_S)$ to be

$$\begin{aligned}\pi_S^\star(-\lambda_S) &= \sup_s \{-s\,\lambda_S - \pi_S(s)\}\\ &= \sup_s \{-s\,\lambda_S \,|\, s \in S = [s_A, s_E]\}\\ &= \begin{cases} -s_A\,\lambda_S & \text{if } -\lambda_S \leq 0,\\ -s_E\,\lambda_S & \text{if } -\lambda_S \geq 0.\end{cases}\end{aligned} \tag{6.26}$$

This positively homogeneous convex function is called the support function of the set $[s_A, s_E]$, where $\pi_S^{\star\star} = \pi_S$ can be easily verified in our case because π_S is closed.

In order to finally derive $\pi_P^\star$ we first compute the conjugates $\pi_{Pi}^\star$ of π_{Pi} from (6.4) by applying (6.22) again. We obtain

$$\begin{aligned}\pi_{Pi}^\star(-\lambda_{Pi}) &= \sup_s \{-s\,\lambda_{Pi} - \pi_{Pi}(s)\}\\ &= \sup_s \left\{\begin{matrix} -s\,\lambda_{Pi} & \text{if } s \leq s_i\\ -s\,\lambda_{Pi} - a_i\,(s - s_i) & \text{if } s \geq s_i\end{matrix}\right\}\\ &= \begin{cases} -s_i\,\lambda_{Pi} & \text{if } 0 \leq -\lambda_{Pi} \leq a_i\\ +\infty & \text{else}\end{cases}\\ &= -s_i\,\lambda_{Pi} + \pi_{Ai}^\star(-\lambda_{Pi}),\end{aligned} \tag{6.27}$$

where $\pi^\star_{Ai}(-\lambda_{Pi})$ in the last line denotes the indicator function of the interval $[0, a_i]$. It is shown in [121] that summation of convex functions corresponds to infimal convolution of the conjugates. With

$$\pi_P(s) = \pi_{P1}(s) + \ldots + \pi_{Pm}(s) \tag{6.28}$$

the conjugate $\pi^\star_P(-\lambda_P)$ is therefore obtained via

$$\pi^\star_P(-\lambda_P) = (\pi^\star_{P1} \square \ldots \square \pi^\star_{Pm})(-\lambda_P) \tag{6.29}$$

with the infimal convolution being defined by

$$\begin{aligned} &\inf_{-\lambda_{Pi}} \{\pi^\star_{P1}(-\lambda_{P1}) + \ldots + \pi^\star_{Pm}(-\lambda_{Pm}) \,|\, \lambda_P = \lambda_{P1} + \ldots + \lambda_{Pm}\} \\ &=: (\pi^\star_{P1} \square \ldots \square \pi^\star_{Pm})(-\lambda_P) \end{aligned} \tag{6.30}$$

Expressing $\pi^\star_{Pi}(-\lambda_{Pi})$ by (6.27) yields

$$\pi^\star_P(-\lambda_P) = \inf_{-\lambda_{Pi}} \left\{ \sum_{i=1}^{m} -s_i\,\lambda_{Pi} + \pi^\star_{Ai}(-\lambda_{Pi}) \,\middle|\, \lambda_P = \sum_{i=1}^{m} \lambda_{Pi} \right\} \tag{6.31}$$

which defines the function $\pi^\star_P(-\lambda_P)$. Suppose that $-\lambda_{Pi} \notin [0, a_i]$ for some i in (6.31). In this case the corresponding indicator functions $\pi^\star_{Ai}(-\lambda_{Pi})$ take values $+\infty$, and $\pi^\star_P(-\lambda_P)$ becomes $+\infty$. This happens for every $-\lambda_P \notin \sum_{i=1}^{m}[0, a_i]$, i.e. for $-\lambda_P < 0$ or $-\lambda_P > \sum_{i=1}^{m} a_i$. In the other case ($0 \le -\lambda_P \le \sum_{i=1}^{m} a_i$) the evaluation of the infimum condition yields

$$\begin{aligned} &\inf_{-\lambda_{Pi}} \left\{ \sum_{i=1}^{m} -s_i\,\lambda_{Pi} \,\middle|\, \lambda_P = \sum_{i=1}^{m} \lambda_{Pi},\ -\lambda_{Pi} \in [0, a_i] \right\} \\ &= \begin{cases} -s_1\,\lambda_P & \text{if } 0 \le -\lambda_P \le a_1 \\ s_1\,a_1 - s_2\,(\lambda_P + a_1) & \text{if } a_1 \le -\lambda_P \le a_1 + a_2 \\ s_1\,a_1 + s_2\,a_2 - s_3\,(\lambda_P + a_1 + a_2) & \text{if } a_1 + a_2 \le -\lambda_P \le a_1 + a_2 + a_3 \\ \vdots & \vdots \end{cases} \end{aligned} \tag{6.32}$$

since $s_1 \le \ldots \le s_m$ by assumption. The conjugate $\pi^\star_P(-\lambda_P)$ therefore writes

$$\pi^\star_P(-\lambda_P) = \begin{cases} +\infty & \text{if } -\lambda_P < 0 \\ \sum_{i=0}^{k-1}(s_i - s_k)\,a_i - s_k\,\lambda_P & \text{if } \sum_{i=0}^{k-1} a_i \le -\lambda_P \le \sum_{i=0}^{k} a_i \\ +\infty & \text{if } \sum_{i=0}^{m} a_i < -\lambda_P \end{cases} \tag{6.33}$$

with k running from 0 to m and $a_0 = 0$ as already defined below equation (6.13). Formula (6.33) may also be obtained in a very straightforward geometrical manner from $\pi^\star_{Pi}(-\lambda_{Pi}) = -s_i\,\lambda_{Pi} + \pi^\star_{Ai}(-\lambda_{Pi})$ in (6.27). Infimal convolution of some functions $\pi^\star_{Pi}$ actually corresponds to the addition of

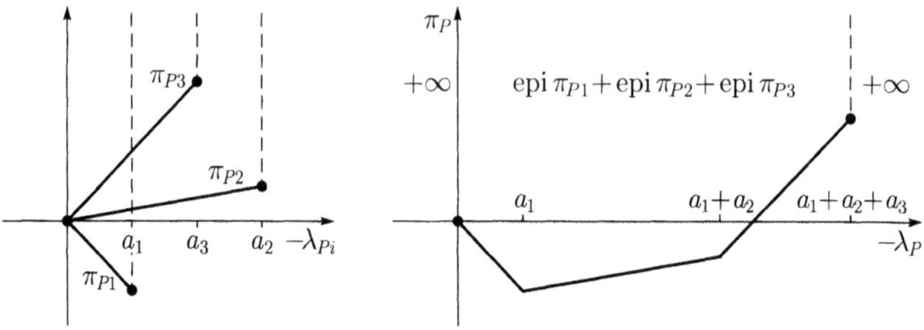

Figure 6.5. Potential π_P obtained by infimal convolution of π_{Pi}

their epigraphs. The result, $\sum_{i=1}^{m} \operatorname{epi} \pi^\star_{Pi}$, is then the epigraph of the desired function $\pi^\star_P$ as it is shown in Figure 6.5 for $m = 3$. As a consequence, the effective domain A of $\pi^\star_P$ is simply the sum of the effective domains $A_i = [0, a_i]$ of $\pi^\star_{Pi}$, i.e. $A = [0, \sum_{i=1}^{m} a_i]$.

With $\pi^\star_D$ from (6.25), $\pi^\star_S$ from (6.26), and $\pi^\star_P$ from (6.33) one is finally able to state the overall conjugate $\pi^\star$ of $\pi = \pi_D + \pi_S + \pi_P$ from (6.2) by infimal convolution as

$$\pi^\star(-\lambda) = (\pi^\star_D \Box \pi^\star_S \Box \pi^\star_P)(-\lambda), \tag{6.34}$$

or, by definition (6.30), more explicitly as

$$\pi^\star(-\lambda) = \inf_{-\lambda_D, -\lambda_S, -\lambda_P} \{\pi^\star_D(-\lambda_D) + \pi^\star_S(-\lambda_S) + \pi^\star_P(-\lambda_P) \mid \lambda = \lambda_D + \lambda_S + \lambda_P\}. \tag{6.35}$$

This function will later be used, in some modified form, to define a convex optimization problem on the forces $-\lambda$ that will turn out to be the dual of the principle of least constraints.

6.4 Force Elements in Engineering Dynamics

Finally we will discuss applications of the aforementioned decomposition $\pi = \pi_D + \pi_S + \pi_P$ with respect to force characteristics that are frequently met in engineering dynamics.

The basic element on the displacement level is the unilateral constraint from the first line in Figure 6.6. It admits relative displacements in only one direction, accompanied by an unbound set of forces that prevent interpenetration ($g < 0$) at $g = 0$ and that also contribute to impacts in the form of impulsions. Arranging two unilateral constraints parallel to each other we obtain a force element that restricts movement to a closed interval $[0, s]$ describing the free play in the mesh of gears or the like. For vanishing play ($s = 0$) this element changes into a classical bilateral constraint, i.e. a set-valued force law that restricts the displacements g to a single point $g = 0$.

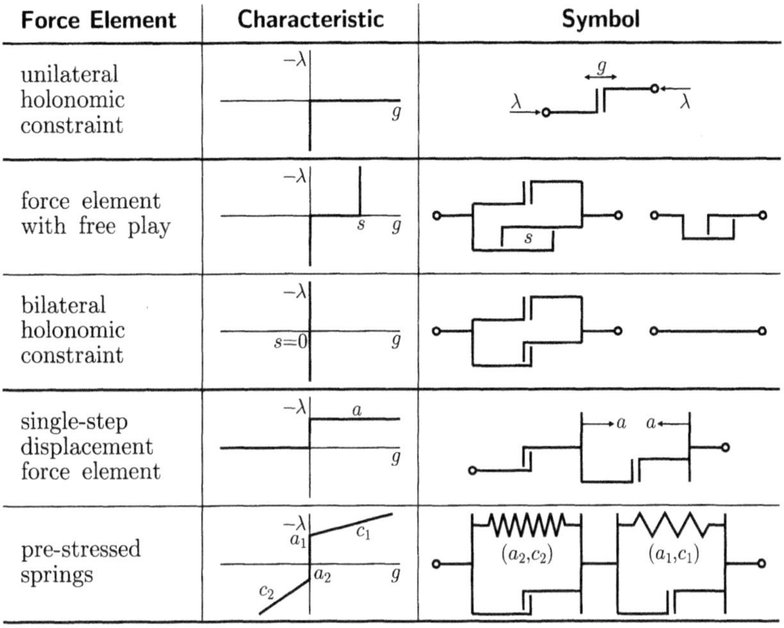

Figure 6.6. Displacement force elements and decompositions

The case $s < 0$ is not of interest since it corresponds to an inconsistent formulation of the constraints with an empty set of admissible displacements. The single-step displacement force element from the fourth line in Figure 6.6 is composed of two unilateral constraints in series together with a constant load a. It occurs in machine dynamics in a slightly modified form as a pre-stressed spring, such as depicted in the lower diagrams of Figure 6.6: The left spring is pre-compressed by an amount $-a_2/c_2$ and is held in that position due to the parallel connected unilateral constraint. Similarly, the right spring is pre-stretched by a_1/c_1. Suppose now that some forces $\lambda \geq 0$ are applied to the ends of the force element. As long as $0 \leq \lambda \leq -a_2$ the displacement g of the force element will remain equal to zero since the right unilateral constraint does not allow movement directed towards $g < 0$, and the left unilateral constraint also stays closed because the load λ is carried by the pre-compressed spring (a_2, c_2). For forces $\lambda > -a_2$, however, the left unilateral constraint will open by displacements $g = -(\lambda + a_2)/c_2$. Similarly, the right unilateral constraint opens if the entire force element is stretched by values $\lambda < -a_1$. Finally, note that force elements of the type shown in the last two lines of Figure 6.6 do not contribute to impacts since the set of transferable forces is bounded.

One-way clutches are the technical realization of unilateral nonholonomic constraints, i.e. unilateral constraints that may only be formulated on the velocity level and that restrict only the velocities to some directions ($\gamma \geq 0$) but not the displacements, see first line of Figure 6.7. They may be found,

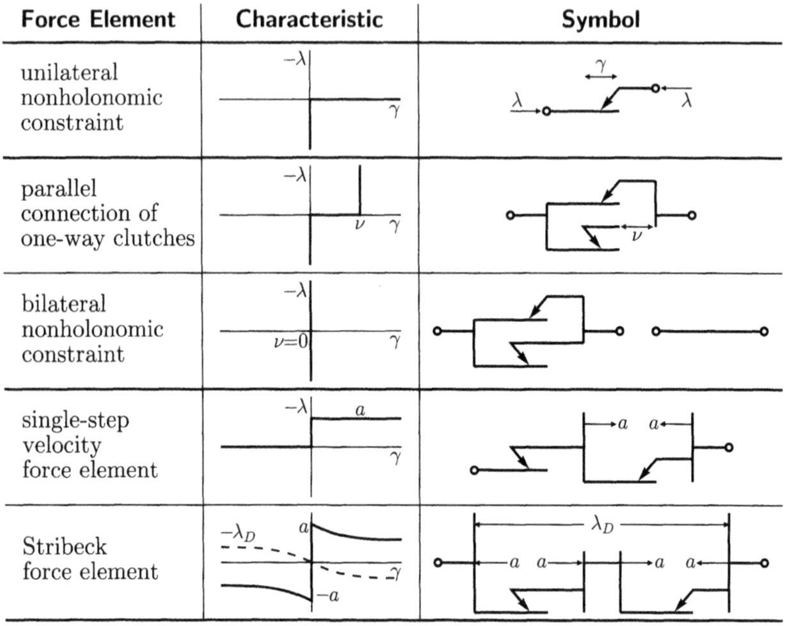

Figure 6.7. Velocity force elements and decompositions

for example, between the pinion and the rear wheel of bicycles and also in automated gears. If a one-way clutch is externally driven with a given velocity ν and connected in parallel to a second clutch, the resulting force element allows only relative velocities between 0 and ν. For $\nu = 0$ we obtain bilateral nonholonomic constraints as depicted in the third line of Figure 6.7. Since the sets of forces are unbound for $\gamma = \{0, \nu\}$, these elements have also to be regarded in the impact equations of the system. Single-step velocity force elements (fourth line in Figure 6.7) characterize dry friction by two unilateral nonholonomic constraints in series and additional constant loads a. The non-monotone set-valued Stribeck curve, for example, is composed in that manner with a smooth part λ_D acting parallel. Dry friction elements do not contribute to impacts since the forces are bounded. This is different to the Coulomb friction force law where the overall contact force has to be taken from a cone, i.e. an unbound set. As a consequence, frictional impacts are only due to Coulomb friction and do *not* occur if dry friction models as above are used.

From the foregoing discussion it is obvious that only two more elements, the unilateral constraint on the displacement and on the velocity level, are needed in order to expand the classical one-dimensional force laws to the set-valued case. All characteristics obtained by any combination represent standard elements in machine dynamics which are met in nearly every technical device.

7. Force Laws on Different Kinematic Levels

After having discussed several representations of the force laws $-\lambda \in \bar{\partial}\pi(s)$ in Chapter 6, we will now come back to our main task, to determine the right and the left accelerations $\dot{\mathbf{u}}^+$ and $\dot{\mathbf{u}}^-$ in the equations of motion (5.19). This might be done easily if the force laws $-\lambda_i \in \bar{\partial}v_i(g_i)$, $-\lambda_j \in \bar{\partial}\varphi_j(\gamma_j)$ are single-valued: For given values of $\mathbf{q}$ and $\mathbf{u}$ one has just to determine the $g_i(\mathbf{q}, t)$ and the $\gamma_j(\mathbf{q}, \mathbf{u}, t)$ in order to obtain the λ_i and λ_j by the corresponding single-valued force laws. The values $\dot{\mathbf{u}}$ may then be uniquely obtained by solving the linear system of equations in the first line of (5.19) with symmetric and positive definite mass matrix $\mathbf{M}(\mathbf{q}, t)$ and given gyroscopical acceleration vector $\mathbf{h}(\mathbf{q}, \mathbf{u}, t)$.

This direct approach does not work for set-valued force laws because the forces $-\lambda_k$ may be taken from an entire interval, depending on the values of g_k and γ_k as we have learned in Chapter 6. This might lead to the wrong impression that also the accelerations are set-valued, and that nothing more specific can be said about their values and about the trajectories of the system. By looking more carefully at equation (5.19) we recognize, however, that the only terms of interest are the left and the right acceleration $\dot{\mathbf{u}}^-$ and $\dot{\mathbf{u}}^+$, being different for only an at most countable set of time points, and that only the left and the right limits of the forces $\lambda_k^{\pm}(t)$ as functions of time have to be regarded.

Moreover, equation (5.19) holds for every time t within the time interval I under consideration when $\{t_{j,\text{right}}\} = \{t_{j,\text{left}}\} = \emptyset$, so we can additionally use some continuity properties of $\mathbf{q}(t)$ and $\mathbf{u}^{\pm}(t)$ in order to obtain more information about the evolution of $\lambda_k^{\pm}(t)$ and $\dot{\mathbf{u}}^{\pm}(t)$. This is – perhaps – the most unaccustomed step in this contribution since we will now reformulate the force laws in a manner that they will also depend on the accelerations of the system. Such, equation (5.19) will change into a quadratic program with optimal solutions $-\lambda_k^{\pm}$ that can also be rewritten as a linear complementarity problem as it will be shown in Chapter 8.

7.1 Continuity Properties of the Trajectories

Let us first study some consequences of the one-sided continuity of the velocities and the displacements, both considered as functions of time. We recall that we assumed $\mathbf{u}(t)$ to be a function of bounded variations, leading to a right-continuous $\mathbf{u}^+(t)$ and an absolutely continuous $\mathbf{q}(t)$ which is obtained via integration (4.11) from $\mathbf{u}(t)$, thus $\dot{\mathbf{q}}(t) = \mathbf{u}(t) = \mathbf{u}^+(t)$ for almost every t. Considered as functions of time, the right velocities $t \to \gamma^+(\mathbf{q}(t), \mathbf{u}^+(t), t)$ and the relative displacements $t \to g(\mathbf{q}(t), t)$ from the third and the second line of (5.19) are therefore also right-continuous and absolutely continuous, respectively, and $g(t) = g(t_A) + \int_{t_A}^{t} \gamma(\tau)\,\mathrm{d}\tau$ with $\dot{g} = \gamma = \gamma^+$ for [dt]-a.e. t. In order to reformulate the displacement force laws on the velocity level which is done in the next section, we have first to answer the following question: Is it possible to deduce from the value of the right velocity γ^+ at time t some information about the evolution of the displacements g for an adjoining time interval?

Proposition 7.1.1. *Let $t \to f^+(t)$ be a right-continuous function, and let*

$$F(t) := F(0) + \int_0^t f^+(\tau)\,\mathrm{d}\tau. \tag{7.1}$$

If $f^+(0) > 0$, there is a time interval $(0, t^\star)$ such that $F(t) > F(0)$ for all $t \in (0, t^\star)$.

Proof. Equation (7.1) can be rewritten as

$$F(t) - F(0) = t\, f^+(0) + \int_0^t (f^+(\tau) - f^+(0))\,\mathrm{d}\tau. \tag{7.2}$$

With the abbreviation

$$r(t) := \int_0^t (f^+(\tau) - f^+(0))\,\mathrm{d}\tau \tag{7.3}$$

equation (7.2) becomes

$$F(t) - F(0) = t\left(f^+(0) + \frac{r(t)}{t}\right). \tag{7.4}$$

Let $\varepsilon > 0$. By the right-continuity of $f^+(t)$ at $t = 0$, there is a $t^\star$ such that

$$\left| f^+(t) - f^+(0) \right| < \varepsilon \quad \text{for } 0 < t < t^\star. \tag{7.5}$$

With the help of (7.5) we may now estimate $r(t)$ from (7.3) to be

$$\begin{aligned} |\,r(t)\,| &= \left| \int_0^t (f^+(\tau) - f^+(0))\,\mathrm{d}\tau \right| \\ &\le \int_0^t |\,f^+(\tau) - f^+(0)\,|\,\mathrm{d}\tau \\ &\le \varepsilon \int_0^t \mathrm{d}\tau \\ &= \varepsilon t. \end{aligned} \tag{7.6}$$

For $f^+(0) > 0$ given, we choose ε such that $0 < \varepsilon < f^+(0)$. By (7.5), (7.6) there is a $t^\star$ such that $|r(t)|/t \le \varepsilon < f^+(0)$ for every $t \in (0, t^\star)$, hence $F(t) - F(0) > 0$ by (7.4). This proves that $f^+(0) > 0$ is sufficient for $F(t) > F(0)$ in some interval $t \in (0, t^\star)$. □

When we now take in Proposition 7.1.1 $f^+(t)$ and $F(t)$ to be the relative velocity $\gamma^+(t)$ and displacement $g(t)$, we have the desired information, i.e. that $g(t) > g(0)$ for some interval $0 < t < t^\star$ when $\gamma^+(0) > 0$. In the same manner we obtain $g(t) < g(0)$ in $(0, t^\star)$ when $\gamma^+(0) < 0$.

Suppose now that we are looking backwards in time, i.e. that $t^\star < t < 0$. In order to obtain the corresponding results for g one has to use, instead of $\gamma^+(t)$, the left velocity $\gamma^-(t)$ which is left-continuous with $\gamma^-(t) = \gamma^+(t)$ a.e. The estimation of $|r(t)|$ in (7.6) becomes $|r(t)| \le -\varepsilon t$ in this case, and $\gamma^-(0) > 0$ is then sufficient for $g(t) < g(0)$ for some interval $t^\star < t < 0$. Also, if $\gamma^-(0) < 0$, then $g(t) > g(0)$ for $t^\star < t < 0$.

Analogous statements on the evolution of the velocities may be derived when (g, γ^+) is replaced by $(\gamma^+, \dot\gamma^+)$, $\dot\gamma^+(t)$ is assumed to be right-continuous at $t = 0$, and the interval $(0, t^\star)$ is impact-free. Then they also apply to $(\gamma^-, \dot\gamma^-)$ in the form discussed above.

7.2 Displacement Force Laws on Acceleration Level

In this section we express the displacement force law $\bar\partial v(g)$ from the left equation in (5.18) (index k omitted) first on velocity level and then even on acceleration level. The latter then involves the second time derivative $\dot\gamma^+$ of its relative displacement $g(t)$ at the points of set-valuedness. We perform this step on the three portions of the force law that have been introduced in (6.1)–(6.6), and we restrict ourselves to the case leading to future events, expressed by the right limits of the forces and the right accelerations in the corresponding directional Newton-Euler equation. The force law under consideration reads

$$-\lambda \in \bar\partial v(g) \quad \text{with } \lambda = \lambda_D + \lambda_S + \lambda_P, \qquad v(g) = v_D(g) + v_S(g) + v_P(g) \tag{7.7}$$

with subscripts D, S and P referring to the differentiable potential, the indicator potential, and the polyhedral potential, respectively. The potentials v_S and v_P considered here are defined in a slightly simpler form than in (6.3), (6.4), describing only *one* unilateral constraint ($g \ge s_A$) and only *one* set-valued single-step at $g = s_P$ as the most essential parts of the force element (6.1), (6.2). The more complicated situation (6.3), (6.4) may be handled in exactly the same manner since it corresponds just to a parallel connection of the force elements considered here, as it has been shown in Section 6.1.

The differentiable portion v_D of the overall force potential v covers classical displacement force laws via

$$-\lambda_D = \nabla v_D(g) \tag{7.8}$$

and requires no further discussion since it is single-valued: The force λ_D may be directly computed from (7.8) for a given relative displacement $g(\mathbf{q}, t)$.

We discuss now the second potential in (7.7), the indicator potential $v_S(g)$ as defined in (7.9), restricting the values of g to $[s_A, +\infty]$ and leading to the force law

$$\begin{aligned} -\lambda_S \in \partial v_S(g) &= \begin{cases} 0 & \text{if } g > s_A \\ [-\infty, 0] & \text{if } g = s_A \end{cases} \\ \text{with } v_S(g) &:= \begin{cases} 0 & \text{if } g \geq s_A \\ +\infty & \text{else.} \end{cases} \end{aligned} \tag{7.9}$$

In the following we will reformulate (7.9), first on the velocity level and then also on the acceleration level, by keeping in mind that only the right limit $\lambda_S^+(t) = \lim_{\tau \downarrow 0} \lambda_S(t+\tau)$ as defined in (4.1) is used in the directional Newton-Euler equation (5.19) for future events. We will now compute this force limit for the different cases contained in the force law (7.9).

Suppose first that, for a given time t, the relative displacement is $g(t) > s_A$. Due to the absolute continuity of $t \to g(t)$ we may find a $\tau^\star$ such that $g(t+\tau) > s_A$ for every $\tau \in (0, \tau^\star)$. The force law (7.9) stated at $t+\tau$ then reads $-\lambda_S(t+\tau) \in \partial v_S(g(t+\tau)) = 0$ since $g(t+\tau) > s_A$. This holds for every $\tau \in (0, \tau^\star)$, hence $\lambda_S^+(t)$ may be computed by the limit $-\lambda_S^+(t) = \lim_{\tau \downarrow 0} -\lambda_S(t+\tau) = 0$. For $g > s_A$ we therefore have $-\lambda_S^+ = 0$ as stated in the first line of (7.10).

Suppose now that $g(t) = s_A$ for some given time t, leading to forces $-\lambda_S(t) \in [-\infty, 0]$ by the second line in (7.9). There are three different cases for the right velocity $\gamma^+(t)$: First, consider $\gamma^+(t) < 0$. By the results of Section 7.1 this would lead to displacements $g(t+\tau) < s_A$ for $\tau \in (0, \tau^\star)$ and some $\tau^\star$, violating the restriction $g \geq s_A$ in (7.9). Hence, $\gamma^+(t) \geq 0$ is a necessary condition on the right velocity at $g(t) = s_A$, usually provided by some impact equations that are not discussed here. Secondly, we consider $\gamma^+(t) > 0$, leading again by the results of Section 7.1 to displacements $g(t+\tau) > s_A$. Hence, $-\lambda_S^+(t) = 0$ may be obtained by the same reasoning as in the case $g(t) > s_A$ discussed first. For $g = s_A$, $\gamma^+ > 0$ the force limit therefore is $-\lambda_S^+ = 0$ as stated in the second line of (7.10). In the last case, $\gamma^+ = 0$, nothing more specific can be said about the values of $-\lambda_S^+(t)$. The only thing we know is that $-\lambda_S^+ \in [-\infty, 0]$ as stated in the third line of (7.10) (note that $g \to \partial v_S(g)$ is an upper semicontinuous map!). Putting together these results we may now rewrite (7.9) in the form

$$-\lambda_S^+ \in \begin{cases} 0 & \text{if } g > s_A \\ 0 & \text{if } g = s_A, \gamma^+ > 0 \\ [-\infty, 0] & \text{if } g = s_A, \gamma^+ = 0 \end{cases} \tag{7.10}$$

which we call the *representation* of the force law *on the velocity level.*

By comparing the last two lines in (7.10) with the force law (7.9) one recognizes a very similar structure: Obviously, the two different cases on γ^+ for $g = s_A$ in (7.10) may also be expressed by the subdifferential of an (artificially introduced) indicator potential $\varphi_S(\gamma^+)$ as defined in (7.11). The force law (7.10) may then be written in a more compact form as

$$-\lambda_S^+ \in \begin{cases} 0 & \text{if } g > s_A \\ \partial\varphi_S(\gamma^+) & \text{if } g = s_A \end{cases}$$
$$\text{with } \varphi_S(\gamma^+) := \begin{cases} 0 & \text{if } \gamma^+ \geq 0 \\ +\infty & \text{else.} \end{cases} \tag{7.11}$$

We call the function $\varphi_S(\gamma^+)$ the *velocity potential associated with* $v_S(g)$, in this context being defined in a different manner than in the paper [38]. Moreover, one observes that the epicontingent derivative of v_S at g in the direction γ^+

$$v_S^{\downarrow}(g, \gamma^+) = \begin{cases} 0 & \text{if } g > s_A \\ 0 & \text{if } g = s_A, \gamma^+ \geq 0 \\ +\infty & \text{if } g = s_A, \gamma^+ < 0 \end{cases} \tag{7.12}$$

coincides with $\varphi_S(\gamma^+)$ in (7.11) when stated at $g = s_A$. The velocity potential is therefore identified to be

$$\gamma^+ \to v_S^{\downarrow}(s_A, \gamma^+) = \varphi_S(\gamma^+). \tag{7.13}$$

Finally, one recognizes from the first line in (7.12) that $\partial_{\gamma^+} v_S^{\downarrow}(g, \gamma^+) = 0$ for $g > s_A$ which covers the first case in (7.10). Since $\partial_{\gamma^+} v_S^{\downarrow}(s_A, \gamma^+) = \partial\varphi_S(\gamma^+)$ takes into account the second and the third case in (7.10) as stated in (7.11), the overall force law on the velocity level may also be expressed in the very compact form

$$-\lambda_S^+ \in \partial_{\gamma^+} v_S^{\downarrow}(g, \gamma^+) \subset \partial v_S(g). \tag{7.14}$$

We therefore have three equivalent representations of the displacement force law (7.9) on the velocity level, given by the equations (7.10), (7.11) and (7.14).

The representation of the force law in (7.14) as well as the definition of the velocity potential $\varphi_S(\gamma^+)$ in (7.11) seems to be very artificially. There is, however, hope for some general rules on how to obtain representations of potential force laws like in (7.14) that do *not* require a decomposition as in (7.10) or (7.11). The underlying idea is to realize that the force law (7.7) has to hold for every time point, then to approximate the potential $v(g(t))$ for succeeding times $t + \tau$ by some Taylor expansion of the form $v(g(t+\tau)) = v(g(t)) + v^{\downarrow}(g(t), \gamma^+(t))\,\tau$, and finally to take the limit (lim sup) of the generalized gradients for $\tau \downarrow 0$. In this way, the use of the epicontingent derivative is required in a quite natural manner when we think in the direction of nonregular potentials. Some attempts to perform these steps on regular potentials in the configuration space may be found in [38].

Now we reformulate the force law (7.10) on the acceleration level by performing the same steps as in (7.10)–(7.14) again. The first two cases in (7.10) remain unchanged since the force law is single-valued in some neighborhood of g when $g > s_A$ and in some neighborhood of γ^+ when $g = s_A$, $\gamma^+ > 0$. In the third case, i.e. $g = s_A$, $\gamma^+ = 0$, the force law is set-valued, and we proceed as in the paragraph preceding equation (7.10). As a necessary condition we have $\dot{\gamma}^+ \geq 0$, because $\dot{\gamma}^+ < 0$ (right-continuity is assumed) would yield $\gamma^+(t+\tau) < 0$ for $\tau \in (0, \tau^*)$ and some τ^*, leading to a constraint violation $g(t+\tau) < s_A$. For $\dot{\gamma}^+ > 0$ we have $g > s_A$ in near future, hence $-\lambda_S^+ = 0$, whereas for $\dot{\gamma}^+ = 0$ no further evaluation of the force law is possible. Altogether we have

$$-\lambda_S^+ \in \begin{cases} 0 & \text{if } g > s_A \\ 0 & \text{if } g = s_A, \gamma^+ > 0 \\ 0 & \text{if } g = s_A, \gamma^+ = 0, \dot{\gamma}^+ > 0 \\ [-\infty, 0] & \text{if } g = s_A, \gamma^+ = 0, \dot{\gamma}^+ = 0 \end{cases} \tag{7.15}$$

which is the *representation* of the displacement force law (7.9) on the *acceleration level.* As in (7.11) we realize that the last two conditions in (7.15) may be alternatively expressed by the subdifferential of an indicator function $\psi_S(\dot{\gamma}^+)$ which we call the *associated acceleration potential.* With its help the force law (7.15) is now stated as

$$-\lambda_S^+ \in \begin{cases} 0 & \text{if } g > s_A \\ 0 & \text{if } g = s_A, \gamma^+ > 0 \\ \partial\psi_S(\dot{\gamma}^+) & \text{if } g = s_A, \gamma^+ = 0 \end{cases}$$
$$\text{with } \psi_S(\dot{\gamma}^+) := \begin{cases} 0 & \text{if } \dot{\gamma}^+ \geq 0 \\ +\infty & \text{else.} \end{cases} \tag{7.16}$$

The epicontingent derivative of the velocity potential φ_S from (7.11) at γ^+ in the direction $\dot{\gamma}^+$ reads

$$\varphi_S^\downarrow(\gamma^+, \dot{\gamma}^+) = \begin{cases} 0 & \text{if } \gamma^+ > 0 \\ 0 & \text{if } \gamma^+ = 0, \dot{\gamma}^+ \geq 0 \\ +\infty & \text{if } \gamma^+ = 0, \dot{\gamma}^+ < 0 \end{cases} \tag{7.17}$$

and coincides for $\gamma^+ = 0$ with the acceleration potential ψ_S as it can be seen when comparing (7.17) for $\gamma^+ = 0$ with the definition of ψ_S in (7.16). We therefore have

$$\dot{\gamma}^+ \to \varphi_S^\downarrow(0, \dot{\gamma}^+) = \psi_S(\dot{\gamma}^+) \tag{7.18}$$

and, since $\partial_{\dot{\gamma}^+}\varphi_S^\downarrow(\gamma^+, \dot{\gamma}^+) = 0$ for $\gamma^+ > 0$, an alternative formulation of the force laws (7.15), (7.16) that reads

$$-\lambda_S^+ \in \begin{cases} 0 & \text{if } g > s_A \\ \partial_{\dot\gamma^+}\varphi_S^\downarrow(\gamma^+,\dot\gamma^+) \subset \partial\varphi_S(\gamma^+) & \text{if } g = s_A \end{cases} \subset \partial v_S(g). \qquad (7.19)$$

From the analytical point of view a more condensed form of the force law (7.19) without any explicit distinction between different cases would be highly desirable. This, however, requires the use of second directional derivatives of non-smooth functions that are quite difficult to handle.

Finally, we investigate the third potential in (7.7) which expresses the single-step displacement force law with the step of height $a > 0$ occurring at a displacement s_P. In order to perform the transition from the displacement level to the acceleration level we have to carry out exactly the same steps as for the unilateral constraint. The only difference is that neither the displacements nor the velocities or accelerations are restricted to any subsets of $\mathbb{R}$ since the single-step displacement force law as introduced in (7.20) is set-valued but not a constraint. On the displacement level the force law reads

$$-\lambda_P \in \partial v_P(g) = \begin{cases} 0 & \text{if } g < s_P, \\ [0,a] & \text{if } g = s_P \\ a & \text{if } g > s_P. \end{cases}$$

$$\text{with } v_P(g) := \begin{cases} 0 & \text{if } g \le s_P \\ a\,(g - s_P) & \text{if } g \ge s_P \end{cases} \qquad (7.20)$$

with the displacement potential $v_P(g)$ being a polyhedral convex function. The representation of (7.20) on the velocity level is given by

$$-\lambda_P^+ \in \begin{cases} 0 & \text{if } g < s_P \\ 0 & \text{if } g = s_P, \gamma^+ < 0 \\ [0,a] & \text{if } g = s_P, \gamma^+ = 0 \\ a & \text{if } g = s_P, \gamma^+ > 0 \\ a & \text{if } g > s_P \end{cases} \qquad (7.21)$$

where the second, third and fourth case may be expressed by the subdifferential of a velocity potential $\varphi_P(\gamma^+)$ being the support function of the set $[0, a]$, i.e.

$$-\lambda_P^+ \in \begin{cases} 0 & \text{if } g < s_P \\ \partial\varphi_P(\gamma^+) & \text{if } g = s_P \\ a & \text{if } g > s_P \end{cases}$$

$$\text{with } \varphi_P(\gamma^+) := \begin{cases} 0 & \text{if } \gamma^+ \le 0 \\ a\gamma^+ & \text{if } \gamma^+ \ge 0. \end{cases} \qquad (7.22)$$

With the help of the epicontingent derivative of v_P,

$$v_P^\downarrow(g,\gamma^+) = \begin{cases} 0 & \text{if } g < s_P \\ 0 & \text{if } g = s_P, \gamma^+ \le 0 \\ a\gamma^+ & \text{if } g = s_P, \gamma^+ \ge 0 \\ a\gamma^+ & \text{if } g > s_P \end{cases} \qquad (7.23)$$

we identify the velocity potential to be

$$\gamma^+ \to v_P^{\downarrow}(s_P, \gamma^+) = \varphi_P(\gamma^+), \tag{7.24}$$

similarly as in (7.13). With $\partial_{\gamma^+} v_P^{\downarrow}(g, \gamma^+)$ taking values 0 and a for $g < s_P$ and $g > s_P$, the force law on the velocity level may also be written as

$$-\lambda_P^+ \in \partial_{\gamma^+} v_P^{\downarrow}(g, \gamma^+) \subset \partial v_P(g). \tag{7.25}$$

From (7.21) we obtain the representation of the force law on the acceleration level,

$$-\lambda_P^+ \in \begin{cases} 0 & \text{if } g < s_P \\ 0 & \text{if } g = s_P, \gamma^+ < 0 \\ 0 & \text{if } g = s_P, \gamma^+ = 0, \dot\gamma^+ < 0 \\ [0, a] & \text{if } g = s_P, \gamma^+ = 0, \dot\gamma^+ = 0 \\ a & \text{if } g = s_P, \gamma^+ = 0, \dot\gamma^+ > 0 \\ a & \text{if } g = s_P, \gamma^+ > 0 \\ a & \text{if } g > s_P \end{cases} \tag{7.26}$$

which may be also stated in an equivalent way by using some acceleration potential $\psi_P(\dot\gamma^+)$, i.e.

$$-\lambda_P^+ \in \begin{cases} 0 & \text{if } g < s_P \\ 0 & \text{if } g = s_P, \gamma^+ < 0 \\ \partial\psi_P(\dot\gamma^+) & \text{if } g = s_P, \gamma^+ = 0 \\ a & \text{if } g = s_P, \gamma^+ > 0 \\ a & \text{if } g > s_P \end{cases}$$

$$\text{with } \psi_P(\dot\gamma^+) := \begin{cases} 0 & \text{if } \dot\gamma^+ \leq 0 \\ a\dot\gamma^+ & \text{if } \dot\gamma^+ \geq 0. \end{cases} \tag{7.27}$$

Again, we identify the potential $\psi_P(\dot\gamma^+)$ from the epicontingent derivative of φ_P that reads

$$\varphi_P^{\downarrow}(\gamma^+, \dot\gamma^+) = \begin{cases} 0 & \text{if } \gamma^+ < 0 \\ 0 & \text{if } \gamma^+ = 0, \dot\gamma^+ \leq 0 \\ a\dot\gamma^+ & \text{if } \gamma^+ = 0, \dot\gamma^+ \geq 0 \\ a\dot\gamma^+ & \text{if } \gamma^+ > 0. \end{cases} \tag{7.28}$$

By comparing (7.28) for $\gamma^+ = 0$ with the definition of ψ_P in (7.27) we recognize that

$$\dot\gamma^+ \to \varphi_P^{\downarrow}(0, \dot\gamma^+) = \psi_P(\dot\gamma^+) \tag{7.29}$$

which finally leads to the third representation of the force law, i.e.

$$-\lambda_P^+ \in \begin{cases} 0 & \text{if } g < s_P \\ \partial_{\dot\gamma^+} \varphi_P^{\downarrow}(\gamma^+, \dot\gamma^+) \subset \partial\varphi_P(\gamma^+) & \text{if } g = s_P \\ a & \text{if } g > s_P \end{cases} \subset \partial v_P(g). \tag{7.30}$$

By the reformulation of the displacement force laws on the acceleration level we have obtained representations that also depend on the relative accelerations. Moreover, we were able to state the set-valued force laws in terms of subdifferentials of certain functions, called the acceleration potentials, that depend on the unknown relative accelerations only, best seen from equations (7.16) and (7.27). These inclusions, together with the directional Newton-Euler equations in (5.19) and the relative accelerations in (5.15), enable us to state an equilibrium problem or an optimization problem on the unknowns $\dot{\gamma}^+$ which will be subject of Chapter 8.

7.3 Velocity Force Laws on Acceleration Level

As we have seen by the examples in Chapter 6, there are also force laws that deny a formulation on the displacement level and that may be only expressed in terms of velocities. Viscous damping, dry friction and unilateral velocity constraints belong to that class. With respect to their representation on the acceleration level they are more easy to handle than the displacement force laws, since only one kinematic level has to be changed. Indeed, the required procedure corresponds exactly to the reformulation of the displacement force laws on the velocity level with the only difference that all terms referring to displacements and velocities have to be replaced by corresponding terms on velocities and accelerations. We will therefore only present the results of this transition process, rather than repeating the entire procedure which is already stated in Section 7.2.

We consider force laws with the same structure as in (7.7) but now depending on the relative velocities $\dot{\gamma}^+$ as variables,

$$-\lambda \in \bar{\partial}\varphi(\gamma^+) \quad \text{with } \lambda = \lambda_D + \lambda_S + \lambda_P, \qquad \varphi(\gamma^+) = \varphi_D(\gamma^+) + \varphi_S(\gamma^+) + \varphi_P(\gamma^+). \tag{7.31}$$

The functions φ denote the velocity potentials, also called the dissipation functions, and the subscripts D, S and P refer to the differential portion (viscous damping), the indicator portion (unilateral velocity constraint) and the single-step portion (dry friction).

As in (7.8), the force law represented by the differentiable potential reads

$$-\lambda_D = \nabla\varphi_D(\gamma^+) \tag{7.32}$$

and is single-valued. No further investigation is necessary since λ_D may be computed uniquely by a direct evaluation of (7.32).

In order to express the indicator-type force law we first define its velocity potential $\varphi_S(\gamma^+)$ and also the associated acceleration potential $\psi_S(\dot{\gamma}^+)$, i.e.

$$\begin{aligned} \varphi_S(\gamma^+) &:= \begin{cases} 0 & \text{if } \gamma^+ \geq \nu_A \\ +\infty & \text{else} \end{cases} \\ \psi_S(\dot\gamma^+) &:= \begin{cases} 0 & \text{if } \dot\gamma^+ \geq 0 \\ +\infty & \text{else.} \end{cases} \end{aligned} \tag{7.33}$$

Hereby, ν_A is a given relative velocity that is equal to zero, for example, in the case of the unilateral nonholonomic constraint in Figure 6.7. By the definition of φ_S, the force law on the velocity level reads

$$-\lambda_S^+ \in \partial\varphi_S(\gamma^+) = \begin{cases} 0 & \text{if } \gamma^+ > \nu_A \\ [-\infty, 0] & \text{if } \gamma^+ = \nu_A \end{cases} \tag{7.34}$$

and may be transferred to the acceleration level by performing the same steps as in (7.10)–(7.14). One obtains

$$\begin{aligned} -\lambda_S^+ &\in \partial_{\dot\gamma^+}\varphi_S^\downarrow(\gamma^+, \dot\gamma^+) \\ &= \begin{cases} 0 & \text{if } \gamma^+ > \nu_A \\ \partial\psi_S(\dot\gamma^+) & \text{if } \gamma^+ = \nu_A \end{cases} \\ &= \begin{cases} 0 & \text{if } \gamma^+ > \nu_A \\ 0 & \text{if } \gamma^+ = \nu_A, \dot\gamma^+ > 0 \\ [-\infty, 0] & \text{if } \gamma^+ = \nu_A, \dot\gamma^+ = 0, \end{cases} \end{aligned} \tag{7.35}$$

which corresponds to the velocity representations (7.14), (7.11) (7.10) of the displacement indicator potential in Section 7.2, and which is the desired representation of the velocity indicator potential on the acceleration level.

Finally, we treat the single-step velocity force law. It is defined via the velocity potential $\varphi_P(\gamma^+)$ and associated acceleration potential $\psi_P(\dot\gamma^+)$

$$\begin{aligned} \varphi_P(\gamma^+) &:= \begin{cases} 0 & \text{if } \gamma^+ \leq \nu_P \\ a\,(\gamma^+ - \nu_P) & \text{if } \gamma^+ \geq \nu_P \end{cases} \\ \psi_P(\dot\gamma^+) &:= \begin{cases} 0 & \text{if } \dot\gamma^+ \leq 0 \\ a\dot\gamma^+ & \text{if } \dot\gamma^+ \geq 0 \end{cases} \end{aligned} \tag{7.36}$$

by $-\lambda_P^+ \in \partial\varphi_P(\gamma^+)$ and may be written down analogously to equation (7.20) in a more explicit form as

$$-\lambda_P^+ \in \partial\varphi_P(\gamma^+) = \begin{cases} 0 & \text{if } \gamma^+ < \nu_P \\ [0, a] & \text{if } \gamma^+ = \nu_P \\ a & \text{if } \gamma^+ > \nu_P. \end{cases} \tag{7.37}$$

The dry friction element with friction bounds $+a/2$, $-a/2$, for example, is obtained from (7.37) by setting $\nu_P = 0$ and by adjusting a differentiable force potential with $-\lambda_D = \nabla\varphi_D(\gamma^+) = -a/2$ parallel to (7.37). Following equations (7.21)–(7.25) the representation of the single-step force law (7.37) on the acceleration level reads

$$
\begin{aligned}
-\lambda_P^+ &\in \partial_{\dot\gamma^+}\varphi_P^{\downarrow}(\gamma^+,\dot\gamma^+) \\
&= \begin{cases} 0 & \text{if } \gamma^+ < \nu_P \\ \partial\psi_P(\dot\gamma^+) & \text{if } \gamma^+ = \nu_P \\ a & \text{if } \gamma^+ > \nu_P \end{cases} \\
&= \begin{cases} 0 & \text{if } \gamma^+ < \nu_P \\ 0 & \text{if } \gamma^+ = \nu_P, \dot\gamma^+ < 0 \\ [0,a] & \text{if } \gamma^+ = \nu_P, \dot\gamma^+ = 0 \\ a & \text{if } \gamma^+ = \nu_P, \dot\gamma^+ > 0 \\ a & \text{if } \gamma^+ > \nu_P . \end{cases}
\end{aligned}
\tag{7.38}
$$

In contrast to the displacement force elements, the formulation of the velocity force laws on the acceleration level (7.35), (7.38) can be done in a very compact form by using the epicontingent derivative. As already pointed out, this fails for the displacement force laws since *two* kinematic levels have to be bridged which requires second directional derivatives.

8. Index Sets and LCP-Formulation

In this chapter we give three equivalent formulations of the problem in terms of the unknown forces λ_{Sk}^{+}, λ_{Pk}^{+} from the unilateral constraints and the single-step force elements ($k = 1, \ldots, n$): A linear complementarity problem with maximal dimension $3n$ corresponding to the maximal number of unilateral primitives in use, i.e. the case that every force law is set-valued when expressed on the acceleration level; a variational inequality in terms of forces which is the dual of Gauß' principle; an associated convex minimization problem which is the dual of the principle of least constraints.

As in Chapter 7, only the evolution problem leading to future events will be discussed, and upper indices "+" will further be omitted. The directional Newton-Euler equations (5.8) with the forces $\mathbf{f}$ split up according to (5.17) and (7.7), (7.31) may therefore be written as

$$\mathbf{M}\dot{\mathbf{u}} - \mathbf{h} - \sum_{k=1}^{n} \mathbf{w}_k \lambda_{Dk} - \sum_{k=1}^{n} \mathbf{w}_k \lambda_{Sk} - \sum_{k=1}^{n} \mathbf{w}_k \lambda_{Pk} = 0, \tag{8.1}$$

where $\mathbf{w}_k(\mathbf{q}, t)$ is the generalized force direction of force element k, being itself composed of three sub-elements connected in parallel. As introduced in (5.14), the numbers k refer to a displacement force element for $k \in \{1, \ldots, l\}$, and to a velocity force element for $k \in \{l+1, \ldots, n\}$. In addition, one has by (5.14) for each $k \in \{1, \ldots, l\}$ an equation

$$g_k = g_k(\mathbf{q}, t), \qquad k \in \{1, \ldots, l\} \tag{8.2}$$

which measures the "elongation" of the displacement force element in the direction of $\mathbf{w}_k$. Furthermore, for each $k \in \{1, \ldots, n\}$, the (one-sided) relative velocities and their (one-sided) time derivatives are given by (5.14), (5.15) as

$$\begin{aligned} \gamma_k &= \mathbf{w}_k^T \mathbf{u} + \tilde{w}_k, \\ \dot{\gamma}_k &= \mathbf{w}_k^T \dot{\mathbf{u}} + \bar{w}_k, \end{aligned} \qquad k \in \{1, \ldots, n\}. \tag{8.3}$$

The problem is completed by the l scalar displacement force laws

$$\begin{aligned} -\lambda_{Dk} &= \nabla v_{Dk}(g_k), \\ -\lambda_{Sk} &\in \partial v_{Sk}(g_k), \\ -\lambda_{Pk} &\in \partial v_{Pk}(g_k), \end{aligned} \qquad k \in \{1, \ldots, l\} \tag{8.4}$$

from (7.8), (7.9), (7.20), where the latter two may also be expressed on the velocity level (7.11), (7.22) or on the acceleration level (7.16), (7.27) by choice, and by the $(n-l)$ scalar velocity force laws

$$\begin{aligned} -\lambda_{Dk} &= \nabla\varphi_{Dk}(\gamma_k), \\ -\lambda_{Sk} &\in \partial\varphi_{Sk}(\gamma_k), \qquad k \in \{l+1,\dots,n\} \\ -\lambda_{Pk} &\in \partial\varphi_{Pk}(\gamma_k), \end{aligned} \tag{8.5}$$

with a representation on the acceleration level according to (7.35) and (7.38).

In Section 8.2 we will state the problem (8.1)–(8.5) in three different ways, with a different dividing up of the force laws in single-valued and set-valued classes, according to the particular kinematic level that has been chosen for their representation. This division requires the introduction of certain index sets in Section 8.1, and will smooth the way to the principles of d'Alembert/Lagrange, Jourdain, and Gauß in Chapter 9 as well as to the linear complementarity problem in Section 8.3 and to some variational force expressions in Section 8.4.

8.1 Index Sets

In this section some *index sets* are introduced which group, on each kinematic level, the force laws into single-valued and set-valued classes. For the indicator displacement potentials we define the three sets

$$\begin{aligned} \mathcal{S}_{vg} &:= \{1,\dots,l\} \\ &\supset \mathcal{S}_{v\gamma} := \{k \in \mathcal{S}_{vg} \mid g_k = s_{Ak}\} \\ &\qquad \supset \mathcal{S}_{v\dot{\gamma}} := \{k \in \mathcal{S}_{v\gamma} \mid \gamma_k = 0\}, \end{aligned} \tag{8.6}$$

which will be used in the next section within the following context:

$\mathcal{S}_{vg}$ contains simply the l numbers of the force elements which are expressed on the displacement level by the indicator potential force law (7.9), i.e. for every $k \in \mathcal{S}_{vg}$ we have $-\lambda_{Sk} \in \partial v_{Sk}(g_k)$.

If the force law is stated on the velocity level in the form (7.11), we may divide up the numbers in $\mathcal{S}_{vg}$ into two sets, $\mathcal{S}_{v\gamma}$ and $\mathcal{S}_{vg}\backslash\mathcal{S}_{v\gamma}$, corresponding to the two cases in (7.11): By the definition of $\mathcal{S}_{v\gamma}$, the force law (7.11) reads $-\lambda_{Sk} \in \partial\varphi_{Sk}(\gamma_k)$ for $k \in \mathcal{S}_{v\gamma}$ and $-\lambda_{Sk} = 0$ for $k \in \mathcal{S}_{vg}\backslash\mathcal{S}_{v\gamma}$. For every $k \in \mathcal{S}_{vg}\backslash\mathcal{S}_{v\gamma}$ the force law (7.11) is therefore just a function of the displacements, $g_k \to -\lambda_{Sk}(g_k) = 0$, and may be directly evaluated if g_k is known. In contrast, for certain g_k, i.e. $g_k = s_{Ak}$, the force law is stated as an inclusion, depending additionally on the relative velocities γ_k. The indices k for which this case applies are collected in $\mathcal{S}_{v\gamma}$.

The third index set, $\mathcal{S}_{v\dot{\gamma}}$, refers to the representation of the indicator displacement force law on the acceleration level (7.16). For every $k \in \mathcal{S}_{v\dot{\gamma}}$ we have by (8.6) that $g_k = s_{Ak}$, $\gamma_k = 0$. Hence, the third line in the force law (7.16) applies, i.e. $-\lambda_{Sk} \in \partial\psi_{Sk}(\dot\gamma_k)$. For $k \in \mathcal{S}_{vg}\backslash\mathcal{S}_{v\dot\gamma} \equiv (\mathcal{S}_{vg}\backslash\mathcal{S}_{v\gamma}) \cup (\mathcal{S}_{v\gamma}\backslash\mathcal{S}_{v\dot\gamma})$ the force law is single-valued and depends on g_k and γ_k as stated in the first two lines of (7.16).

In the same manner we define now the index sets for the single-step displacement force element,

$$\begin{aligned}\mathcal{P}_{vg} :=& \{1,\dots,l\}\\ &\supset \mathcal{P}_{v\gamma} := \{k \in \mathcal{P}_{vg} \mid g_k = s_{Pk}\}\\ &\qquad\supset \mathcal{P}_{v\dot\gamma} := \{k \in \mathcal{P}_{v\gamma} \mid \gamma_k = 0\}\end{aligned} \tag{8.7}$$

where the three sets refer to the representation of the force law on displacement level (7.20), velocity level (7.22), and acceleration level (7.27), respectively.

The index sets for the velocity force elements of Section 7.3 may be stated analogously. They are

$$\begin{aligned}\mathcal{S}_{\varphi\gamma} :=& \{l+1,\dots,n\}\\ &\supset \mathcal{S}_{\varphi\dot\gamma} := \{k \in \mathcal{S}_{\varphi\gamma} \mid \gamma_k = \nu_{Ak}\}\end{aligned} \tag{8.8}$$

for the non-holonomic indicator-type force law (7.34) and its representation on the acceleration level (7.35), and

$$\begin{aligned}\mathcal{P}_{\varphi\gamma} :=& \{l+1,\dots,n\}\\ &\supset \mathcal{P}_{\varphi\dot\gamma} := \{k \in \mathcal{P}_{\varphi\gamma} \mid \gamma_k = \nu_{Pk}\}\end{aligned} \tag{8.9}$$

for the single-step velocity force law (7.37), (7.38). Note that only two index sets each are defined, since these force laws can not be stated on the displacement level.

8.2 Formulation on Different Kinematic Levels

By using the index sets as introduced in the last section, we are going to formulate the entire problem (8.1)–(8.5) again, but on each kinematic level separately by choosing the corresponding representations of the force laws from Chapter 7.

Assume first that all force laws are stated on the displacement level. This is only possible if there are no velocity force laws (8.5), which means in particular that $\mathcal{S}_{\varphi\gamma} = \mathcal{P}_{\varphi\gamma} = \emptyset$. The numbers k in the Newton-Euler equations (8.1) are then just running from 1 to l, with each k being also associated with a kinematic equation (8.2) and a force law (8.4), i.e.

$$\begin{aligned}&\mathbf{M}\dot{\mathbf{u}} - \mathbf{h}_g - \sum_{k\in\mathcal{S}_{vg}} \mathbf{w}_k\lambda_{Sk} - \sum_{k\in\mathcal{P}_{vg}} \mathbf{w}_k\lambda_{Pk} = 0\\ &-\lambda_{Sk} \in \partial v_{Sk}(g_k), \quad g_k = g_k(\mathbf{q},t) \qquad \text{for } k \in \mathcal{S}_{vg}\\ &-\lambda_{Pk} \in \partial v_{Pk}(g_k), \quad g_k = g_k(\mathbf{q},t) \qquad \text{for } k \in \mathcal{P}_{vg},\end{aligned} \tag{8.10}$$

where the forces λ_{Dk} from the differentiable force potentials have been collected in $\mathbf{h}_g$,

$$\mathbf{h}_g = \mathbf{h} + \sum_{k=1}^{l} \mathbf{w}_k \, \lambda_{Dk}(g_k). \tag{8.11}$$

The relative velocities γ_k and their time derivatives $\dot{\gamma}_k$ from (8.4) are not needed in this case, since they do not enter the representation of the force laws on the displacement level. The set of equations (8.10) will be used in Chapter 9 in order to discuss the principle of d'Alembert/Lagrange. Finally, note that $v_{Sk}(g_k)$ and $v_{Pk}(g_k)$ are convex *but not* positively homogeneous functions by (7.9) and (7.20), and that $\mathbf{q} \to g_k(\mathbf{q}, t)$ is generally assumed to be *non-convex* but smooth.

Assume now that all set-valued force laws are stated on the velocity level. Equations (8.1)–(8.5) then apply with the only difference that $\dot{\gamma}_k$ is not yet needed, and that the force laws for λ_{Sk} and λ_{Pk} in (8.4) are re-written as in (7.11) and (7.22). Our problem may therefore be stated as

$$\begin{aligned} &\mathbf{M}\dot{\mathbf{u}} - \mathbf{h}_\gamma - \sum_{k \in \mathcal{S}_{v\gamma} \cup \mathcal{S}_{\varphi\gamma}} \mathbf{w}_k \lambda_{Sk} - \sum_{k \in \mathcal{P}_{v\gamma} \cup \mathcal{P}_{\varphi\gamma}} \mathbf{w}_k \lambda_{Pk} = 0 \\ &-\lambda_{Sk} \in \partial\varphi_{Sk}(\gamma_k), \quad \gamma_k = \mathbf{w}_k^T \mathbf{u} + \tilde{w}_k \qquad \text{for } k \in \mathcal{S}_{v\gamma} \cup \mathcal{S}_{\varphi\gamma} \\ &-\lambda_{Pk} \in \partial\varphi_{Pk}(\gamma_k), \quad \gamma_k = \mathbf{w}_k^T \mathbf{u} + \tilde{w}_k \qquad \text{for } k \in \mathcal{P}_{v\gamma} \cup \mathcal{P}_{\varphi\gamma}, \end{aligned} \tag{8.12}$$

where all the force elements that are single-valued as functions of γ_k or g_k have been collected in $\mathbf{h}_\gamma$,

$$\mathbf{h}_\gamma = \mathbf{h}_g + \sum_{k=l+1}^{n} \mathbf{w}_k \, \lambda_{Dk}(\gamma_k) + \sum_{k \in \mathcal{S}_{vg} \backslash \mathcal{S}_{v\gamma}} \mathbf{w}_k \, \lambda_{Sk}(g_k) + \sum_{k \in \mathcal{P}_{vg} \backslash \mathcal{P}_{v\gamma}} \mathbf{w}_k \, \lambda_{Pk}(g_k) \tag{8.13}$$

with $\mathbf{h}_g$ from (8.11). The potentials φ_{Sk} and φ_{Pk} are convex. They are, in addition, positively homogeneous for $k \in \mathcal{S}_{v\gamma}$ and $k \in \mathcal{P}_{v\gamma}$, respectively, as it can be seen from their definition in (7.11) and (7.22), or by their identification with the corresponding epicontingent derivatives (7.13) and (7.24). Equations (8.12) will be used in the next chapter to state the principle of Jourdain, a variational formulation in terms of the generalized velocities.

Finally, we rewrite equations (8.1)–(8.5) for the case that all the set-valued force laws (8.4), (8.5) are stated on the acceleration level, according to equations (7.16), (7.27) for the displacement force elements and (7.35), (7.38) for the velocity force elements. Again, we group together all the single-valued portions in a vector $\mathbf{h}_{\dot{\gamma}}$ which yields

$$\begin{aligned} &\mathbf{M}\dot{\mathbf{u}} - \mathbf{h}_{\dot{\gamma}} - \sum_{k \in \mathcal{S}_{v\dot{\gamma}} \cup \mathcal{S}_{\varphi\dot{\gamma}}} \mathbf{w}_k \lambda_{Sk} - \sum_{k \in \mathcal{P}_{v\dot{\gamma}} \cup \mathcal{P}_{\varphi\dot{\gamma}}} \mathbf{w}_k \lambda_{Pk} = 0 \\ &-\lambda_{Sk} \in \partial\psi_{Sk}(\dot{\gamma}_k), \quad \dot{\gamma}_k = \mathbf{w}_k^T \dot{\mathbf{u}} + \bar{w}_k \qquad \text{for } k \in \mathcal{S}_{v\dot{\gamma}} \cup \mathcal{S}_{\varphi\dot{\gamma}} \\ &-\lambda_{Pk} \in \partial\psi_{Pk}(\dot{\gamma}_k), \quad \dot{\gamma}_k = \mathbf{w}_k^T \dot{\mathbf{u}} + \bar{w}_k \qquad \text{for } k \in \mathcal{P}_{v\dot{\gamma}} \cup \mathcal{P}_{\varphi\dot{\gamma}} \end{aligned} \tag{8.14}$$

with

$$\mathbf{h}_{\dot{\gamma}} = \mathbf{h}_\gamma + \sum_{\substack{k\in\mathcal{S}_{v\gamma}\setminus\mathcal{S}_{v\dot{\gamma}}\\ \cup\mathcal{S}_{\varphi\gamma}\setminus\mathcal{S}_{\varphi\dot{\gamma}}}} \mathbf{w}_k\,\lambda_{Sk}(g_k,\gamma_k) + \sum_{\substack{k\in\mathcal{P}_{v\gamma}\setminus\mathcal{P}_{v\dot{\gamma}}\\ \cup\mathcal{P}_{\varphi\gamma}\setminus\mathcal{P}_{\varphi\dot{\gamma}}}} \mathbf{w}_k\,\lambda_{Pk}(g_k,\gamma_k) \tag{8.15}$$

and $\mathbf{h}_\gamma$ from (8.13). Note that $\mathbf{h}_{\dot{\gamma}}$ may be directly computed for a given $(\mathbf{q},\mathbf{u},t)$, since $g_k = g_k(\mathbf{q},t)$, $\gamma_k = \gamma_k(\mathbf{q},\mathbf{u},t)$, $\mathbf{w}_k = \mathbf{w}_k(\mathbf{q},t)$, and $\mathbf{h} = \mathbf{h}(\mathbf{q},\mathbf{u},t)$, see e.g. (5.14), (5.15), and (5.8). Furthermore, note that every force in (8.4) and (8.5) has been taken into account in (8.14) exactly once. This may be seen from (8.15) by re-substituting $\mathbf{h}_\gamma$ from (8.13) and $\mathbf{h}_g$ from (8.11),

$$\mathbf{h}_{\dot{\gamma}} = \mathbf{h} + \sum_{k=1}^{n} \mathbf{w}_k\lambda_{Dk} + \sum_{\substack{k\in\mathcal{S}_{vg}\setminus\mathcal{S}_{v\dot{\gamma}}\\ \cup\mathcal{S}_{\varphi\gamma}\setminus\mathcal{S}_{\varphi\dot{\gamma}}}} \mathbf{w}_k\,\lambda_{Sk} + \sum_{\substack{k\in\mathcal{P}_{vg}\setminus\mathcal{P}_{v\dot{\gamma}}\\ \cup\mathcal{P}_{\varphi\gamma}\setminus\mathcal{P}_{\varphi\dot{\gamma}}}} \mathbf{w}_k\,\lambda_{Pk}. \tag{8.16}$$

The excluded sets $(\mathcal{S}_{v\dot{\gamma}} \cup \mathcal{S}_{\varphi\dot{\gamma}})$ and $(\mathcal{P}_{v\dot{\gamma}} \cup \mathcal{P}_{\varphi\dot{\gamma}})$ in (8.16) are covered instead by the force laws in (8.14). Since $(\mathcal{S}_{vg} \cup \mathcal{S}_{\varphi\gamma})$ and $(\mathcal{P}_{vg} \cup \mathcal{P}_{\varphi\gamma})$ are $\{1,\dots,n\}$ each, every force in (8.1) has been regarded. Note that *all* the potentials ψ_{Sk} and ψ_{Pk} in (8.14) are positively homogeneous, see e.g. (7.16), (7.27), (7.33), (7.36), a nice property that turns their conjugates into indicator functions, indicators of some closed intervals in our case, which makes it very easy to treat the problem by complementarity conditions. We will use equation (8.14) in Chapter 9 in order to state Gauß' principle and the associated principle of least constraints, as well as in the next two sections when we re-formulate it as a linear complementarity problem on the one hand, and as a minimization problem on the scalar force values on the other hand.

8.3 The Linear Complementarity Problem

In this section we rewrite equations (8.14) as a linear complementarity problem. There are many approaches spread through the literature on how to perform this step, more or less suited for contact problems in multibody dynamics. In our opinion a good formulation should fulfill at least the following three demands: In the first place, any inversion of an expression $\mathbf{W}^T\mathbf{M}^{-1}\mathbf{W}$ has to be avoided. For multi-contact problems the matrix $\mathbf{W}$ consists of a selection of generalized force directions $\mathbf{w}_k$ depending on the actual contact state which changes during time. In multibody dynamics there is nearly no way to avoid contact situations where these vectors become linearly dependent. Even for the simple case of a planar rectangular block resting on a surface this situation occurs when the unilateral frictional constraints are modeled by two contact points at the block's lower corners, see e.g. [37] and [116]. Thus, formulations as presented in [64] and [101] usually fail for dynamic multibody problems. This situation has been overcome in [37], [116] by introducing additional variables and thus enlarging the dimension of the LCP which leads to our second demand: Since the LCP has to be solved at every time step during numerical integration, its dimension should be kept as small as possible. In our case an optimal LCP would have a dimension equal to the number of unilateral constraints plus twice the number of single-step force laws, corresponding to the overall number of unilateral primitives in use. Thirdly, a standard LCP formulation is desirable.

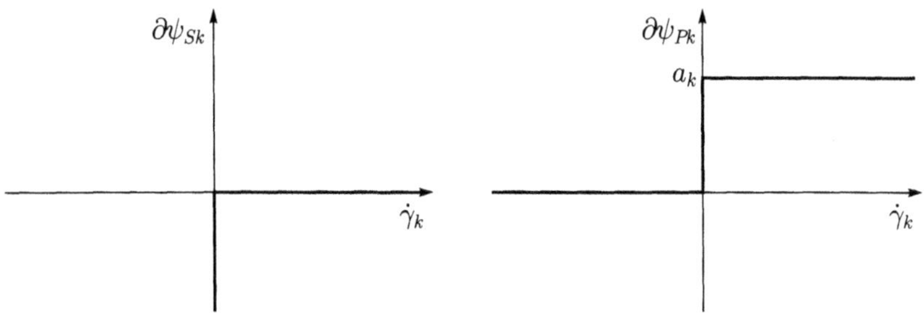

Figure 8.1. Subdifferentials of the acceleration potentials

In order to formulate (8.14) as a linear complementarity problem we state in a first step the force laws $-\lambda_{Sk} \in \partial\psi_{Sk}(\dot{\gamma}_k)$ and $-\lambda_{Pk} \in \partial\psi_{Pk}(\dot{\gamma}_k)$ by complementarity conditions. The acceleration potentials ψ_{Sk} and ψ_{Pk} are defined according to equations (7.16), (7.33) and (7.27), (7.36) with associated subdifferential mappings as depicted in Figure 8.1. The complementarities describing the indicator force law $-\lambda_{Sk} \in \partial\psi_{Sk}(\dot{\gamma}_k)$ are simply

$$\lambda_{Sk} \geq 0, \quad \dot{\gamma}_k \geq 0, \quad \lambda_{Sk}\,\dot{\gamma}_k = 0 \qquad \text{for } k \in \mathcal{S}_{v\dot{\gamma}} \cup \mathcal{S}_{\varphi\dot{\gamma}}, \tag{8.17}$$

as it can be seen from the corresponding graph in the left diagram of Figure 8.1. The single-step force law from the right diagram may be regarded as two unilateral primitives connected in series, see Figure 6.3 and equation (6.8) with $s_i = 0$ (replace index i by k, s by $\dot{\gamma}_k$ and ξ by κ), and may therefore be expressed via

$$\begin{aligned}
&\dot{\gamma}_k = \kappa_{Rk} - \kappa_{Lk} \\
&\kappa_{Rk} \geq 0, \quad (\lambda_{Pk} + a_k) \geq 0, \quad \kappa_{Rk}\,(\lambda_{Pk} + a_k) = 0 \qquad \text{for } k \in \mathcal{P}_{v\dot{\gamma}} \cup \mathcal{P}_{\varphi\dot{\gamma}}, \\
&\kappa_{Lk} \geq 0, \quad -\lambda_{Pk} \geq 0, \quad \kappa_{Lk}\,\lambda_{Pk} = 0
\end{aligned} \tag{8.18}$$

according to the representation that was already given in (6.8). We denote by σ_{Rk} and σ_{Lk} the two expressions

$$\sigma_{Rk} := \lambda_{Pk} + a_k, \qquad \sigma_{Lk} := -\lambda_{Pk}, \tag{8.19}$$

and may then rewrite the force laws (8.18) as standard complementarities,

$$
\begin{aligned}
&\dot{\gamma}_k = \kappa_{Rk} - \kappa_{Lk}, \\
&\kappa_{Rk} \geq 0, \quad \sigma_{Rk} \geq 0, \quad \kappa_{Rk}\,\sigma_{Rk} = 0, \\
&\kappa_{Lk} \geq 0, \quad \sigma_{Lk} \geq 0, \quad \kappa_{Lk}\,\sigma_{Lk} = 0.
\end{aligned} \tag{8.20}
$$

Equations (8.17) and (8.19), (8.20) replace now the two inclusions $-\lambda_{Sk} \in \partial\psi_{Sk}(\dot{\gamma}_k)$ and $-\lambda_{Pk} \in \partial\psi_{Pk}(\dot{\gamma}_k)$ in (8.14).

In order to proceed with matrix-vector notation we introduce the following abbreviations:

$$
\begin{aligned}
&\text{For } k \in \mathcal{S}_{v\dot{\gamma}} \cup \mathcal{S}_{\varphi\dot{\gamma}}: \\
&\boldsymbol{\lambda}_S = (\lambda_{Sk}), \quad \dot{\boldsymbol{\gamma}}_S = (\dot{\gamma}_k), \quad \mathbf{W}_S = (\mathbf{w}_k), \quad \bar{\mathbf{w}}_S = (\bar{w}_k). \\
&\text{For } k \in \mathcal{P}_{v\dot{\gamma}} \cup \mathcal{P}_{\varphi\dot{\gamma}}: \\
&\boldsymbol{\lambda}_P = (\lambda_{Pk}), \quad \dot{\boldsymbol{\gamma}}_P = (\dot{\gamma}_k), \quad \mathbf{W}_P = (\mathbf{w}_k), \quad \bar{\mathbf{w}}_P = (\bar{w}_k), \\
&\boldsymbol{\kappa}_R = (\kappa_{Rk}), \quad \boldsymbol{\kappa}_L = (\kappa_{Lk}), \quad \boldsymbol{\sigma}_R = (\sigma_{Rk}), \quad \boldsymbol{\sigma}_L = (\sigma_{Lk}), \quad \mathbf{a} = (a_k).
\end{aligned} \tag{8.21}
$$

Thus, equations (8.14) together with the alternative representations of the force laws (8.17) and (8.19), (8.20) become

$$
\begin{aligned}
&\mathbf{M}\dot{\mathbf{u}} - \mathbf{h}_{\dot{\gamma}} - \mathbf{W}_S\boldsymbol{\lambda}_S - \mathbf{W}_P\boldsymbol{\lambda}_P = 0, \\
&\dot{\boldsymbol{\gamma}}_S = \mathbf{W}_S^T\dot{\mathbf{u}} + \bar{\mathbf{w}}_S, \\
&\dot{\boldsymbol{\gamma}}_P = \mathbf{W}_P^T\dot{\mathbf{u}} + \bar{\mathbf{w}}_P, \\
&\boldsymbol{\lambda}_S \geq 0 \quad \dot{\boldsymbol{\gamma}}_S \geq 0 \quad \boldsymbol{\lambda}_S^T\dot{\boldsymbol{\gamma}}_S = 0, \\
&\boldsymbol{\sigma}_R = \boldsymbol{\lambda}_P + \mathbf{a}, \\
&\boldsymbol{\sigma}_L = -\boldsymbol{\lambda}_P, \\
&\dot{\boldsymbol{\gamma}}_P = \boldsymbol{\kappa}_R - \boldsymbol{\kappa}_L, \\
&\boldsymbol{\kappa}_R \geq 0 \quad \boldsymbol{\sigma}_R \geq 0 \quad \boldsymbol{\kappa}_R^T\boldsymbol{\sigma}_R = 0, \\
&\boldsymbol{\kappa}_L \geq 0 \quad \boldsymbol{\sigma}_L \geq 0 \quad \boldsymbol{\kappa}_L^T\boldsymbol{\sigma}_L = 0,
\end{aligned} \tag{8.22}
$$

a set of six equations and three complementarities. Note that there is no complementarity between the single-step forces $\boldsymbol{\lambda}_P$ and the associated "relative accelerations" $\dot{\boldsymbol{\gamma}}_P$. Both magnitudes may therefore be eliminated, which can be done with the help of the sixth and seventh line in (8.22). The remaining four equations and three complementarities then are

$$
\begin{aligned}
&\mathbf{M}\dot{\mathbf{u}} - \mathbf{h}_{\dot{\gamma}} - \mathbf{W}_S\boldsymbol{\lambda}_S + \mathbf{W}_P\boldsymbol{\sigma}_L = 0, \\
&\dot{\boldsymbol{\gamma}}_S = \mathbf{W}_S^T\dot{\mathbf{u}} + \bar{\mathbf{w}}_S, \\
&\boldsymbol{\kappa}_L = -\mathbf{W}_P^T\dot{\mathbf{u}} - \bar{\mathbf{w}}_P + \boldsymbol{\kappa}_R, \\
&\boldsymbol{\sigma}_R = -\boldsymbol{\sigma}_L + \mathbf{a}, \\
&\dot{\boldsymbol{\gamma}}_S \geq 0 \quad \boldsymbol{\lambda}_S \geq 0 \quad \dot{\boldsymbol{\gamma}}_S^T\boldsymbol{\lambda}_S = 0, \\
&\boldsymbol{\kappa}_L \geq 0 \quad \boldsymbol{\sigma}_L \geq 0 \quad \boldsymbol{\kappa}_L^T\boldsymbol{\sigma}_L = 0, \\
&\boldsymbol{\sigma}_R \geq 0 \quad \boldsymbol{\kappa}_R \geq 0 \quad \boldsymbol{\sigma}_R^T\boldsymbol{\kappa}_R = 0.
\end{aligned} \tag{8.23}
$$

In a final step we solve the Newton-Euler equations for $\dot{\mathbf{u}}$ and substitute the resulting expression into the kinematic equations from the second and the third line. This gives

$$\begin{pmatrix} \dot{\gamma}_S \\ \kappa_L \\ \sigma_R \end{pmatrix} = \begin{pmatrix} \mathbf{W}_S^T\mathbf{M}^{-1}\mathbf{W}_S & -\mathbf{W}_S^T\mathbf{M}^{-1}\mathbf{W}_P & 0 \\ -\mathbf{W}_P^T\mathbf{M}^{-1}\mathbf{W}_S & \mathbf{W}_P^T\mathbf{M}^{-1}\mathbf{W}_P & \mathbf{E} \\ 0 & -\mathbf{E} & 0 \end{pmatrix} \begin{pmatrix} \lambda_S \\ \sigma_L \\ \kappa_R \end{pmatrix} + \begin{pmatrix} \mathbf{W}_S^T\mathbf{M}^{-1}\mathbf{h}_{\dot{\gamma}} + \bar{\mathbf{w}}_S \\ -\mathbf{W}_P^T\mathbf{M}^{-1}\mathbf{h}_{\dot{\gamma}} - \bar{\mathbf{w}}_P \\ \mathbf{a} \end{pmatrix} \tag{8.24}$$

which constitutes, together with the complementarities from the last three lines in (8.23), a *linear complementarity problem* (LCP) in standard form, i.e. a problem

$$\mathbf{y} = \mathbf{A}\mathbf{x} + \mathbf{b}, \qquad \mathbf{y} \geq 0, \quad \mathbf{x} \geq 0, \quad \mathbf{y}^T\mathbf{x} = 0 \tag{8.25}$$

with unknowns $\mathbf{y}$ and $\mathbf{x}$, and given quadratic matrix $\mathbf{A}$ and vector $\mathbf{b}$. For a theoretical treatment of linear complementarity problems as well as for numerical algorithms we refer especially to the books [95], [27]. In our case, the LCP (8.24) has the special structure

$$\begin{pmatrix} \mathbf{u} \\ \mathbf{v} \end{pmatrix} = \begin{pmatrix} \mathbf{G} & \mathbf{I} \\ -\mathbf{I}^T & 0 \end{pmatrix} \begin{pmatrix} \mathbf{p} \\ \mathbf{q} \end{pmatrix} + \begin{pmatrix} \mathbf{c} \\ \mathbf{a} \end{pmatrix}, \qquad 0 \leq \begin{pmatrix} \mathbf{u} \\ \mathbf{v} \end{pmatrix} \perp \begin{pmatrix} \mathbf{p} \\ \mathbf{q} \end{pmatrix} \geq 0, \tag{8.26}$$

with $\mathbf{G} = \mathbf{W}^T\mathbf{M}^{-1}\mathbf{W}$, $\mathbf{W} = (\mathbf{W}_S \; -\mathbf{W}_P)$ and $\mathbf{I}^T = (0 \; \mathbf{E})$, where $\mathbf{E}$ denotes the identity matrix. Furthermore, due to $\bar{\mathbf{w}}_S$ and $\bar{\mathbf{w}}_P$ in (8.24), $\mathbf{c}$ is any arbitrary (given) vector, and $\mathbf{a} > 0$. Note that $\mathbf{G}$ is symmetric and positive semidefinite. The first property is obvious due to the symmetry of the positive definite mass matrix $\mathbf{M}$, whereas the second property may be seen from $\mathbf{p}^T\mathbf{G}\mathbf{p} = \mathbf{p}^T\mathbf{W}^T\mathbf{L}^T\mathbf{L}\mathbf{W}\mathbf{p} = ||\mathbf{L}\mathbf{W}\mathbf{p}||_2^2 \geq 0 \; \forall \mathbf{p}$, where $\mathbf{L}^T\mathbf{L} := \mathbf{M}^{-1}$ denotes the Cholesky factorization of $\mathbf{M}^{-1}$. If the generalized force directions $\mathbf{w}_k$ are in addition linearly independent, $\mathbf{G}$ becomes positive definite. Note also that the LCP-matrix in (8.26) is not symmetric but positive semidefinite, since $\mathbf{x}^T\mathbf{A}\mathbf{x} = \mathbf{p}^T\mathbf{G}\mathbf{p} \geq 0 \; \forall \mathbf{x}^T = (\mathbf{p}^T \quad \mathbf{q}^T)$, and it is never positive definite since the latter inequality holds in particular for every $\mathbf{q} \neq 0$. Moreover, this matrix is never a P-matrix since there are principal minors equal to zero as it can be seen from the zero entries in its lower right corner. This matrix belongs to the so-called bisymmetric matrices [27], and there is an associated quadratic program with Karush-Kuhn-Tucker conditions as in (8.26) which we will discuss in the next section.

8.4 The Dual Principle of Least Constraints

Let us look again at equation (8.14). In order to rewrite it as a minimization problem on λ_{Sk} and λ_{Pk}, we have first to invert each of the force laws and then to define an overall force potential such that all force laws together are stated by a single inclusion. The latter may be done due to the following property:

Proposition 8.4.1. *Let* $f_i : \mathbb{R}^{m_i} \to \bar{\mathbb{R}}$ *be convex functions with effective domains* $\operatorname{dom} f_i$ *and* $f(\mathbf{x}) := \sum_i f_i(\mathbf{x}_i)$ *with* $\operatorname{dom} f = \operatorname{dom} f_1 \times \operatorname{dom} f_2 \times \ldots$. *Then*

$$\mathbf{y} \in \partial f(\mathbf{x}) \quad \Leftrightarrow \quad \mathbf{y}_i \in \partial f_i(\mathbf{x}_i) \text{ for every index } i \tag{8.27}$$

with $\mathbf{x}^T = (\mathbf{x}_1^T, \mathbf{x}_2^T, \ldots)$ *and* $\mathbf{y}^T = (\mathbf{y}_1^T, \mathbf{y}_2^T, \ldots)$.

Proof. We first show that (8.27) holds in the direction "⇒". By the definition of the subdifferential [121], $\mathbf{y}$ belongs to $\partial f(\mathbf{x})$ if

$$\mathbf{y}^T(\mathbf{x}^\star - \mathbf{x}) \leq f(\mathbf{x}^\star) - f(\mathbf{x}), \quad \mathbf{x} \in \operatorname{dom} f, \quad \forall \mathbf{x}^\star \tag{8.28}$$

or, written as a sum,

$$\sum_i \mathbf{y}_i^T(\mathbf{x}_i^\star - \mathbf{x}_i) \leq \sum_i f_i(\mathbf{x}_i^\star) - \sum_i f_i(\mathbf{x}_i), \quad \mathbf{x}_i \in \operatorname{dom} f_i, \quad \forall \mathbf{x}_i^\star. \tag{8.29}$$

Choose an index k and set $\mathbf{x}_i^\star = \mathbf{x}_i$ for every $i \neq k$. Equation (8.29) then becomes

$$\mathbf{y}_k^T(\mathbf{x}_k^\star - \mathbf{x}_k) \leq f_k(\mathbf{x}_k^\star) - f_k(\mathbf{x}_k), \quad \mathbf{x}_k \in \operatorname{dom} f_k, \quad \forall \mathbf{x}_k^\star \tag{8.30}$$

which is

$$\mathbf{y}_k \in \partial f_k(\mathbf{x}_k). \tag{8.31}$$

On the other hand, suppose that (8.31) or, equivalently, (8.30) holds for every index k. Then (8.29) holds also by summation, hence $\mathbf{y} \in \partial f(\mathbf{x})$. □

We are now going to derive from (8.14) a variational inequality on the forces λ_{Sk} and λ_{Pk}. The potentials ψ_{Sk} of the unilateral constraints (7.16), (7.33) as well as the potentials ψ_{Pk} of the single step force laws (7.27), (7.36) are depicted in Figure 8.2 together with their conjugates $\psi_{Sk}^\star$ and $\psi_{Pk}^\star$. Note that both, $\psi_{Sk}^\star$ and $\psi_{Pk}^\star$, are indicator functions, with effective domains $\operatorname{dom} \psi_{Sk}^\star = (-\infty, 0] = \mathbb{R}^-$ and $\operatorname{dom} \psi_{Pk}^\star = [0, a_k]$ respectively, as a consequence on the positive homogeneity of ψ_{Sk} and ψ_{Pk}. According to equation (6.24) we may now "invert" the force laws $-\lambda_{Sk} \in \partial \psi_{Sk}(\dot\gamma_k)$ and $-\lambda_{Pk} \in \partial \psi_{Pk}(\dot\gamma_k)$ in (8.14) which yields

$$\begin{aligned}
&\mathbf{M}\dot{\mathbf{u}} - \mathbf{h}_{\dot\gamma} - \sum_{k \in \mathcal{S}_{\dot\gamma}} \mathbf{w}_k \lambda_{Sk} - \sum_{k \in \mathcal{P}_{\dot\gamma}} \mathbf{w}_k \lambda_{Pk} = 0 \\
&\dot\gamma_k \in \partial \psi_{Sk}^\star(-\lambda_{Sk}), \quad \dot\gamma_k = \mathbf{w}_k^T \dot{\mathbf{u}} + \bar{w}_k \qquad \text{for } k \in \mathcal{S}_{\dot\gamma} \\
&\dot\gamma_k \in \partial \psi_{Pk}^\star(-\lambda_{Pk}), \quad \dot\gamma_k = \mathbf{w}_k^T \dot{\mathbf{u}} + \bar{w}_k \qquad \text{for } k \in \mathcal{P}_{\dot\gamma},
\end{aligned} \tag{8.32}$$

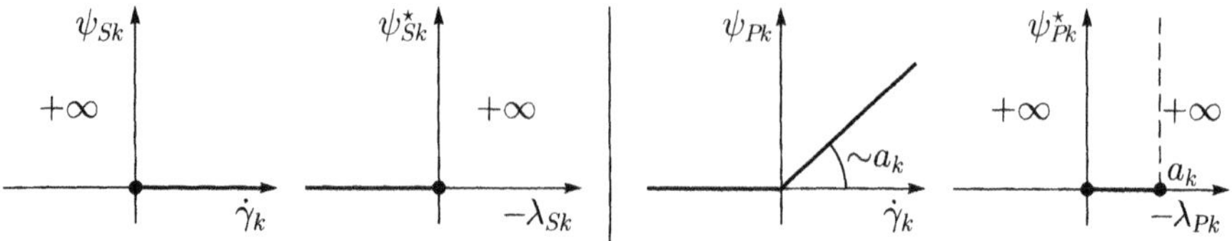

$$\psi_{Sk}(\dot\gamma_k) = \begin{cases} 0 & \text{if } \dot\gamma_k \geq 0 \\ +\infty & \text{else} \end{cases} \qquad \psi_{Pk}(\dot\gamma_k) = \begin{cases} 0 & \text{if } \dot\gamma_k \leq 0 \\ a_k\dot\gamma_k & \text{if } \dot\gamma_k \geq 0 \end{cases}$$

$$\psi^\star_{Sk}(-\lambda_{Sk}) = \begin{cases} 0 & \text{if } -\lambda_{Sk} \leq 0 \\ +\infty & \text{else} \end{cases} \qquad \psi^\star_{Pk}(-\lambda_{Pk}) = \begin{cases} 0 & \text{if } 0 \leq -\lambda_{Pk} \leq a_k \\ +\infty & \text{else} \end{cases}$$

Figure 8.2. Potentials and conjugates

where $\mathcal{S}_{\dot\gamma} := \mathcal{S}_{v\dot\gamma} \cup \mathcal{S}_{\varphi\dot\gamma}$ and $\mathcal{P}_{\dot\gamma} := \mathcal{P}_{v\dot\gamma} \cup \mathcal{P}_{\varphi\dot\gamma}$ have been used as abbreviations.

We further introduce $\boldsymbol{\lambda}_S$, $\boldsymbol{\lambda}_P$, $\dot{\boldsymbol{\gamma}}_S$, $\dot{\boldsymbol{\gamma}}_P$, $\mathbf{W}_S$, $\mathbf{W}_P$, $\bar{\mathbf{w}}_S$, $\bar{\mathbf{w}}_P$ as in (8.21) and define

$$\begin{aligned} \psi^\star_S(-\boldsymbol{\lambda}_S) &:= \sum_{k\in\mathcal{S}_{\dot\gamma}} \psi^\star_{Sk}(-\lambda_{Sk}) \quad \text{with dom}\,\psi^\star_S = \underbrace{\mathbb{R}^- \times \mathbb{R}^- \times \ldots}_{n_S} \\ \psi^\star_P(-\boldsymbol{\lambda}_P) &:= \sum_{k\in\mathcal{P}_{\dot\gamma}} \psi^\star_{Pk}(-\lambda_{Pk}) \quad \text{with dom}\,\psi^\star_P = \underbrace{[0,a_1] \times [0,a_2] \times \ldots}_{n_P} \end{aligned} \tag{8.33}$$

with n_S and n_P denoting the numbers of elements in $\mathcal{S}_{\dot\gamma}$ and $\mathcal{P}_{\dot\gamma}$. Note that both, $\psi^\star_S$ and $\psi^\star_P$, are indicator functions of their effective domains,

$$\begin{aligned} \psi^\star_S(-\boldsymbol{\lambda}_S) &= \begin{cases} 0 & \text{if } -\boldsymbol{\lambda}_S \in \text{dom}\,\psi^\star_S \\ +\infty & \text{else} \end{cases} \\ \psi^\star_P(-\boldsymbol{\lambda}_P) &= \begin{cases} 0 & \text{if } -\boldsymbol{\lambda}_P \in \text{dom}\,\psi^\star_P \\ +\infty & \text{else} \end{cases} \end{aligned} \tag{8.34}$$

restricting the values of $-\boldsymbol{\lambda}_S$ and $-\boldsymbol{\lambda}_P$ to certain polyhedral convex subsets of $\mathbb{R}^{n_S}$ and $\mathbb{R}^{n_P}$, i.e. to the non-positive orthant $\mathbb{R}^- \times \mathbb{R}^- \times \ldots$ of $\mathbb{R}^{n_S}$ and to the closed parallelepiped $[0,a_1] \times [0,a_2] \times \ldots$ in $\mathbb{R}^{n_P}$. With the result (8.27) equations (8.32) may now be expressed as

$$\begin{aligned} &\mathbf{M}\dot{\mathbf{u}} - \mathbf{h}_{\dot\gamma} - \mathbf{W}_S\boldsymbol{\lambda}_S - \mathbf{W}_P\boldsymbol{\lambda}_P = 0 \\ &\dot{\boldsymbol{\gamma}}_S \in \partial\psi^\star_S(-\boldsymbol{\lambda}_S), \quad \dot{\boldsymbol{\gamma}}_S = \mathbf{W}_S^T\dot{\mathbf{u}} + \bar{\mathbf{w}}_S \\ &\dot{\boldsymbol{\gamma}}_P \in \partial\psi^\star_P(-\boldsymbol{\lambda}_P), \quad \dot{\boldsymbol{\gamma}}_P = \mathbf{W}_P^T\dot{\mathbf{u}} + \bar{\mathbf{w}}_P \end{aligned} \tag{8.35}$$

with $\psi^\star_S$ and $\psi^\star_P$ according to (8.33), (8.34).

By doing the same steps as in (8.33)–(8.35) again, we may rewrite (8.35) in an even more compact form: Denote by

$$\dot{\gamma} := \begin{pmatrix} \dot{\gamma}_S \\ \dot{\gamma}_P \end{pmatrix}, \quad \boldsymbol{\lambda} := \begin{pmatrix} \boldsymbol{\lambda}_S \\ \boldsymbol{\lambda}_P \end{pmatrix}, \quad \bar{\mathbf{w}} := \begin{pmatrix} \bar{\mathbf{w}}_S \\ \bar{\mathbf{w}}_P \end{pmatrix}, \quad \mathbf{W} := (\mathbf{W}_S \ \ \mathbf{W}_P)$$

$$\begin{aligned} &\text{and } \psi^\star(-\boldsymbol{\lambda}) := \psi_S^\star(-\boldsymbol{\lambda}_S) + \psi_P^\star(-\boldsymbol{\lambda}_P) \\ &\text{with } \operatorname{dom}\psi^\star = \operatorname{dom}\psi_S^\star \times \operatorname{dom}\psi_P^\star \end{aligned} \tag{8.36}$$

with the overall force potential $\psi^\star$ being also an indicator,

$$\psi^\star(-\boldsymbol{\lambda}) = \begin{cases} 0 & \text{if } -\boldsymbol{\lambda} \in \operatorname{dom}\psi^\star \\ +\infty & \text{else} \end{cases} \tag{8.37}$$

and apply property (8.27) again. Equations (8.35) then become

$$\begin{aligned} &\mathbf{M}\dot{\mathbf{u}} - \mathbf{h}_{\dot{\gamma}} - \mathbf{W}\boldsymbol{\lambda} = 0 \\ &\dot{\gamma} \in \partial\psi^\star(-\boldsymbol{\lambda}), \quad \dot{\gamma} = \mathbf{W}^T\dot{\mathbf{u}} + \bar{\mathbf{w}}, \end{aligned} \tag{8.38}$$

consisting of the directional Newton-Euler equations, the set-valued force law, and the kinematics equation on the acceleration level.

The generalized accelerations $\dot{\mathbf{u}}$ may be eliminated from (8.38) by solving the Newton-Euler equations for $\dot{\mathbf{u}}$ and substituting the result into the kinematics equation. With

$$\mathbf{G} := \mathbf{W}^T\mathbf{M}^{-1}\mathbf{W}, \qquad \bar{\mathbf{g}} := \mathbf{W}^T\mathbf{M}^{-1}\mathbf{h}_{\dot{\gamma}} + \bar{\mathbf{w}} \tag{8.39}$$

equation (8.38) becomes

$$\dot{\gamma} \in \partial\psi^\star(-\boldsymbol{\lambda}), \quad \dot{\gamma} = \mathbf{G}\boldsymbol{\lambda} + \bar{\mathbf{g}}, \tag{8.40}$$

which describes completely the original problem (8.14). Note that the matrix $\mathbf{G}$ is symmetric and at least positive semidefinite, since $\mathbf{M} = \mathbf{M}^T$ is positive definite.

The inclusion in (8.40) may be also expressed by the variational inequality

$$\dot{\gamma}^T(-\boldsymbol{\lambda}^\star + \boldsymbol{\lambda}) \le \psi^\star(-\boldsymbol{\lambda}^\star) - \psi^\star(-\boldsymbol{\lambda}), \quad -\boldsymbol{\lambda} \in \operatorname{dom}\psi^\star, \quad \forall -\boldsymbol{\lambda}^\star \tag{8.41}$$

defining the subdifferential, and may be further simplified: The second term in the right-hand side of (8.41), $\psi^\star(-\boldsymbol{\lambda})$, vanishes identically, since $-\boldsymbol{\lambda} \in \operatorname{dom}\psi^\star$ and $\psi^\star$ is the indicator function from (8.37). Furthermore, the variational inequality has to hold for every $-\boldsymbol{\lambda}^\star$, with the most restrictive case $-\boldsymbol{\lambda}^\star \in \operatorname{dom}\psi^\star$, since then $\psi^\star(-\boldsymbol{\lambda}^\star) = 0$, whereas $\psi^\star(-\boldsymbol{\lambda}^\star) = +\infty$ for $-\boldsymbol{\lambda}^\star \notin \operatorname{dom}\psi^\star$. Equation (8.41) may thus be rewritten as

$$(-\boldsymbol{\lambda}^\star + \boldsymbol{\lambda})^T(\mathbf{G}\boldsymbol{\lambda} + \bar{\mathbf{g}}) \le 0, \quad -\boldsymbol{\lambda} \in \operatorname{dom}\psi^\star, \quad \forall -\boldsymbol{\lambda}^\star \in \operatorname{dom}\psi^\star, \tag{8.42}$$

where $\dot{\gamma}$ in (8.41) has been replaced by $\mathbf{G}\boldsymbol{\lambda} + \bar{\mathbf{g}}$ from the kinematics equation in (8.40). The expression (8.42), the *dual principle of Gauß*, is a variational inequality on the forces $\boldsymbol{\lambda}$ and contains all information needed to solve (8.14): For a given $\mathbf{G}$ and $\bar{\mathbf{g}}$, find $-\boldsymbol{\lambda} \in \operatorname{dom}\psi^\star$ such that the inequality (8.42) is satisfied for every $-\boldsymbol{\lambda}^\star \in \operatorname{dom} \psi^\star$. This $-\boldsymbol{\lambda}$, not necessarily unique, is then a solution of (8.40) and may be used to compute $\dot{\mathbf{u}}$ in (8.38).

There is an optimization problem associated with the variational inequality (8.42). In order to derive it we return to equation (8.40) and eliminate $\dot{\boldsymbol{\gamma}}$, which yields the inclusion

$$0 \in \partial\psi^\star(-\boldsymbol{\lambda}) - \mathbf{G}\boldsymbol{\lambda} - \bar{\mathbf{g}}. \tag{8.43}$$

Let us denote by $k(-\boldsymbol{\lambda})$ the convex function

$$k(-\boldsymbol{\lambda}) := \tfrac{1}{2}\,\boldsymbol{\lambda}^T\mathbf{G}\boldsymbol{\lambda} + \bar{\mathbf{g}}^T\boldsymbol{\lambda} \tag{8.44}$$

with gradient

$$\nabla k(-\boldsymbol{\lambda}) = -\mathbf{G}\boldsymbol{\lambda} - \bar{\mathbf{g}}. \tag{8.45}$$

Substituting (8.45) into (8.43) yields

$$0 \in \nabla k(-\boldsymbol{\lambda}) + \partial\psi^\star(-\boldsymbol{\lambda}), \tag{8.46}$$

which are the necessary and sufficient conditions on $-\boldsymbol{\lambda}$ to belong to the minimum set of $k(-\boldsymbol{\lambda})+\psi^\star(-\boldsymbol{\lambda})$. Since $\psi^\star$ is an indicator function ($\psi^\star(-\boldsymbol{\lambda}) = +\infty$ for $-\boldsymbol{\lambda} \notin \operatorname{dom}\psi^\star$), minimization of $k(-\boldsymbol{\lambda})+\psi^\star(-\boldsymbol{\lambda})$ over $\mathbb{R}^{n_S+n_P}$ is the same as minimization of $k(-\boldsymbol{\lambda})$ over $\operatorname{dom}\psi^\star$. The forces $-\boldsymbol{\lambda}$ are therefore the optimal solutions of a constrained minimization problem with convex cost function $k(-\boldsymbol{\lambda}^\star)$ and feasible solutions $-\boldsymbol{\lambda}^\star \in \operatorname{dom}\psi^\star$, i.e.

$$-\boldsymbol{\lambda} \in \arg\min\,\{k(-\boldsymbol{\lambda}^\star) \mid -\boldsymbol{\lambda}^\star \in \operatorname{dom}\psi^\star\}. \tag{8.47}$$

This is the *dual principle of least constraints.* Actually, (8.47) is a quadratic program with affine inequality constraints as it can be seen by the definition of $k(-\boldsymbol{\lambda})$ in (8.44) and $\operatorname{dom}\psi^\star$ from (8.36), (8.33). Note that $\operatorname{dom}\psi^\star$ is never empty, i.e. there are always feasible solutions. Note also that $k(-\boldsymbol{\lambda})$ is not necessarily strictly convex, since $\mathbf{G} = \mathbf{W}^T\mathbf{M}^{-1}\mathbf{W}$ may be positive semidefinite: This happens when the generalized force directions $\mathbf{w}_k$ contained in $\mathbf{W}$ are linearly dependent. In such a situation, the problem (8.46) has either one solution or infinitely many solutions or no solution. In the latter case, the function $k(-\boldsymbol{\lambda}) + \psi^\star(-\boldsymbol{\lambda})$ is not bounded from below, hence the minimum set becomes $\emptyset$. This situation, however, does not occur when the function $k(-\boldsymbol{\lambda}) + \psi^\star(-\boldsymbol{\lambda})$ is constant in every of its directions of recession, see e.g. [121], which applies always for "reasonable" multibody systems. If this is not the case, one has to look for inconsistencies in the kinematics of the system which cause the aforementioned problem, and which, equivalently, lead to an empty set of feasible solutions of the dual of (8.46), see e.g. Chapter 9.

Non-unique forces $\boldsymbol{\lambda}$ might also occur. They are, however, *very natural* for rigid bodies: Any imaginable distribution of the contact forces between a four-legged rigid table and a rigid floor, for example, which is in accordance with the unilateral constraints, may be gained in this way. These force distributions do *not* affect the uniqueness of the accelerations $\dot{\mathbf{u}}$, as we will see in the next chapter. This is in analogy to the classical principle that "constraint forces do not change the dynamics" (Note: Coulomb friction is *not* considered, but only dry friction!).

Unique forces are obtained as soon as the force directions $\mathbf{w}_k$ in $\mathbf{W}$ are no longer linearly dependent. The matrix $\mathbf{G}$ then becomes positive definite, and $k(-\boldsymbol{\lambda})$ is strictly convex with a unique optimal solution.

We finally note that the Karush-Kuhn-Tucker conditions of the problem (8.46) correspond exactly to the LCP-formulation of the foregoing section.

9. Principles in Dynamics

In equations (8.10), (8.12) and (8.14) we have formulated the set-valued force laws on three different kinematic levels. For each of them, there is an associated variational formulation in the configuration space with a counterpart in classical mechanics, best known as the principles of d'Alembert/Lagrange, Jourdain and Gauß, respectively. Within the next sections we are going to put our problem into this framework with special emphasis on the question of how to choose appropriate sets for the variations in use, i.e. on the choice of the virtual displacements, velocities, and accelerations. First of all, however, we discuss the meaning of Gauß' principle of least constraints, a strictly convex minimization problem in terms of the generalized accelerations, when it is extended to arbitrary subdifferential mappings which are representing the set-valued forces.

9.1 The Principle of Least Constraints

In this section we derive a strictly convex minimization problem in terms of $\dot{\mathbf{u}}$, with associated dual as already stated in equation (8.47). We start again with the representation of the problem on the acceleration level (8.14), with the positively homogeneous convex potentials ψ_{Sk} and ψ_{Pk} as depicted in Figure 9.1. By using the same nomenclature as in (8.21), (8.36) we introduce the overall potential $\psi(\dot{\gamma})$ to be

$$\begin{aligned}
&\psi(\dot{\gamma}) := \sum_{k \in \mathcal{S}_{\dot{\gamma}}} \psi_{Sk}(\dot{\gamma}_k) + \sum_{k \in \mathcal{P}_{\dot{\gamma}}} \psi_{Pk}(\dot{\gamma}_k) \\
&\text{with } \operatorname{dom}\psi = \underbrace{\mathrm{I\!R}^+ \times \mathrm{I\!R}^+ \times \ldots}_{n_S} \times \underbrace{\mathrm{I\!R} \times \mathrm{I\!R} \times \ldots}_{n_P},
\end{aligned} \tag{9.1}$$

where $\mathcal{S}_{\dot{\gamma}} := \mathcal{S}_{v\dot{\gamma}} \cup \mathcal{S}_{\varphi\dot{\gamma}}$ with n_S elements and $\mathcal{P}_{\dot{\gamma}} := \mathcal{P}_{v\dot{\gamma}} \cup \mathcal{P}_{\varphi\dot{\gamma}}$ with n_P elements. Applying the composition rule (8.27), equation (8.14) becomes

$$\begin{aligned}
&\mathbf{M}\dot{\mathbf{u}} - \mathbf{h}_{\dot{\gamma}} - \mathbf{W}\boldsymbol{\lambda} = 0 \\
&-\boldsymbol{\lambda} \in \partial\psi(\dot{\gamma}), \quad \dot{\gamma} = \mathbf{W}^T\dot{\mathbf{u}} + \bar{\mathbf{w}}
\end{aligned} \tag{9.2}$$

which is the same as (8.38) but with inverted force laws.

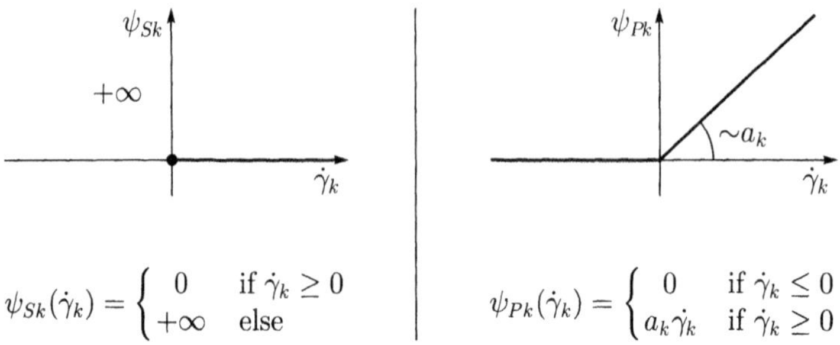

$$\psi_{Sk}(\dot\gamma_k) = \begin{cases} 0 & \text{if } \dot\gamma_k \geq 0 \\ +\infty & \text{else} \end{cases} \qquad \psi_{Pk}(\dot\gamma_k) = \begin{cases} 0 & \text{if } \dot\gamma_k \leq 0 \\ a_k \dot\gamma_k & \text{if } \dot\gamma_k \geq 0 \end{cases}$$

Figure 9.1. Potentials ψ_{Sk} and ψ_{Pk}

By the last equation in (9.2) we may regard $\psi(\dot{\boldsymbol\gamma})$ as a function of $\dot{\mathbf{u}}$, i.e.

$$\begin{aligned} &\Psi(\dot{\mathbf{u}}) := \psi(\dot{\boldsymbol\gamma}(\dot{\mathbf{u}})) \\ &\operatorname{dom}\Psi = \{\dot{\mathbf{u}} \,|\, \mathbf{W}^T\dot{\mathbf{u}} + \bar{\mathbf{w}} = \dot{\boldsymbol\gamma} \in \operatorname{dom}\psi\}. \end{aligned} \tag{9.3}$$

We call Ψ the *acceleration potential* of the system. Obviously, Ψ is a polyhedral convex function in our case, but no longer positively homogeneous, and $\operatorname{dom}\Psi$ is a convex set, not empty, at least as long as the kinematics of the system have been derived in a consistent way. Applying the chain rule which has already been assumed in Chapter 5 to hold as an equality, the subdifferential of Ψ becomes

$$\partial\Psi(\dot{\mathbf{u}}) = \mathbf{W}\,\partial\psi(\dot{\boldsymbol\gamma}), \tag{9.4}$$

and we obtain from (9.4), (9.2) by setting $\mathbf{f}_\Psi := \mathbf{W}\boldsymbol\lambda$ the inclusion

$$-\mathbf{f}_\Psi = -\mathbf{W}\boldsymbol\lambda \in \mathbf{W}\,\partial\psi(\dot{\boldsymbol\gamma}) = \partial\Psi(\dot{\mathbf{u}}), \tag{9.5}$$

and therefore the problem (9.2) in the form

$$\mathbf{M}\dot{\mathbf{u}} - \mathbf{h}_{\dot\gamma} - \mathbf{f}_\Psi = 0, \quad -\mathbf{f}_\Psi \in \partial\Psi(\dot{\mathbf{u}}). \tag{9.6}$$

Let $\dot{\mathbf{u}}_R$ denote the accelerations associated with the system without any set-valued forces $\mathbf{f}_\Psi$, i.e. the accelerations obtained from

$$\mathbf{M}\dot{\mathbf{u}}_R - \mathbf{h}_{\dot\gamma} = 0, \tag{9.7}$$

and recall that $\mathbf{M}$ denotes the mass matrix which is symmetric and positive definite. The quadratic form

$$Z(\dot{\mathbf{u}}) := \tfrac{1}{2}\,\|\dot{\mathbf{u}}_R - \dot{\mathbf{u}}\|^2_{\mathbf{M}} \tag{9.8}$$

is therefore strictly convex with gradient

$$\nabla Z(\dot{\mathbf{u}}) = \mathbf{M}(\dot{\mathbf{u}} - \dot{\mathbf{u}}_R) = \mathbf{M}\dot{\mathbf{u}} - \mathbf{h}_{\dot\gamma}, \tag{9.9}$$

and may hence be used to rewrite (9.6) as

$$0 \in \nabla Z(\dot{\mathbf{u}}) + \partial \Psi(\dot{\mathbf{u}}). \tag{9.10}$$

These, however, are the necessary and sufficient optimality conditions of the following problem: Minimize over $\mathbb{R}^f$ the strictly convex cost function

$$P(\dot{\mathbf{u}}) := Z(\dot{\mathbf{u}}) + \Psi(\dot{\mathbf{u}}) \tag{9.11}$$

in order to obtain, as the unique optimal solution, the unknown accelerations $\dot{\mathbf{u}}$,

$$\dot{\mathbf{u}} = \arg\min\{P(\dot{\mathbf{u}}^\star)\}. \tag{9.12}$$

Note that the cost function P is strictly convex, since it is the sum of a convex function Ψ and a strictly convex function Z. Moreover, P has obviously no directions of recession, which follows immediately from the additional properties that Z is quadratic with minimum at $\dot{\mathbf{u}} = \dot{\mathbf{u}}_R$, and Ψ is by (9.1)–(9.3) the sum of $(n_S + n_P)$ convex functions $\Psi_k(\dot{\mathbf{u}}) := \psi_k(\dot{\gamma}_k(\dot{\mathbf{u}}))$ which are the translates of the positively homogeneous convex functions $\tilde{\Psi}_k(\dot{\mathbf{u}}) := \psi_k(\dot{\tilde{\gamma}}_k(\dot{\mathbf{u}}))$ with $\dot{\tilde{\gamma}}_k(\dot{\mathbf{u}}) = \dot{\gamma}_k(\dot{\mathbf{u}}) - \bar{w}_k = \mathbf{w}_k^T\dot{\mathbf{u}}$. This means that the infimum of P is finite and attained, and that the minimum set of P consists of a unique acceleration $\dot{\mathbf{u}}$ as stated in equation (9.12). The uniqueness of $\dot{\mathbf{u}}$ is therefore *not* affected by linearly dependent force directions $\mathbf{w}_k$ which was already mentioned in Section 8.4.

The quadratic form $Z(\dot{\mathbf{u}})$ in (9.8) is sometimes called the *Zwang*[1] of the system, since it rates the distance $d(\dot{\mathbf{u}}_R, \dot{\mathbf{u}})_{\mathbf{M}} = ||\dot{\mathbf{u}}_R - \dot{\mathbf{u}}||_{\mathbf{M}}$ of the accelerations $\dot{\mathbf{u}}_R$ without any set-valued force laws from the accelerations $\dot{\mathbf{u}}$ under the restrictions expressed by Ψ. The minimization problem (9.11) should therefore be regarded as an *extension* of Gauß' classical *principle of least constraints.* Note that more is contained in (9.11) than unilaterally or bilaterally constrained motion: Even single-step force laws with a potential as depicted in the right diagram of Figure 9.1 are captured, which obviously do not restrict the values of the accelerations to any subset of $\mathbb{R}^f$.

As already mentioned, the acceleration potential $\Psi(\dot{\mathbf{u}})$ is composed of (n_S+n_P) potentials $\Psi_k(\dot{\mathbf{u}})$ which are the *translates* of positively homogeneous polyhedral convex functions $\tilde{\Psi}_k(\dot{\mathbf{u}})$. This does not mean that $\Psi(\dot{\mathbf{u}})$ itself is the translate of a positively homogeneous convex function, but only that $\Psi(\dot{\mathbf{u}})$ is polyhedral, as it can be seen from Figure 9.2. Only in the case that $\dot{\gamma}_k(\dot{\mathbf{u}}) = 0$ is admissible for every $k \in \mathcal{S}_{\dot{\gamma}} \cup \mathcal{P}_{\dot{\gamma}}$ for the *same* $\dot{\mathbf{u}}$, we may gain this additional property: $\Psi(\dot{\mathbf{u}})$ is then the translate of the positively homogeneous convex function $\tilde{\Psi}(\dot{\mathbf{u}}) = \psi(\dot{\tilde{\gamma}}(\dot{\mathbf{u}}))$ with $\dot{\tilde{\gamma}}(\dot{\mathbf{u}}) = \dot{\gamma}(\dot{\mathbf{u}}) - \bar{\mathbf{w}} = \mathbf{W}^T\dot{\mathbf{u}}$, since then $\{\dot{\mathbf{u}} \,|\, \mathbf{W}^T\dot{\mathbf{u}} + \bar{\mathbf{w}} = 0\} \neq \emptyset$ by assumption, and hence $\Psi(\dot{\mathbf{u}}) = \tilde{\Psi}(\dot{\mathbf{u}} - \mathbf{a})$ with $\mathbf{a}$ such that $\mathbf{W}^T\mathbf{a} + \bar{\mathbf{w}} = 0$, as direct calculation shows. From the physical point of view, this situation applies if constant relative velocities γ_k are at the

[1] German word for "constraint"

same time kinematically accessible for all force elements $k \in \mathcal{S}_{\dot\gamma} \cup \mathcal{P}_{\dot\gamma}$ whose force directions $\mathbf{w}_k$ lie in ker $\mathbf{W}^T$. This holds, in particular, for *independent* force directions, as well as for systems which are *not* driven kinematically from outside. Furthermore, note that $\bar{\mathbf{w}}$ vanishes for scleronomic systems if, in addition, $\mathbf{W}$ is constant. Although being of restricted practical relevance, we may mention this case, since then $\Psi(\dot{\mathbf{u}})$ itself is positively homogeneous. The results concerning Gauß' principle as presented in [38] do only apply for this situation, since the more complicated structure of Ψ as discussed above has not been regarded by mistake.

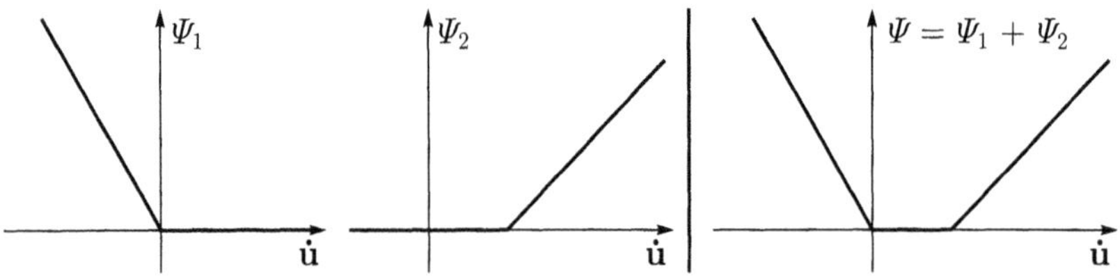

Figure 9.2. Sum of translates of positively homogeneous convex functions

Gauß' principle of least constraints may be more appropriately stated in terms of proximations: As a consequence of Moreau's Theorem, see [121], each $\dot{\mathbf{u}}_R \in \mathbb{R}^f$ can be decomposed uniquely into a sum $\dot{\mathbf{u}} - \mathbf{M}^{-1}\mathbf{f}_\Psi$ with respect to some closed proper convex function Ψ such that $(\dot{\mathbf{u}}, -\mathbf{f}_\Psi)$ belongs to the graph of $\partial\Psi$. This situation applies, because we have by (9.6), (9.7)

$$\dot{\mathbf{u}}_R = \dot{\mathbf{u}} - \mathbf{M}^{-1}\mathbf{f}_\Psi, \qquad -\mathbf{f}_\Psi \in \partial\Psi(\dot{\mathbf{u}}). \tag{9.13}$$

The unique component $\dot{\mathbf{u}}$ in this decomposition is then denoted by

$$\dot{\mathbf{u}} = \operatorname{prox}_\Psi(\dot{\mathbf{u}}_R), \tag{9.14}$$

and the mapping $\dot{\mathbf{u}}_R \to \operatorname{prox}_\Psi(\dot{\mathbf{u}}_R)$ is called the proximation corresponding to Ψ. Clearly,

$$\operatorname{prox}_\Psi(\dot{\mathbf{u}}_R) = \arg\min_{\dot{\mathbf{u}}^\star}\{\tfrac{1}{2}\,\|\dot{\mathbf{u}}_R - \dot{\mathbf{u}}^\star\|^2_{\mathbf{M}} + \Psi(\dot{\mathbf{u}}^\star)\} \tag{9.15}$$

by (9.14) and (9.12). For Ψ being an indicator function, for example, the proximation (9.15) becomes

$$\operatorname{prox}_\Psi(\dot{\mathbf{u}}_R) = \arg\min_{\dot{\mathbf{u}}^\star}\{\tfrac{1}{2}\,\|\dot{\mathbf{u}}_R - \dot{\mathbf{u}}^\star\|^2_{\mathbf{M}} \mid \dot{\mathbf{u}}^\star \in \operatorname{dom}\Psi\}. \tag{9.16}$$

The acceleration $\dot{\mathbf{u}} = \operatorname{prox}_\Psi(\dot{\mathbf{u}}_R)$ is therefore the nearest point to $\dot{\mathbf{u}}_R$ in the convex set dom Ψ, i.e. the point in dom Ψ which minimizes the distance to $\dot{\mathbf{u}}_R$. In our case, Ψ becomes an indicator when $\mathcal{P}_{\dot\gamma} = \emptyset$, defining a polyhedral convex set in $\mathbb{R}^f$.

Let us now discuss the meaning of proximations in the special case that $\varPsi$ is the translate of a closed convex positively homogeneous function $\tilde{\varPsi}$, say $\varPsi(\dot{\mathbf{u}}) = \tilde{\varPsi}(\dot{\mathbf{u}} - \mathbf{a})$. Note that $\tilde{\varPsi}(0) = 0$, and that $\operatorname{dom}\tilde{\varPsi} = \operatorname{dom}\varPsi - \mathbf{a}$ is a closed convex cone. After exchange of variables, $\dot{\tilde{\mathbf{u}}} := \dot{\mathbf{u}} - \mathbf{a}$, $\dot{\tilde{\mathbf{u}}}_R := \dot{\mathbf{u}}_R - \mathbf{a}$, the proximation may be written as in (9.13), (9.14),

$$\dot{\tilde{\mathbf{u}}} = \operatorname{prox}_{\tilde{\varPsi}}(\dot{\tilde{\mathbf{u}}}_R) \quad \Leftrightarrow \quad \dot{\tilde{\mathbf{u}}}_R = \dot{\tilde{\mathbf{u}}} - \mathbf{M}^{-1}\mathbf{f}_\varPsi, \quad -\mathbf{f}_\varPsi \in \partial\tilde{\varPsi}(\dot{\tilde{\mathbf{u}}}). \tag{9.17}$$

Suppose now that $\tilde{\varPsi}$ is, in addition, an indicator function, which reflects unilaterally constrained motion. By the corresponding minimization problem (9.16), the proximation yields $(\dot{\mathbf{u}} - \mathbf{a})$ to be the nearest point to $(\dot{\mathbf{u}}_R - \mathbf{a})$ in the closed convex cone $\operatorname{dom}\varPsi - \mathbf{a}$. In this case, the proximation (9.17) might be equivalently expressed by the extended complementarity conditions

$$\dot{\tilde{\mathbf{u}}}_R = \dot{\tilde{\mathbf{u}}} - \mathbf{M}^{-1}\mathbf{f}_\varPsi, \quad \dot{\tilde{\mathbf{u}}} \in \operatorname{dom}\tilde{\varPsi}, \quad -\mathbf{f}_\varPsi \in (\operatorname{dom}\tilde{\varPsi})^\circ, \quad -\mathbf{f}_\varPsi^T\dot{\tilde{\mathbf{u}}} = 0, \tag{9.18}$$

where C° denotes the polar of a closed convex cone C defined by $C^\circ = \{\mathbf{y} \,|\, \mathbf{x}^T\mathbf{y} \le 0,\ \forall\mathbf{x} \in C\}$. We will prove this result:

Proof. Suppose first that (9.17) holds. Since $\tilde{\varPsi}$ is an indicator, we have $-\mathbf{f}_\varPsi \in \partial\tilde{\varPsi}(\dot{\tilde{\mathbf{u}}}) = N_{\operatorname{dom}\tilde{\varPsi}}(\dot{\tilde{\mathbf{u}}})$, where the latter denotes the normal cone to the closed convex cone $\operatorname{dom}\tilde{\varPsi}$ at the point $\dot{\tilde{\mathbf{u}}}$. By the definition of the normal cone, this inclusion may also be written as $-\mathbf{f}_\varPsi \in N_{\operatorname{dom}\tilde{\varPsi}}(\dot{\tilde{\mathbf{u}}}) \Leftrightarrow -\mathbf{f}_\varPsi^T(\dot{\tilde{\mathbf{u}}}^\star - \dot{\tilde{\mathbf{u}}}) \le 0$, $\dot{\tilde{\mathbf{u}}} \in \operatorname{dom}\tilde{\varPsi}$, $\forall\dot{\tilde{\mathbf{u}}}^\star \in \operatorname{dom}\tilde{\varPsi}$. The set $\operatorname{dom}\tilde{\varPsi}$ is a closed convex cone, i.e. $\lambda\dot{\tilde{\mathbf{u}}} \in \operatorname{dom}\tilde{\varPsi}$ holds whenever $\dot{\tilde{\mathbf{u}}} \in \operatorname{dom}\tilde{\varPsi}$ and $\lambda \ge 0$. We therefore choose $\dot{\tilde{\mathbf{u}}}^\star = \lambda\dot{\tilde{\mathbf{u}}}$ with $\lambda = 0$ and $\lambda = 2$, and obtain from the variational inequality $-\mathbf{f}_\varPsi^T\dot{\tilde{\mathbf{u}}} \ge 0$ and $-\mathbf{f}_\varPsi^T\dot{\tilde{\mathbf{u}}} \le 0$, which proves the last condition in (9.18). The validity of $-\mathbf{f}_\varPsi \in (\operatorname{dom}\tilde{\varPsi})^\circ$ is obvious, since $(\operatorname{dom}\tilde{\varPsi})^\circ = N_{\operatorname{dom}\tilde{\varPsi}}(0) \supset N_{\operatorname{dom}\tilde{\varPsi}}(\dot{\tilde{\mathbf{u}}}) = \partial\tilde{\varPsi}(\dot{\tilde{\mathbf{u}}})$. Suppose now that (9.18) holds. The inclusion $-\mathbf{f}_\varPsi \in (\operatorname{dom}\tilde{\varPsi})^\circ$ may be written as $-\mathbf{f}_\varPsi^T\dot{\tilde{\mathbf{u}}}^\star \le 0, \forall\dot{\tilde{\mathbf{u}}}^\star \in \operatorname{dom}\tilde{\varPsi}$. Together with the second and the last expression in (9.18), this inequality becomes $-\mathbf{f}_\varPsi^T(\dot{\tilde{\mathbf{u}}}^\star - \dot{\tilde{\mathbf{u}}}) \le 0$, $\dot{\tilde{\mathbf{u}}} \in \operatorname{dom}\tilde{\varPsi}$, $\forall\dot{\tilde{\mathbf{u}}}^\star \in \operatorname{dom}\tilde{\varPsi}$, which means that $-\mathbf{f}_\varPsi \in N_{\operatorname{dom}\tilde{\varPsi}}(\dot{\tilde{\mathbf{u}}}) = \partial\tilde{\varPsi}(\dot{\tilde{\mathbf{u}}})$ as in (9.17). □

The geometric meaning of (9.18) is as follows: With respect to the Euclidean space $\mathbb{R}^f$ with inner product $(\mathbf{a}, \mathbf{b})_\mathbf{M} := \mathbf{a}^T\mathbf{M}\mathbf{b}$, the proximation (9.17) applied to the indicator of a closed convex cone (9.18) yields the unique orthogonal decomposition of $\dot{\tilde{\mathbf{u}}}_R$ into $\dot{\tilde{\mathbf{u}}}$ and $-\mathbf{M}^{-1}\mathbf{f}_\varPsi$, i.e. $(\dot{\tilde{\mathbf{u}}}, -\mathbf{M}^{-1}\mathbf{f}_\varPsi)_\mathbf{M} = 0$, where both members are taken from closed convex cones that are polar to each other, that means that for each $\dot{\tilde{\mathbf{u}}}^\star \in \operatorname{dom}\tilde{\varPsi}$ and $(-\mathbf{M}^{-1}\mathbf{f}_\varPsi)^\star \in \mathbf{M}^{-1}(\operatorname{dom}\tilde{\varPsi})^\circ$ one has $(\dot{\tilde{\mathbf{u}}}^\star, (-\mathbf{M}^{-1}\mathbf{f}_\varPsi)^\star)_\mathbf{M} \le 0$. In the case of classical bilaterally constrained motion, the cone $\operatorname{dom}\tilde{\varPsi}$ becomes a linear subspace of $\mathbb{R}^f$ with $\mathbf{M}^{-1}(\operatorname{dom}\tilde{\varPsi})^\circ$ being its orthogonal complement, and the proximation of $\dot{\tilde{\mathbf{u}}} \in \operatorname{dom}\tilde{\varPsi} \oplus \mathbf{M}^{-1}(\operatorname{dom}\tilde{\varPsi})^\circ$ reduces to the orthogonal projection of $\dot{\tilde{\mathbf{u}}}_R$ on $\operatorname{dom}\tilde{\varPsi}$, i.e. $\dot{\tilde{\mathbf{u}}} = \operatorname{prox}_{\tilde{\varPsi}}(\dot{\tilde{\mathbf{u}}}_R) = \operatorname{proj}_{\operatorname{dom}\tilde{\varPsi}}(\dot{\tilde{\mathbf{u}}}_R)$.

9.2 The Principle of Gauß

In a similar fashion as for the quadratic program (8.47) with variational inequality (8.42), there are some variational formulations related to the principle of least constraints (9.12), which we are going to discuss in this section. These variational expressions involve the difference of certain accelerations, called the virtual accelerations, and appear already in classical mechanics in some slightly modified form, denoted here as the principle of Gauß.

We start our investigations with equation (9.6) which we rewrite as a variational inequality. By the definition of the subdifferential, the inclusion $-\mathbf{f}_\Psi \in \partial\Psi(\dot{\mathbf{u}})$ holds for $\dot{\mathbf{u}} \in \operatorname{dom}\Psi$ iff $-\mathbf{f}_\Psi^T(\dot{\mathbf{u}}^\star - \dot{\mathbf{u}}) \leq \Psi(\dot{\mathbf{u}}^\star) - \Psi(\dot{\mathbf{u}})$ for all $\dot{\mathbf{u}}^\star$. Moreover, since the generalized gradient and the subdifferential agree for convex functions, we may use even further inequalities to describe the aforementioned inclusion. This step has already been done in equation (6.18) for scalar functions; the expressions presented there, however, are general. Thus, by setting $\delta\dot{\mathbf{u}} := \dot{\mathbf{u}}^\star - \dot{\mathbf{u}}$ and after elimination of $\mathbf{f}_\Psi$ we obtain from (9.6)

$$
\begin{aligned}
-(\mathbf{M}\dot{\mathbf{u}} - \mathbf{h}_{\dot{\gamma}})^T\delta\dot{\mathbf{u}} \leq \Psi^\uparrow(\dot{\mathbf{u}}, \delta\dot{\mathbf{u}}) \leq \Psi'(\dot{\mathbf{u}}, \delta\dot{\mathbf{u}}) \leq \Psi(\dot{\mathbf{u}} + \delta\dot{\mathbf{u}}) - \Psi(\dot{\mathbf{u}}),& \\
\dot{\mathbf{u}} \in \operatorname{dom}\Psi, \quad \forall\,\delta\dot{\mathbf{u}},&
\end{aligned}
\tag{9.19}
$$

where any of the three inequalities between the left hand member in (9.19) and one of the three Ψ–expressions reflects equivalently (9.6), as long as it is regarded to hold for *every* $\delta\dot{\mathbf{u}}$. We call $\delta\dot{\mathbf{u}}$ a *virtual acceleration*, and the variational formulation (9.19) of (9.6) the *principle of Gauß*. Note again that Gauß' principle may be stated by either one out of three variational inequalities which are all equivalent, since the virtual accelerations $\delta\dot{\mathbf{u}}$ are arbitrary. Note also that more is contained in (9.19) than just (impact-free) unilaterally constrained motion, because the elements of $\mathcal{P}_{\dot{\gamma}}$ contributing to Ψ reflect single-steps and hence unconstrained motion under set-valued force laws.

Most suitable for further investigations is the epiderivative $\Psi^\uparrow$ in (9.19) due to its natural linking to the tangent cone, and the difference in the acceleration potentials since it may conveniently be used to express properties of convexity. We therefore skip the representation via the directional derivative Ψ' in (9.19) and proceed with Gauß' principle in the form

$$
\begin{aligned}
-(\mathbf{M}\dot{\mathbf{u}} - \mathbf{h}_{\dot{\gamma}})^T\delta\dot{\mathbf{u}} \leq \Psi^\uparrow(\dot{\mathbf{u}}, \delta\dot{\mathbf{u}}) \leq \Psi(\dot{\mathbf{u}} + \delta\dot{\mathbf{u}}) - \Psi(\dot{\mathbf{u}}),& \\
\dot{\mathbf{u}} \in \operatorname{dom}\Psi, \quad \forall\,\delta\dot{\mathbf{u}}.&
\end{aligned}
\tag{9.20}
$$

Equation (9.20) is required to hold for every $\delta\dot{\mathbf{u}}$, in particular for $\dot{\mathbf{u}}^\star = \dot{\mathbf{u}} + \delta\dot{\mathbf{u}}$ even outside of $\operatorname{dom}\Psi$. In mechanics, however, one usually thinks in terms of physically admissible variations. We are thus going to rewrite Gauß' principle with variations taken from certain smaller sets which are closer related to the physical problem.

We first restrict the variations $\delta\dot{\mathbf{u}}$ to belong to the tangent cone to $\operatorname{dom}\Psi$ at $\dot{\mathbf{u}}$. Gauß' principle then reads

$$\begin{aligned} -(\mathbf{M}\dot{\mathbf{u}} - \mathbf{h}_{\dot{\gamma}})^T \delta\dot{\mathbf{u}} \leq \Psi^{\uparrow}(\dot{\mathbf{u}}, \delta\dot{\mathbf{u}}) \leq \Psi(\dot{\mathbf{u}} + \delta\dot{\mathbf{u}}) - \Psi(\dot{\mathbf{u}}), \\ \dot{\mathbf{u}} \in \operatorname{dom}\Psi, \quad \forall\, \delta\dot{\mathbf{u}} \in T_{\operatorname{dom}\Psi}(\dot{\mathbf{u}}), \end{aligned} \tag{9.21}$$

and we may prove equivalence with (9.20):

Proof. It is evident that (9.21) holds when (9.20) is presupposed. To show the reverse we first note that

$$\Psi^{\uparrow}(\dot{\mathbf{u}}, \delta\dot{\mathbf{u}}) = +\infty \quad \text{for } \delta\dot{\mathbf{u}} \notin T_{\operatorname{dom}\Psi}(\dot{\mathbf{u}}), \tag{9.22}$$

since Ψ is convex and hence regular in the sense that $\operatorname{epi}\Psi$ does not contain any re-intrant corners, and that

$$\operatorname{dom}\Psi - \dot{\mathbf{u}} \subset T_{\operatorname{dom}\Psi}(\dot{\mathbf{u}}) \tag{9.23}$$

which is a consequence on the convexity of Ψ, see e.g. [9] and Figure 9.3. In order to deduce (9.20) from (9.21) it is sufficient to show that (9.20) still holds for any $\delta\dot{\mathbf{u}}$ not considered in (9.21), i.e. for any $\delta\dot{\mathbf{u}} \notin T_{\operatorname{dom}\Psi}(\dot{\mathbf{u}})$. In this case, the first inequality in (9.20) is still valid, since $\Psi^{\uparrow}(\dot{\mathbf{u}}, \delta\dot{\mathbf{u}}) = +\infty$ by (9.22). The second inequality also holds. In particular, it holds as an equality of the form $+\infty = +\infty$, since $\Psi(\dot{\mathbf{u}})$ is finite due to $\dot{\mathbf{u}} \in \operatorname{dom}\Psi$, and $\Psi(\dot{\mathbf{u}}+\delta\dot{\mathbf{u}}) = +\infty$, because $\dot{\mathbf{u}} + \delta\dot{\mathbf{u}} \notin \dot{\mathbf{u}} + T_{\operatorname{dom}\Psi}(\dot{\mathbf{u}}) \supset \operatorname{dom}\Psi$ by (9.23). □

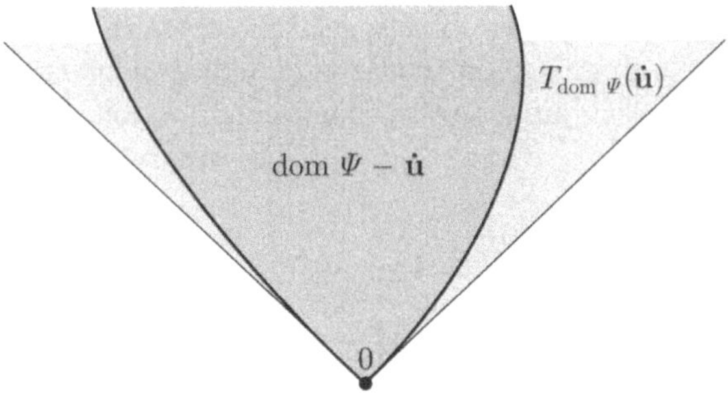

Figure 9.3. Convex set $\operatorname{dom}\Psi$ with tangent cone at $\dot{\mathbf{u}}$

We may even further restrict the variations $\delta\dot{\mathbf{u}}$ such that they become physically admissible. This is to allow values of $\dot{\mathbf{u}}^{\star} = \dot{\mathbf{u}} + \delta\dot{\mathbf{u}}$ only in the set $\operatorname{dom}\Psi$, which yields Gauß' principle in the form

$$\begin{aligned} -(\mathbf{M}\dot{\mathbf{u}} - \mathbf{h}_{\dot{\gamma}})^T \delta\dot{\mathbf{u}} \leq \Psi^{\uparrow}(\dot{\mathbf{u}}, \delta\dot{\mathbf{u}}) \leq \Psi(\dot{\mathbf{u}} + \delta\dot{\mathbf{u}}) - \Psi(\dot{\mathbf{u}}), \\ \dot{\mathbf{u}} \in \operatorname{dom}\Psi, \quad \forall\, \delta\dot{\mathbf{u}} \in \operatorname{dom}\Psi - \dot{\mathbf{u}}. \end{aligned} \tag{9.24}$$

Note that the set $\operatorname{dom}\Psi$ is polyhedral in our case, that means even simpler than the more general situation depicted in Figure 9.3 and used throughout the proofs. For (9.24), equivalence to (9.21) and hence to (9.20) can be shown, again:

Proof. As before, equation (9.24) follows immediately from (9.21), since, by (9.23), the variations in (9.24) belong to a subset of the variations in (9.21). To prove the reverse we have to show that (9.21) is still true for any $\delta\dot{\mathbf{u}} \in T_{\mathrm{dom}\,\Psi}(\dot{\mathbf{u}})\backslash(\mathrm{dom}\,\Psi - \dot{\mathbf{u}})$, i.e. for any $(\dot{\mathbf{u}}+\delta\dot{\mathbf{u}}) \in (\dot{\mathbf{u}} + T_{\mathrm{dom}\,\Psi}(\dot{\mathbf{u}}))\backslash\,\mathrm{dom}\,\Psi$. In this case we thus have $(\dot{\mathbf{u}}+\delta\dot{\mathbf{u}}) \notin \mathrm{dom}\,\Psi$, hence $\Psi(\dot{\mathbf{u}}+\delta\dot{\mathbf{u}}) = +\infty$, which proves already the second inequality in (9.21), since $\Psi(\dot{\mathbf{u}})$ is finite. We have still to show the first inequality in (9.21), that is, we have to show that

$$-(\mathbf{M}\dot{\mathbf{u}} - \mathbf{h}_{\dot{\gamma}})^T\delta\dot{\mathbf{u}} \le \Psi^{\uparrow}(\dot{\mathbf{u}}, \delta\dot{\mathbf{u}}) \tag{9.25}$$

holds for every $\delta\dot{\mathbf{u}} \in T_{\mathrm{dom}\,\Psi}(\dot{\mathbf{u}})$ when it applies for arbitrary $\delta\dot{\mathbf{u}} \in \mathrm{dom}\,\Psi - \dot{\mathbf{u}}$. In order to do this we use a characterization of the contingent cone $K_{\mathrm{dom}\,\Psi}(\dot{\mathbf{u}})$ which is, in our case, the same as the tangent cone $T_{\mathrm{dom}\,\Psi}(\dot{\mathbf{u}})$, since the set $\mathrm{dom}\,\Psi$ is convex and hence tangentially regular: An element $\delta\dot{\mathbf{u}}$ belongs to the contingent cone to $\mathrm{dom}\,\Psi$ at $\dot{\mathbf{u}}$ and hence to $T_{\mathrm{dom}\,\Psi}(\dot{\mathbf{u}})$, iff there exist sequences $\mu_n \downarrow 0$ and sequences $\delta\dot{\mathbf{u}}_n \to \delta\dot{\mathbf{u}}$ satisfying $\dot{\mathbf{u}} + \mu_n\,\delta\dot{\mathbf{u}}_n \in \mathrm{dom}\,\Psi$ for all n, see [9]. For any $\delta\dot{\mathbf{u}} \in T_{\mathrm{dom}\,\Psi}(\dot{\mathbf{u}})$ we may thus find sequences $(\mu_n\,\delta\dot{\mathbf{u}}_n) \in \mathrm{dom}\,\Psi - \dot{\mathbf{u}}$ for which (9.24) applies by presupposition, i.e.

$$-(\mathbf{M}\dot{\mathbf{u}} - \mathbf{h}_{\dot{\gamma}})^T(\mu_n\,\delta\dot{\mathbf{u}}_n) \le \Psi^{\uparrow}(\dot{\mathbf{u}}, \mu_n\,\delta\dot{\mathbf{u}}_n), \qquad \forall n \quad (\text{and } \mu_n > 0). \tag{9.26}$$

Since the epiderivative is positively homogeneous in its second argument, (9.26) simplifies to $-(\mathbf{M}\dot{\mathbf{u}} - \mathbf{h}_{\dot{\gamma}})^T\delta\dot{\mathbf{u}}_n \le \Psi^{\uparrow}(\dot{\mathbf{u}}, \delta\dot{\mathbf{u}}_n)$ $\forall n$, and the limit for $n \to \infty$ yields (9.25) with $\delta\dot{\mathbf{u}} \in T_{\mathrm{dom}\,\Psi}(\dot{\mathbf{u}})$, which completes the proof. □

Let us finally discuss Gauß' principle for some special choices of the acceleration potential Ψ. We assume from now on that Ψ is an indicator function ($\mathcal{P}_{\dot{\gamma}} = \emptyset$ in our case), which reflects unilaterally constrained motion. With $\Psi(\dot{\mathbf{u}}) = 0$ for $\dot{\mathbf{u}} \in \mathrm{dom}\,\Psi$ and

$$\Psi^{\uparrow}(\dot{\mathbf{u}}, \delta\dot{\mathbf{u}}) = \begin{cases} 0 & \text{if } \delta\dot{\mathbf{u}} \in T_{\mathrm{dom}\,\Psi}(\dot{\mathbf{u}}) \\ +\infty & \text{else} \end{cases} \tag{9.27}$$

the three versions (9.20), (9.21), (9.24) of Gauß' principle become

$$\begin{aligned} &-(\mathbf{M}\dot{\mathbf{u}} - \mathbf{h}_{\dot{\gamma}})^T\delta\dot{\mathbf{u}} \le \Psi^{\uparrow}(\dot{\mathbf{u}}, \delta\dot{\mathbf{u}}) \le \Psi(\dot{\mathbf{u}}+\delta\dot{\mathbf{u}}) && \dot{\mathbf{u}} \in \mathrm{dom}\,\Psi,\ \forall\,\delta\dot{\mathbf{u}} \\ &-(\mathbf{M}\dot{\mathbf{u}} - \mathbf{h}_{\dot{\gamma}})^T\delta\dot{\mathbf{u}} \le 0 \le \Psi(\dot{\mathbf{u}}+\delta\dot{\mathbf{u}}) && \dot{\mathbf{u}} \in \mathrm{dom}\,\Psi,\ \forall\,\delta\dot{\mathbf{u}} \in T_{\mathrm{dom}\,\Psi}(\dot{\mathbf{u}}) \\ &-(\mathbf{M}\dot{\mathbf{u}} - \mathbf{h}_{\dot{\gamma}})^T\delta\dot{\mathbf{u}} \le 0 && \dot{\mathbf{u}} \in \mathrm{dom}\,\Psi,\ \forall\,\delta\dot{\mathbf{u}} \in \mathrm{dom}\,\Psi - \dot{\mathbf{u}}. \end{aligned} \tag{9.28}$$

If bilaterally constrained motion is considered, $\mathrm{dom}\,\Psi$ becomes an affine subspace of $\mathbb{R}^f$, and $T_{\mathrm{dom}\,\Psi}(\dot{\mathbf{u}})$ is the associated linear subspace obtained from a translation of $\mathrm{dom}\,\Psi$ such that $\dot{\mathbf{u}}$ is moved to the origin. Equation (9.23) thus holds as an equality,

$$\mathrm{dom}\,\Psi - \dot{\mathbf{u}} \equiv T_{\mathrm{dom}\,\Psi}(\dot{\mathbf{u}}). \tag{9.29}$$

Moreover, since the epiderivative $\Psi^{\uparrow}(\dot{\mathbf{u}},\cdot)$ in (9.27) may be regarded as the indicator of the set $T_{\mathrm{dom}\,\Psi}(\dot{\mathbf{u}})$ and hence of the set $(\mathrm{dom}\,\Psi - \dot{\mathbf{u}})$ by (9.29), we have $\Psi^{\uparrow}(\dot{\mathbf{u}},\delta\dot{\mathbf{u}}) = \Psi(\dot{\mathbf{u}}+\delta\dot{\mathbf{u}})$. Note also that $\Psi^{\uparrow}(\dot{\mathbf{u}},\delta\dot{\mathbf{u}}) = \Psi^{\uparrow}(\dot{\mathbf{u}},-\delta\dot{\mathbf{u}})$, since $T_{\mathrm{dom}\,\Psi}(\dot{\mathbf{u}})$ is a linear space. Together, we therefore have

$$\Psi^{\uparrow}(\dot{\mathbf{u}},\delta\dot{\mathbf{u}}) \equiv \Psi(\dot{\mathbf{u}}+\delta\dot{\mathbf{u}}) \equiv \Psi(\dot{\mathbf{u}}-\delta\dot{\mathbf{u}}) \equiv \Psi^{\uparrow}(\dot{\mathbf{u}},-\delta\dot{\mathbf{u}}) \tag{9.30}$$

which will be used to further simplify equations (9.28). Note that (9.28) holds for $-\delta\dot{\mathbf{u}}$ when it does for $+\delta\dot{\mathbf{u}}$, since all the variations in (9.28) are now restricted to linear spaces. Together with (9.30), all the inequalities in (9.28) thus change to equalities, and the second and the third line become identical due to (9.29). We obtain Gauß' variational principle in equality form,

$$\begin{aligned} -(\mathbf{M}\dot{\mathbf{u}}-\mathbf{h}_{\dot\gamma})^T\delta\dot{\mathbf{u}} &= \Psi^{\uparrow}(\dot{\mathbf{u}},\delta\dot{\mathbf{u}}) = \Psi(\dot{\mathbf{u}}+\delta\dot{\mathbf{u}}) && \dot{\mathbf{u}} \in \mathrm{dom}\,\Psi,\ \forall\,\delta\dot{\mathbf{u}} \\ -(\mathbf{M}\dot{\mathbf{u}}-\mathbf{h}_{\dot\gamma})^T\delta\dot{\mathbf{u}} &= 0 = \Psi(\dot{\mathbf{u}}+\delta\dot{\mathbf{u}}) && \dot{\mathbf{u}} \in \mathrm{dom}\,\Psi,\ \forall\,\delta\dot{\mathbf{u}} \in \mathrm{dom}\,\Psi - \dot{\mathbf{u}}, \end{aligned} \tag{9.31}$$

with the expression $-(\mathbf{M}\dot{\mathbf{u}}-\mathbf{h}_{\dot\gamma})^T\delta\dot{\mathbf{u}} = 0$ for $\dot{\mathbf{u}} \in \mathrm{dom}\,\Psi$ and every $\delta\dot{\mathbf{u}} \in \mathrm{dom}\,\Psi - \dot{\mathbf{u}}$ best known from classical mechanics.

9.3 The Principle of Jourdain

Jourdain's principle in classical mechanics is a variational equation which considers holonomic and non-holonomic bilaterally constrained motion by a virtual power expression. This expression with values equal to zero has to be understood as the natural pairing of the constraint forces and the virtual velocities and reflects the physical behavior of what is called an *perfect* bilateral constraint. We may easily carry over this concept to our problem when starting from its velocity representation (8.12) and performing the same steps as for Gauß' variational principle in Section 9.2.

Thus, the main goal of this section is to state (8.12) as a variational inequality which, again, might be immediately done via the definition of the subdifferential. We therefore abbreviate in (8.12) the generalized forces by $\mathbf{f}_\Phi$ and introduce the overall velocity potential $\Phi(\mathbf{u})$ as

$$\begin{aligned} \mathbf{f}_\Phi &:= \sum_{k\in\mathcal{S}_\gamma} \mathbf{w}_k\lambda_{Sk} + \sum_{k\in\mathcal{P}_\gamma} \mathbf{w}_k\lambda_{Pk} \\ \Phi(\mathbf{u}) &:= \sum_{k\in\mathcal{S}_\gamma} \varphi_{Sk}(\gamma_k(\mathbf{u})) + \sum_{k\in\mathcal{P}_\gamma} \varphi_{Pk}(\gamma_k(\mathbf{u})) \end{aligned} \tag{9.32}$$

with $\mathcal{S}_\gamma = \mathcal{S}_{v\gamma} \cup \mathcal{S}_{\varphi\gamma}$ and $\mathcal{P}_\gamma = \mathcal{P}_{v\gamma} \cup \mathcal{P}_{\varphi\gamma}$. By the same reasoning as in equations (9.1)–(9.6) one may now rewrite (8.12) together with (9.32) as

$$\mathbf{M}\dot{\mathbf{u}} - \mathbf{h}_\gamma - \mathbf{f}_\Phi = 0, \quad -\mathbf{f}_\Phi \in \partial\Phi(\mathbf{u}) \tag{9.33}$$

in order to eliminate $\mathbf{f}_\Phi$ and to express the remaining inclusion by either one of the two variational inequalities

$$-(\mathbf{M}\dot{\mathbf{u}} - \mathbf{h}_\gamma)^T \delta\mathbf{u} \le \Phi^\uparrow(\mathbf{u}, \delta\mathbf{u}) \le \Phi(\mathbf{u} + \delta\mathbf{u}) - \Phi(\mathbf{u}), \quad \mathbf{u} \in \operatorname{dom}\Phi. \quad (9.34)$$

The variations $\delta\mathbf{u}$, called the *virtual velocities*, have to be chosen in the same fashion as in (9.20), (9.21) or (9.24); this means that (9.34) has to hold

$$\forall\, \delta\mathbf{u} \quad \text{or} \quad \forall\, \delta\mathbf{u} \in T_{\operatorname{dom}\Phi}(\mathbf{u}) \quad \text{or} \quad \forall\, \delta\mathbf{u} \in \operatorname{dom}\Phi - \mathbf{u}. \quad (9.35)$$

As in (9.28), equations (9.34) and (9.35) express pure unilaterally constrained motion when the velocity potential is an indicator function. If the effective domain of $\Phi(\mathbf{u})$ is, in addition, an affine subspace, one obtains classical holonomic and non-holonomic bilaterally constrained motion. The variational formulation (9.34), (9.35) then becomes $-(\mathbf{M}\dot{\mathbf{u}} - \mathbf{h}_\gamma)^T \delta\mathbf{u} = 0$ for $\mathbf{u} \in \operatorname{dom}\Phi$ and every $\delta\mathbf{u} \in \operatorname{dom}\Phi - \mathbf{u}$ which is the classical principle of Jourdain.

9.4 The Principle of d'Alembert/Lagrange

Similarly to the previous two sections we derive finally a variational formulation of our problem which is stated on the displacement level. This expression is usually called the *principle of d'Alembert/Lagrange* which generalizes the virtual work principle from statics to the dynamic case, see Chapter 3. We start with the representation of the problem on displacement level (8.10) and note again that only displacement force laws such as bilateral and unilateral displacement constraints or pre-stressed springs are considered, whereas any force laws depending also on velocities have to be excluded ($\mathcal{S}_{\varphi\gamma} = \mathcal{P}_{\varphi\gamma} = \emptyset$). In order to rewrite (8.10) as a differential inclusion we introduce the generalized forces $\mathbf{f}_V$ and the overall displacement potential $V(\mathbf{q})$ as

$$\begin{aligned} \mathbf{f}_V &:= \sum_{k \in \mathcal{S}_{vg}} \mathbf{w}_k \lambda_{Sk} + \sum_{k \in \mathcal{P}_{vg}} \mathbf{w}_k \lambda_{Pk} \\ V(\mathbf{q}) &:= \sum_{k \in \mathcal{S}_{vg}} v_{Sk}(g_k(\mathbf{q})) + \sum_{k \in \mathcal{P}_{vg}} v_{Pk}(g_k(\mathbf{q})). \end{aligned} \quad (9.36)$$

Note that $V(\mathbf{q})$ *must not be expected to be convex*, although convexity does hold for the scalar potentials $v_{Sk}(g_k)$ and $v_{Pk}(g_k)$ by assumption: In multibody dynamics, the mappings $\mathbf{q} \to g_k(\mathbf{q})$ usually involve sine and cosine expressions leading immediately to *non-convexity*, which is then transferred to $V(\mathbf{q})$ via (9.36). In most situations, convexity of $V(\mathbf{q})$ is thus lacking, which makes it mandatory to replace the subdifferential ∂ by the generalized gradient $\bar{\partial}$. We are thus forced to express (8.10) with the help of (9.36) as

$$\mathbf{M}\dot{\mathbf{u}} - \mathbf{h}_g - \mathbf{f}_V = 0, \quad -\mathbf{f}_V \in \bar{\partial} V(\mathbf{q}), \quad (9.37)$$

in contrast to the corresponding terms (9.33) on the velocity level, where the subdifferential is still meaningful. Eliminating $\mathbf{f}_V$ from (9.37) and rewriting the generalized gradient with the help of the upper subderivative yields the variational inequality

$$-(\mathbf{M}\dot{\mathbf{u}} - \mathbf{h}_g)^T\delta\mathbf{q} \leq V^{\uparrow}(\mathbf{q}, \delta\mathbf{q}), \quad \mathbf{q} \in \operatorname{dom} V \tag{9.38}$$

with virtual displacements $\delta\mathbf{q}$ such that (9.38) has to hold

$$\forall\, \delta\mathbf{q} \quad \text{or} \quad \forall\, \delta\mathbf{q} \in T_{\operatorname{dom} V}(\mathbf{q}). \tag{9.39}$$

Note that in (9.38) a variational formulation in terms of differences in the displacement potential is *no longer possible*, since it would require the convexity of V, in contrast to the corresponding expression on the velocity level (9.34). For the same reason, a restriction of the virtual displacements to physically admissible sets as in (9.35) also fails for (9.39).

9.5 Remarks on d'Alembert/Lagrange's Principle

Let us finally look at some special cases contained in the variational inequality (9.38), (9.39) or, equivalently, in the differential inclusion (9.37) when non-smooth non-convex potentials V are used. We start with the aforementioned equations

$$\begin{aligned} &-(\mathbf{M}\dot{\mathbf{u}} - \mathbf{h})^T\delta\mathbf{q} \leq V^{\uparrow}(\mathbf{q}, \delta\mathbf{q}), \quad \mathbf{q} \in C, \quad \forall\, \delta\mathbf{q} \\ &-(\mathbf{M}\dot{\mathbf{u}} - \mathbf{h}) \in \bar{\partial} V(\mathbf{q}) \end{aligned} \tag{9.40}$$

to narrow them down to differentiable potentials as well as to perfect unilaterally and bilaterally constrained motion, but we do not assume the special structure of V from the composition (9.36).

Differentiable potentials: We consider unconstrained motion ($C = \mathrm{I\!R}^f$) and assume the function $\mathbf{q} \to V(\mathbf{q})$ to be differentiable. In this case, the generalized gradient of V consists of one element only which is the gradient of V, $\bar{\partial} V(\mathbf{q}) = \{\nabla V(\mathbf{q})\}$, and the upper subderivative becomes $V^{\uparrow}(\mathbf{q}, \delta\mathbf{q}) = \nabla V(\mathbf{q})\, \delta\mathbf{q}$. By taking these two results, (9.40) changes to

$$\begin{aligned} &-(\mathbf{M}\dot{\mathbf{u}} - \mathbf{h})^T\delta\mathbf{q} = \nabla V(\mathbf{q})\, \delta\mathbf{q}, \quad \forall\, \delta\mathbf{q}, \\ &-(\mathbf{M}\dot{\mathbf{u}} - \mathbf{h}) = \nabla V(\mathbf{q}), \end{aligned} \tag{9.41}$$

where the variational inequality $-(\mathbf{M}\dot{\mathbf{u}} - \mathbf{h})^T\delta\mathbf{q} \leq \nabla V(\mathbf{q})\, \delta\mathbf{q}$ is actually an equality, because it applies for $\delta\mathbf{q}$ as well as for $-\delta\mathbf{q}$. The first equation in (9.41) balances the virtual work, whereas the former differential inclusion from the second line has turned into the classical second order differential equation with $\dot{\mathbf{q}} = \mathbf{u}$.

Perfect unilateral constraints: Suppose now in (9.40) that V is the indicator of the set C. With $\bar{\partial}V(\mathbf{q}) = N_C(\mathbf{q})$ equations (9.40) become

$$\begin{aligned} &-(\mathbf{M}\dot{\mathbf{u}} - \mathbf{h})^T \delta\mathbf{q} \leq 0, \quad \mathbf{q} \in C, \quad \forall\, \delta\mathbf{q} \in T_C(\mathbf{q}), \\ &-(\mathbf{M}\dot{\mathbf{u}} - \mathbf{h}) \in N_C(\mathbf{q}). \end{aligned} \tag{9.42}$$

The variational inequality in the first line may either be obtained from the inclusion in the second line together with the polarity of $N_C(\mathbf{q})$ and $T_C(\mathbf{q})$, or by taking into account (9.39) and that $V^{\uparrow}(\mathbf{q}, \delta\mathbf{q}) = 0$ for $\delta\mathbf{q} \in T_C(\mathbf{q})$ when V is the indicator of C. This inequality states that the virtual work done by the constraint forces $\mathbf{f}$ of perfect unilateral constraints ($-\mathbf{f} \in N_C(\mathbf{q})$ with $\mathbf{f} = \mathbf{M}\dot{\mathbf{u}} - \mathbf{h}$) is less or equal to zero, *if the virtual displacements* $\delta\mathbf{q}$ *are taken from the tangent cone* $T_C(\mathbf{q})$. The way in which $\delta\mathbf{q} \in T_C(\mathbf{q})$ are related to physically admissible virtual displacements will be discussed at the end of this section.

Perfect classical bilateral constraints: This is a special case of (9.42) which applies when C is a smooth submanifold: The tangent cone $T_C(\mathbf{q})$ then reduces to the tangent space, thus $-\delta\mathbf{q} \in T_C(\mathbf{q})$ whenever $\delta\mathbf{q} \in T_C(\mathbf{q})$, and the variational inequality holds as an equality. Thus, (9.40) simplifies to

$$\begin{aligned} &-(\mathbf{M}\dot{\mathbf{u}} - \mathbf{h})^T \delta\mathbf{q} = 0, \quad \mathbf{q} \in C, \quad \forall\, \delta\mathbf{q} \in T_C(\mathbf{q}), \\ &-(\mathbf{M}\dot{\mathbf{u}} - \mathbf{h}) \in N_C(\mathbf{q}), \end{aligned} \tag{9.43}$$

which is the classical principle of d'Alembert/Lagrange as already discussed in equations (3.58)–(3.63) of Chapter 3.

Some constraint configurations are depicted in Figure 9.4. The left diagram shows classical bilaterally constrained motion according to equations (9.43) with the virtual displacements $\delta\mathbf{q} = \mathbf{q}^\star - \mathbf{q}$ taken from the tangent space $T_C(\mathbf{q})$.

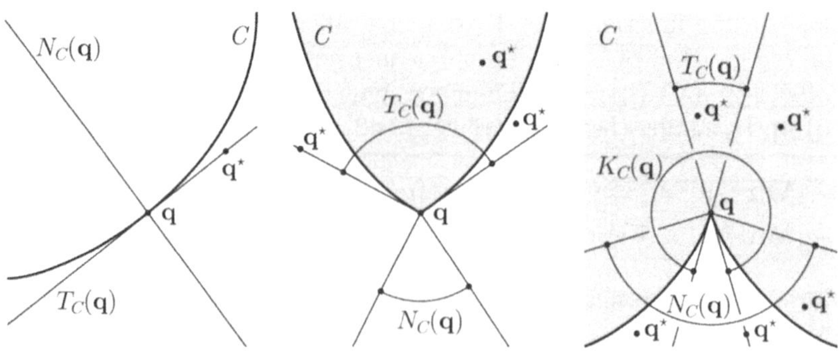

Figure 9.4. Different constraints with cones

In the middle diagram tangentially regular unilateral constraints are considered, i.e. constraints for which the tangent cone $T_C(\mathbf{q})$ and the contingent cone $K_C(\mathbf{q})$ coincide. Tangential regularity usually applies as long as the admissible displacements are defined as the *intersection* of smooth sets of constraints. In this case, the variations $\delta\mathbf{q}$ in (9.42) as members of the tangent cone $T_C(\mathbf{q})$ reflect the small virtual motions the system may take. Thus, polarity holds between all physically admissible virtual motions and constraint forces $-\mathbf{f}$ as elements of the normal cone $N_C(\mathbf{q})$.

Up to now we have excluded the so-called reintrant corners; these are configurations at which the tangent cone and the contingent cone to some points $(\mathbf{q}, V(\mathbf{q}))$ of the set epi V do not coincide. There are reasonable mechanical models of physical devices at which such situations turn up, even in the case of bilaterally constrained motion. For instance, the constrained manifold may be obtained by the *union* of some sub-manifolds with boundaries pasted together such that corners appear. A practical application problem of this kind is treated in Chapter 11, see also [41]. Applied to unilaterally constrained motion, one ends up with configurations as depicted in the right diagram of Figure 9.4. Even this case is contained in equations (9.42), but requires some additional remarks: In order to get reasonable mechanical models, one must think about the forces which might act as reactions, as well as about the choice of the virtual displacements. The inclusion $-\mathbf{f} \in N_C(\mathbf{q})$ seems still to properly express the physics of the reactions: From the mechanical point of view, the set of admissible forces should contain at least the convex hull of the limits of the reactions at points $\mathbf{q}_i$ when approaching $\mathbf{q}$. With respect to the variations $\delta\mathbf{q} = \mathbf{q}^\star - \mathbf{q}$, there is obviously no doubt that they must be taken from the contingent cone $K_C(\mathbf{q})$ when *all* physically admissible virtual motions have to be considered as, for example, in perturbation theory. The tangent cone $T_C(\mathbf{q})$ as a proper subset of $K_C(\mathbf{q})$ is *not* suited for this task, because it describes only a subset of them. In contrast, exactly the elements $\delta\mathbf{q}$ of $T_C(\mathbf{q})$ are needed to express the inclusion $-\mathbf{f} \in N_C(\mathbf{q})$ as a variational inequality, because it requires the polarity of $N_C(\mathbf{q})$ and $T_C(\mathbf{q})$. This is in remarkable contrast to tangentially regular systems, for which the physically admissible virtual displacements $\delta\mathbf{q} \in K_C(\mathbf{q})$ can always be identified with the elements of the polar $T_C(\mathbf{q})$ of $N_C(\mathbf{q})$. Thus, every formulation of the force law as a variational expression involving *arbitrary* physically admissible virtual displacements must fail, a fact which was stated for the first time in [80]. For such systems, convexification fails and, as a consequence, non-uniqueness of solutions has to be expected. This, however, should not lead to a general rejection of models containing reintrant corners, because often physical behavior might be reasonably approximated in that manner.

10. Spatial Coulomb Friction

As already discussed in Chapter 1, Coulomb friction does not fit in with the framework of force potentials. Because of its enormous importance for practical applications, however, we will present at least some mathematical formulations of the spatial problem. As we have seen, set-valued force laws may be conveniently stated in terms of complementarities which express the complementary behavior of forces and displacements or velocities, and which finally lead to a linear complementarity problem in standard form as discussed in Chapter 8. Linear complementarity is still available for the planar friction problem and for spatial friction when the friction cone is approximated by a pyramid, as one might easily imagine. If, however, the conic friction law is considered, nonlinearities are involved in the formulation of the complementarity conditions. Thus, one cannot expect linear complementarity to be available. In most cases, this situation is treated with special algorithms such as in [91], [101] and [3], but no formulation as a standard problem is given.

The main objective of this chapter is therefore the formulation of the conic Coulomb friction law as a nonlinear complementarity problem in standard form. In addition, we will provide some linear complementarity formulations of approximations of the friction cone that are also used by other authors and that may probably turn out to be the most suitable. Within this chapter, only one contact will be considered which is sufficient to demonstrate the structure of the underlying LCP's and NCP's. The extension to multiple contacts is trivial since only certain matrices have to be enlarged and rearranged. This procedure can be found in [126], for example, however only for the friction pyramid model. Impacts are also not treated. Since the structure of the impact equations is very close to the Newton-Euler equations, see e.g. [37], [116], one might immediately transfer the results if the impact laws are chosen as in [91].

The chapter is organized as follows: In the first section some basic geometric and kinematic relations of single rigid bodies are put together and are then used in Section 10.2 to formulate the kinematic equations of spatial contact problems. The distance, the relative velocities, and their changes over time which we call relative accelerations, are defined there. Section 10.3 starts with Newton's second law and Euler's axiom and describes how the contact forces may be considered in the equations of motion for the spatial contact case. In Section 10.4 we will then introduce the contact laws, consisting of the impenetrability condition in the normal direction and Coulomb friction in the tangential directions. Both contact laws are first stated in their original form at the level of displacements and velocities, respectively, and then transferred to the acceleration level, following the same procedure as presented in Chapter 7. Using the results obtained so far, the sliding contact problem is then treated in Section 10.5 and formulated as a linear complementarity problem. Section 10.6 deals with the rolling contact problem by using friction pyramid approximations of the Coulomb friction law. Based on the approximation techniques presented in [101], [64], we pass the LCP formulation process [103] and generalize this approach to arbitrary polytopes. This leads to LCP's with a dimension that is equal to the number of inequalities used for the approximation of the friction cone. In Sections 10.7 and 10.8 the conic friction law is discussed. We present standard NCP formulations for the conic friction law which are obtained by the introduction of additional inequalities on the contact forces. This is done in Section 10.7 with respect to a minimal number of equations in use, whereas a continuous and differentiable formulation is given in Section 10.8.

We are now going to put Coulomb's friction law in the framework already provided by equations (5.19). In order to avoid unessential and cumbersome indexing only one contact is considered. The blow up to multi-contact configurations is straightforward and consists of nothing more than a rearrangement of certain matrices, see e.g. [126] for spatial problems, similar as it has been already done for the one-dimensional case in the previous chapters. Furthermore, we restrict ourselves only to the evolution problem leading to future events, such as it has been stated in the left column of equation (5.19). All superscripts "+ " are omitted for brevity, and we understand, for example, by $\mathbf{u}$ the right velocity and by $\dot{\mathbf{u}}$ the right acceleration.

10.1 Geometry of Surfaces

In this section some basic kinematic relations are put together which will later be needed when contact kinematical magnitudes such as relative velocities and distances have to be derived. The approach is straightforward and does not use the tensor calculus notation from differential geometry.

We consider two bodies moving with $\mathbf{v}_{Pi}$, $\mathbf{\Omega}_i$ $(i = 1, 2)$ where $\mathbf{v}_{Pi}$ denotes the velocity of the body-fixed point P of body i and $\mathbf{\Omega}_i$ its angular velocity. Both bodies are assumed to be convex and smooth, at least in a neighborhood of a potential contact point, and may be represented by a parametrization of their surfaces in the form $\mathbf{r}_{P\Sigma i} = \mathbf{r}_{P\Sigma i}(\mathbf{q}, \xi_i, \eta_i)$, see Figure 10.1 [81].

Note that $\mathbf{r}_{P\Sigma}$ only depends on the surface parameters ξ and η when stated in the corresponding body-fixed frame whereas an additional dependence on the generalized coordinates $\mathbf{q}$ is observed when using arbitrary

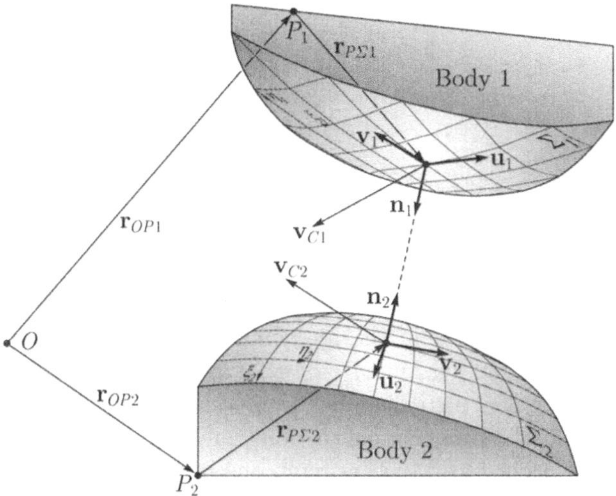

Figure 10.1. The spatial contact problem

frames. For each of the bodies one defines an orthonormal basis $(\mathbf{t}, \mathbf{s}, \mathbf{n})$ via the tangents $\mathbf{u} := \partial_\xi \mathbf{r}_{P\Sigma}$ and $\mathbf{v} := \partial_\eta \mathbf{r}_{P\Sigma}$ in order to obtain

$$\mathbf{n} := \frac{\mathbf{u} \times \mathbf{v}}{|\mathbf{u} \times \mathbf{v}|}, \quad \mathbf{t} := \frac{\mathbf{u}}{|\mathbf{u}|}, \quad \mathbf{s} := \mathbf{n} \times \frac{\mathbf{u}}{|\mathbf{u}|} \tag{10.1}$$

where we agree that (ξ, η) are ordered such that $\mathbf{n}$ becomes the outward normal. The (absolute) changes in time of the contour vector $\dot{\mathbf{r}}_{P\Sigma}$ and the trihedral $(\dot{\mathbf{t}}, \dot{\mathbf{s}}, \dot{\mathbf{n}})$ are

$$\begin{aligned} \dot{\mathbf{r}}_{P\Sigma} &= \boldsymbol{\Omega} \times \mathbf{r}_{P\Sigma} + \mathbf{R}\dot{\boldsymbol{\zeta}} \quad &&\text{with } \mathbf{R} = \partial_\zeta \mathbf{r}_{P\Sigma} = (\mathbf{u}, \mathbf{v}) \\ \dot{\mathbf{k}} &= \boldsymbol{\Omega} \times \mathbf{k} + \mathbf{K}\dot{\boldsymbol{\zeta}} &&\text{with } \mathbf{K} = \partial_\zeta \mathbf{k} \end{aligned} \tag{10.2}$$

where $\boldsymbol{\zeta} = (\xi, \eta)^T$ and $(\mathbf{k}, \mathbf{K})$ is any of $(\mathbf{t}, \mathbf{T})$, $(\mathbf{s}, \mathbf{S})$, $(\mathbf{n}, \mathbf{N})$. The velocity $\mathbf{v}_\Sigma$ of a point moving on the surface in terms of the rigid body motion $(\mathbf{v}_P, \boldsymbol{\Omega})$ is thus given by $\mathbf{v}_\Sigma = \mathbf{v}_P + \dot{\mathbf{r}}_{P\Sigma}$ which yields

$$\mathbf{v}_\Sigma = \mathbf{v}_C + \mathbf{R}\dot{\boldsymbol{\zeta}} \quad \text{with } \mathbf{v}_C = \mathbf{v}_P + \boldsymbol{\Omega} \times \mathbf{r}_{P\Sigma} \tag{10.3}$$

Here, $\mathbf{v}_C$ denotes the rigid body velocity portion of $\mathbf{v}_\Sigma$. Analogously we may find the acceleration $\mathbf{a}_C = \dot{\mathbf{v}}_C$ by differentiation of the second equation in (10.3),

$$\mathbf{a}_C = \mathbf{a}_Q + \boldsymbol{\Omega} \times \mathbf{R}\dot{\boldsymbol{\zeta}} \quad \text{with } \mathbf{a}_Q = \mathbf{a}_P + \dot{\boldsymbol{\Omega}} \times \mathbf{r}_{P\Sigma} + \boldsymbol{\Omega} \times (\boldsymbol{\Omega} \times \mathbf{r}_{P\Sigma}). \tag{10.4}$$

Finally we express $\mathbf{v}_C$ and $\mathbf{a}_C$ by the generalized magnitudes $\mathbf{u}$ and $\dot{\mathbf{u}}$,

$$\begin{aligned} \mathbf{v}_C &= \mathbf{J}_C\,\mathbf{u} + \tilde{\iota}_C, \\ \mathbf{a}_C &= \mathbf{J}_C\,\dot{\mathbf{u}} + \bar{\iota}_C, \end{aligned} \tag{10.5}$$

where $\bar{\iota}_C$ is given by $\bar{\iota}_C = \dot{\mathbf{J}}_C\,\mathbf{u} + \dot{\tilde{\iota}}_C$, and $\mathbf{J}_C$ is the *Jacobian of the contact point* as it will turn out later.

10.2 Contact Kinematics

Within the framework as presented in the foregoing section one is able to derive the main equations of contact kinematics in dynamics which are: The definition of the contact points, their distance, their relative velocities with respect to the normal and the tangential directions, and the corresponding relative accelerations. In order to do the first step one introduces a *distance vector*

$$\mathbf{r}_D := \mathbf{r}_{O\Sigma 2} - \mathbf{r}_{O\Sigma 1}, \tag{10.6}$$

cp. [37], where $\mathbf{r}_{O\Sigma i}$ are vectors pointing from a common inertially fixed point O to the surfaces under consideration.

Next, the two *tangent planes* spanned by $(\mathbf{t}_i, \mathbf{s}_i)$ are adjusted parallel to each other such that the distance vector becomes parallel to each of the normals. The surface parameters $(\zeta_1^\star, \zeta_2^\star)$ corresponding with that situation solve the four nonlinear equations

$$\mathbf{n}_1^T\mathbf{t}_2 = 0, \quad \mathbf{n}_1^T\mathbf{s}_2 = 0, \quad \mathbf{r}_D^T\mathbf{t}_1 = 0, \quad \mathbf{r}_D^T\mathbf{s}_1 = 0 \tag{10.7}$$

and are called the *contact parameters*; the corresponding points at the surfaces given by $\mathbf{r}_{P\Sigma i}(\zeta_i^\star)$ are called the *contact point.* When the bodies are moving the conditions (10.7) for the contact points should not change [37], [81], thus

$$\begin{aligned} \dot{\mathbf{n}}_1^T\mathbf{t}_2 + \mathbf{n}_1^T\dot{\mathbf{t}}_2 &= 0, & \dot{\mathbf{r}}_D^T\mathbf{t}_1 + \mathbf{r}_D^T\dot{\mathbf{t}}_1 &= 0 \\ \dot{\mathbf{n}}_1^T\mathbf{s}_2 + \mathbf{n}_1^T\dot{\mathbf{s}}_2 &= 0, & \dot{\mathbf{r}}_D^T\mathbf{s}_1 + \mathbf{r}_D^T\dot{\mathbf{s}}_1 &= 0 \end{aligned} \tag{10.8}$$

must hold which constitutes a system of four linear equations in $\dot{\zeta}_i^\star$ when substituting $(\dot{\mathbf{n}}_i, \dot{\mathbf{t}}_i, \dot{\mathbf{s}}_i)$ from (10.2) and $\mathbf{v}_{\Sigma i}$ from (10.3) with $\dot{\mathbf{r}}_D = \mathbf{v}_{\Sigma 2} - \mathbf{v}_{\Sigma 1}$.

After the solution $(\zeta_1^\star, \zeta_2^\star)$ of (10.7) has been found the *distance* g_N of the contact points can be calculated as

$$g_N = \mathbf{n}_1^T\mathbf{r}_D \tag{10.9}$$

with values greater than zero for separation and values less than zero for overlapping. The *relative velocities* of the contact points with respect to the three directions $(\mathbf{t}_1, \mathbf{s}_1, \mathbf{n}_1)$ are then given by

$$\gamma_K = (\mathbf{v}_{C2} - \mathbf{v}_{C1})^T \mathbf{k}_1, \qquad ({\scriptstyle K}, \mathbf{k}_1) = ({\scriptstyle T}, \mathbf{t}_1), ({\scriptstyle S}, \mathbf{s}_1), ({\scriptstyle N}, \mathbf{n}_1) \tag{10.10}$$

where it can be easily verified that differentiation of (10.9) with respect to time results in (10.10) for $({\scriptstyle K}, \mathbf{k}_1) = ({\scriptstyle N}, \mathbf{n}_1)$: Just the same steps as for the variations δg in (5.27), (5.28) have to be performed. The temporal changes of the relative velocities γ_K in (10.10) are given by

$$\dot{\gamma}_K = (\mathbf{a}_{C2} - \mathbf{a}_{C1})^T \mathbf{k}_1 + (\mathbf{v}_{C2} - \mathbf{v}_{C1})^T \dot{\mathbf{k}}_1, \tag{10.11}$$

where $\mathbf{a}_{Ci}$ is known from (10.4), $\dot{\mathbf{k}}_1$ may be taken from (10.2), and $\dot{\boldsymbol{\zeta}}_i^\star$ involved in $\mathbf{a}_{Ci}$ and $\dot{\mathbf{k}}_1$ is the solution of (10.8).

Substituting $\mathbf{v}_{Ci}$ and $\mathbf{a}_{Ci}$ from (10.5) into (10.10), (10.11) yields a representation of the relative velocities and accelerations in the form

$$\begin{aligned} \gamma_K &= \mathbf{w}_K^T \mathbf{u} + \tilde{w}_K, \\ \dot{\gamma}_K &= \mathbf{w}_K^T \dot{\mathbf{u}} + \bar{w}_K \end{aligned} \tag{10.12}$$

with $\mathbf{w}_K = (\mathbf{J}_{C2} - \mathbf{J}_{C1})^T \mathbf{k}_1$, $\tilde{w}_K = (\tilde{\iota}_{C2} - \tilde{\iota}_{C1})^T \mathbf{k}_1$, and $\bar{w}_K = \dot{\mathbf{w}}_K^T \mathbf{u} + \dot{\tilde{w}}_K$. Finally, we arrange the two tangential relative velocities in the vector $\boldsymbol{\gamma}_T$, i.e. $\boldsymbol{\gamma}_T^T := (\gamma_T, \gamma_S)$. This $\boldsymbol{\gamma}_T$ tells us the sliding direction of the contact points in the common tangent plane, whereas γ_N denotes their relative velocity in the normal direction. With the abbreviations $\mathbf{W}_T := (\mathbf{w}_T, \mathbf{w}_S)$, $\tilde{\mathbf{w}}_T^T = (\tilde{w}_T, \tilde{w}_S)$, $\bar{\mathbf{w}}_T^T = (\bar{w}_T, \bar{w}_S)$ we rewrite (10.12) in the form

$$\begin{aligned} \gamma_N &= \mathbf{w}_N^T \mathbf{u} + \tilde{w}_N, \qquad & \boldsymbol{\gamma}_T &= \mathbf{W}_T^T \mathbf{u} + \tilde{\mathbf{w}}_T \\ \dot{\gamma}_N &= \mathbf{w}_N^T \dot{\mathbf{u}} + \bar{w}_N, \qquad & \dot{\boldsymbol{\gamma}}_T &= \mathbf{W}_T^T \dot{\mathbf{u}} + \bar{\mathbf{w}}_T \end{aligned} \tag{10.13}$$

which will serve as the contact kinematic equations in standard form in all of the following.

10.3 Kinetics

The dynamics equations of the holonomic multibody system may be derived by the projected Newton-Euler equations (3.40) of Chapter 3. For a system consisting of n bodies and one contact they read

$$\sum_{i=1}^{n} \left\{ \mathbf{J}_S^T (\dot{\mathbf{p}} - \mathbf{F}^{(a)}) + \mathbf{J}_R^T (\dot{\mathbf{L}}_S - \mathbf{M}_S^{(a)}) \right\}^{(B_i)} - \mathbf{J}_{C1}^T \mathbf{F}_{C1} - \mathbf{J}_{C2}^T \mathbf{F}_{C2} = 0, \tag{10.14}$$

when for each body B_i the point of reference P is chosen as the center of mass S, i.e. $\mathbf{r}_{PS}^{(B_i)} = 0$ in (3.40), see also [37], [116]. The terms $\mathbf{p}^{(B_i)}$ and $\mathbf{L}_S^{(B_i)}$ denote the linear momentum and the angular momentum with respect to the center of mass of body B_i, and $\mathbf{F}^{(a,B_i)}$ and $\mathbf{M}_S^{(a,B_i)}$ are the external active forces and moments up to the contact forces, respectively. The *contact forces* acting at the contact points C_1 and C_2 are denoted by $\mathbf{F}_{C1}$ and $\mathbf{F}_{C2}$. In order to perform the projection into the configuration space all terms are premultiplied by the corresponding Jacobians. In particular, $\mathbf{J}_{C1}$ and $\mathbf{J}_{C2}$ are the Jacobians of the contact points which have already been introduced in equation (10.5).

The formulation of the generalized contact forces, required for the system (10.14) to fit in with the description given in (5.19), follows precisely the steps performed in (5.21)–(5.26). In particular, a decomposition of the contact forces into components corresponding to the trihedral $(\mathbf{t}_1, \mathbf{s}_1, \mathbf{n}_1)$ admits, together with the cutting principle, a unique representation in the form

$$\mathbf{F}_{C2} = -\mathbf{F}_{C1} =: \mathbf{n}_1 \lambda_N + \mathbf{t}_1 \lambda_T + \mathbf{s}_1 \lambda_S. \tag{10.15}$$

Putting now (10.15) into (10.14) and using the abbreviations $\mathbf{w}_K = (\mathbf{J}_{C2} - \mathbf{J}_{C1})^T \mathbf{k}_1$ and $\mathbf{W}_T = (\mathbf{w}_T, \mathbf{w}_S)$ from equation (10.12) and below, we arrive with

$$\mathbf{M}\dot{\mathbf{u}} - \mathbf{h} - \mathbf{w}_N \lambda_N - \mathbf{W}_T \boldsymbol{\lambda}_T = 0, \tag{10.16}$$

which are the directional Newton-Euler equations for one spatial contact with Coulomb friction. The sum in (10.14) is represented by the term $(\mathbf{M}\dot{\mathbf{u}} - \mathbf{h})$ with $\mathbf{M}$ as the symmetric and positive definite mass matrix of the system and $\mathbf{h}$ as the vector of the gyroscopical accelerations and the remaining external active forces. Furthermore, λ_N denotes the scalar value of the normal force with force direction $\mathbf{w}_N \in \mathbb{R}^f$ corresponding to the contact normal $\mathbf{n} \in \mathbb{R}^3$, see Figure 10.1. The matrix $\mathbf{W}_T = (\mathbf{w}_T, \mathbf{w}_S)$ consists of the two linearly independent force directions $\mathbf{w}_T, \mathbf{w}_S \in \mathbb{R}^f$ that are obtained from the two orthogonal unit vectors $\mathbf{t}, \mathbf{s} \in \mathbb{R}^3$ which are defining the common tangent plane of the two bodies in possible contact, and $\boldsymbol{\lambda}_T^T = (\lambda_T, \lambda_S)$ are the two corresponding scalars of the overall tangential force with absolute value $\sqrt{\lambda_T^2 + \lambda_S^2}$. The contact laws are still missing and have to be chosen according to the physical properties of the contacts.

10.4 Contact Laws

In order to get a complete description of the dynamic Coulomb friction problem we must finally introduce the contact laws. With g_N being the distance of the contact points and λ_N the corresponding normal force one might express impenetrability of rigid bodies by the *Signorini-Fichera condition* [101], [37], [103], [70], [64], [3]

$$g_N \geq 0, \quad \lambda_N \geq 0, \quad g_N \lambda_N = 0. \tag{10.17}$$

This condition may be transferred to the velocity level and, in a second step, to the acceleration level, exactly as it has been done in Chapter 7, equation (7.15). The resulting expression is

$$\begin{array}{ll} \text{if } g_N > 0 & \text{then } \lambda_N = 0 \\ \text{if } g_N = 0 \wedge \gamma_N > 0 & \text{then } \lambda_N = 0 \\ \text{if } g_N = 0 \wedge \gamma_N = 0 & \text{then } \lambda_N \geq 0,\ \dot{\gamma}_N \geq 0,\ \lambda_N \dot{\gamma}_N = 0 \end{array} \tag{10.18}$$

which describes the detachment of a closed contact and also the behavior of open contacts. Note that the occurrence of a new closed contact is usually described by the conditions $g_N = 0$, $\gamma_N < 0$ leading to an impact which is *not* covered by (10.18). Impacts are already excluded by the formulation of the equations of motion (10.16), because a proper formulation of impacts must be done in terms of equation (5.6).

In the tangential directions we assume *Coulomb friction* given by the friction law [101], [37], [103], [70], [3]

$$\begin{array}{ll} \text{if } \boldsymbol{\gamma}_T \neq 0 & \text{then } |\boldsymbol{\lambda}_T| = \mu\lambda_N,\ \boldsymbol{\lambda}_T = -\beta\,\boldsymbol{\gamma}_T,\ \beta \geq 0 \\ \text{if } \boldsymbol{\gamma}_T = 0 & \text{then } |\boldsymbol{\lambda}_T| \leq \mu_0\lambda_N \end{array} \tag{10.19}$$

where μ denotes the coefficient of friction which depends smoothly on $\boldsymbol{\gamma}_T$, and $\mu_0 = \mu(\boldsymbol{\gamma}_T = 0)$. Similarly we obtain from (10.19) the friction law on the acceleration level in the form [103]

$$\begin{array}{ll} \text{if } \boldsymbol{\gamma}_T \neq 0 & \text{then } \boldsymbol{\lambda}_T = -\mathbf{e}\,\mu\lambda_N \\ \text{if } \boldsymbol{\gamma}_T = 0 \wedge \dot{\boldsymbol{\gamma}}_T \neq 0 & \text{then } |\boldsymbol{\lambda}_T| = \mu_0\lambda_N,\ \boldsymbol{\lambda}_T = -\beta\,\dot{\boldsymbol{\gamma}}_T,\ \beta \geq 0 \\ \text{if } \boldsymbol{\gamma}_T = 0 \wedge \dot{\boldsymbol{\gamma}}_T = 0 & \text{then } |\boldsymbol{\lambda}_T| \leq \mu_0\lambda_N \end{array} \tag{10.20}$$

with $\mathbf{e}$ being the unit vector of the sliding direction, i.e. $\mathbf{e} = \boldsymbol{\gamma}_T/|\boldsymbol{\gamma}_T|$. Note that even in the case of impact free motion, Coulomb friction obviously causes acceleration jumps for transitions from sliding to stiction. Thus, even in this impact free case, the accelerations $\dot{\mathbf{u}}$ are *not* defined at the transition points but only their left and right limits. The first line in equation (10.20) describes sliding of the contact up to velocities $\boldsymbol{\gamma}_T = 0$, whereas the second condition takes into account stiction as well as the transitions from stiction to sliding, and is valid for the right derivatives of the velocity functions.

The Newton-Euler equations (10.16), the contact laws in the normal (10.18) and the tangential (10.20) directions, and the contact kinematic terms from (10.13) as far as needed provide a complete set of equations in order to finally solve (10.16) for the unknowns $\dot{\mathbf{u}}$. In the case of closed contacts this is usually done by two steps [37], [103], see Chapter 8: First, the accelerations $\dot{\mathbf{u}}$ are eliminated with the help of (10.16) from the relative accelerations (10.13). Together with the contact laws they constitute a system of inequalities which has to be solved for the relative accelerations $\dot{\boldsymbol{\gamma}}$ and the contact forces $\boldsymbol{\lambda}$. When this solution has been found, $\boldsymbol{\lambda}$ is re-substituted into the Newton-Euler equations (10.16) which permits finally the calculation of $\dot{\mathbf{u}}$.

In the remaining part of this chapter we concentrate on the formulation of the inequality system for sliding and rolling contact situations. We present LCP and NCP formulations in standard form for friction pyramids and for the friction cone, respectively.

10.5 Sliding Contacts

In this section we derive the linear complementarity problem for a closed contact ($g_N = 0$, $\gamma_N = 0$) which is *sliding* ($\boldsymbol{\gamma}_T \neq 0$). This situation is described by the Newton-Euler equations (10.16), the normal contact law for $g_N = 0$, $\gamma_N = 0$ from the third line in (10.18), and the tangential contact law for sliding contacts $\boldsymbol{\gamma}_T \neq 0$ from the first equation in (10.20). Additionally, the lower left equation in (10.13) is needed as a kinematical relation, since $\dot{\gamma}_N$ is involved in the normal contact law. Altogether these equations write

$$
\begin{aligned}
&\mathbf{M}\dot{\mathbf{u}} - \mathbf{h} - \mathbf{w}_N \lambda_N - \mathbf{W}_T \boldsymbol{\lambda}_T = 0 \\
&\dot{\gamma}_N = \mathbf{w}_N^T \dot{\mathbf{u}} + \bar{w}_N \\
&\lambda_N \geq 0,\ \dot{\gamma}_N \geq 0,\ \lambda_N \, \dot{\gamma}_N = 0 \\
&\boldsymbol{\lambda}_T = -\mathbf{e}\,\mu \lambda_N .
\end{aligned}
\tag{10.21}
$$

Using the fourth equation $\boldsymbol{\lambda}_T$ is substituted into the first one which is then solved for $\dot{\mathbf{u}}$. Substituting $\dot{\mathbf{u}}$ into the second equation we obtain the LCP for the sliding contact [103], [126]

$$
\begin{aligned}
&\dot{\gamma}_N = \mathbf{w}_N^T \mathbf{M}^{-1} (\mathbf{w}_N - \mathbf{W}_T \mathbf{e}\,\mu)\, \lambda_N + (\mathbf{w}_N^T \mathbf{M}^{-1} \mathbf{h} + \bar{w}_N) \\
&\dot{\gamma}_N \geq 0,\ \lambda_N \geq 0,\ \dot{\gamma}_N \, \lambda_N = 0.
\end{aligned}
\tag{10.22}
$$

Obviously, friction may cause the LCP-matrix (for one contact being a scalar) to become indefinite.

Note especially that the force direction $\mathbf{w}_S := \mathbf{w}_N - \mathbf{W}_T \mathbf{e}\,\mu$ that occurs in the first line of (10.21) after having eliminated $\boldsymbol{\lambda}_T$ does *not* coincide with the gradient $\mathbf{w}_N$ of g_N in the second line. This difference is due to Coulomb's frictional law and thus physically based. It may serve as an example that the classical treatment of constraints by the use of Lagrange multipliers ($\mathbf{w}_S = \mathbf{w}_N$) is not capable of describing all situations but restricted to potential forces only.

10.6 Friction Pyramid for Rolling Contacts

Rolling contacts are characterized by vanishing tangential velocities $\boldsymbol{\gamma}_T = 0$. In contrast to the previous section, the tangential contact law is now described by the second and third line in (10.20) which holds for stiction as well as for the transition to sliding. Before discussing approximations of this friction law by polytopes we will rewrite (10.20) for $\gamma_T = 0$ in a more suitable form which will also be used in the following sections. Let

$$D_T := \{\boldsymbol{\lambda}_T \mid |\boldsymbol{\lambda}_T| \leq \mu_0 \lambda_N\} \tag{10.23}$$

be the (convex) set of the admissible tangential contact forces, called the *friction disk.* The tangential contact law is then given by

$$-\dot{\gamma}_T \in N_{D_T}(\boldsymbol{\lambda}_T), \tag{10.24}$$

see e.g. [101], [64], where $N_{D_T}(\boldsymbol{\lambda}_T)$ denotes the normal cone to D_T at $\boldsymbol{\lambda}_T$,

$$N_{D_T}(\boldsymbol{\lambda}_T) = \begin{cases} \kappa\mathbf{e} & \text{for } 0 < |\boldsymbol{\lambda}_T| = \mu_0\lambda_N \\ 0 & \text{for } 0 \leq |\boldsymbol{\lambda}_T| < \mu_0\lambda_N \\ \mathbb{R}^2 & \text{for } 0 = |\boldsymbol{\lambda}_T| = \mu_0\lambda_N \end{cases} \tag{10.25}$$

with $\mathbf{e} = \boldsymbol{\lambda}_T/|\boldsymbol{\lambda}_T|$ and $\kappa \geq 0$ as depicted in Figure 10.2.

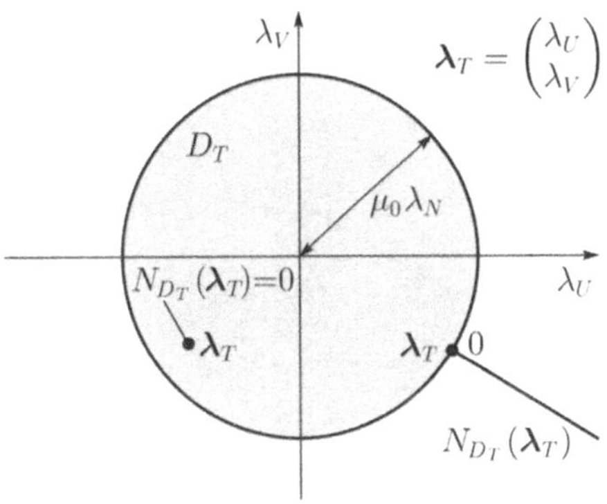

Figure 10.2. Admissible tangential forces for the friction cone

Proposition 10.6.1. *The contact laws (10.20) for $\gamma_T = 0$ and (10.24) are equivalent.*

Proof. By inversion, (10.24) may be also written as $\boldsymbol{\lambda}_T \in \partial I^{\star}_{D_T}(-\dot{\gamma}_T)$, where ∂ denotes the subdifferential from convex analysis and $I^{\star}_{D_T}(-\dot{\gamma}_T)$ is the support function of D_T defined by $I^{\star}_{D_T}(-\dot{\gamma}_T) = \mu_0\lambda_N|-\dot{\gamma}_T|$. Solving this inclusion for $\dot{\gamma}_T = 0$ we obtain $\boldsymbol{\lambda}_T \in \partial I^{\star}_{D_T}(0) = B(0, \mu_0\lambda_N)$, where $B(0,r)$ is the closed disk with center 0 and radius r. Hence, $|\boldsymbol{\lambda}_T| \leq \mu_0\lambda_N$ which is the first condition in (10.20). For $\dot{\gamma}_T \neq 0$ the support function is differentiable. Thus, $\boldsymbol{\lambda}_T = -\mu_0\lambda_N\, \dot{\gamma}_T/|\dot{\gamma}_T|$, i.e. $|\boldsymbol{\lambda}_T| = \mu_0\lambda_N$ and $\boldsymbol{\lambda}_T = -\beta\dot{\gamma}_T$ with $\beta = \mu_0\lambda_N/|\dot{\gamma}_T| \geq 0$ since $\mu_0 \geq 0$, $\lambda_N \geq 0$. This is the second condition in (10.20). □

Following [64] we may approximate the set D_T by a polytope generated by the intersection of n closed half-spaces C_{Ti}, each of which being represented by an affine inequality $\sigma_i(\boldsymbol{\lambda}_T)$,

$$C_T := \bigcap_{i=1}^{n} C_{Ti} \quad \text{with } C_{Ti} = \{\boldsymbol{\lambda}_T \mid \sigma_i(\boldsymbol{\lambda}_T) \geq 0\} \quad \text{and } \sigma_i(\boldsymbol{\lambda}_T) = \mu_0 \lambda_N - \mathbf{e}_i^T \boldsymbol{\lambda}_T \tag{10.26}$$

where $\mathbf{e}_i$ denote the outward unit vectors to D_T at some chosen points on the boundary ∂D_T corresponding to angles α_i, see Figure 10.3. We now claim the contact law to be of the form

$$-\dot{\boldsymbol{\gamma}}_T \in N_{C_T}(\boldsymbol{\lambda}_T). \tag{10.27}$$

This is the friction cone approximation presented in [64]. For inner approximations of the friction cone we refer to [101]. Note that the original friction law (10.24) and its approximation (10.27) differ only in the definition of the sets D_T and C_T. We call the σ_i the *friction saturation* with respect to C_{Ti} because it determines the distance between a given tangential force $\boldsymbol{\lambda}_T$ and the boundary of C_{Ti}. It therefore describes the amount of forces which might additionally act on the contact in the direction $\mathbf{e}_i$ until sliding starts.

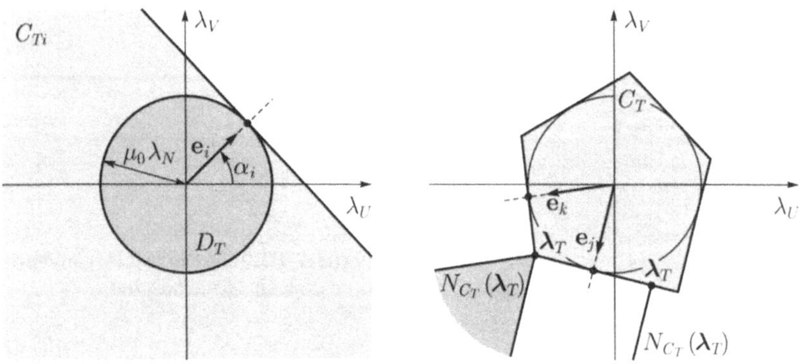

Figure 10.3. Approximation of the friction disk by polytopes

In the following we will derive a system of complementarity conditions that describe the friction law (10.27). The half-spaces C_{Ti} defined in (10.26) have at least the point 0 in common since $\sigma_i(0) \geq 0$ because $\lambda_N \geq 0$, $\mu_0 \geq 0$. Thus, the normal cone to $C_T = \bigcap_{i=1}^n C_{Ti}$ is given by

$$N_{C_T}(\boldsymbol{\lambda}_T) = \sum_{i=1}^{n} N_{C_{Ti}}(\boldsymbol{\lambda}_T), \tag{10.28}$$

cp. [121], and the contact law (10.27) may be expressed by the normal cone conditions

$$-\dot{\gamma}_T \in \sum_{i=1}^{n} -\kappa_i \, \partial\sigma_i(\boldsymbol{\lambda}_T); \quad \sigma_i \geq 0, \; \kappa_i \geq 0, \; \sigma_i\kappa_i = 0. \tag{10.29}$$

Since $\sigma_i(\boldsymbol{\lambda}_T)$ in (10.26) is differentiable this inclusion may be further reduced to the equation

$$-\dot{\gamma}_T = \sum_{i=1}^{n} \mathbf{e}_i\kappa_i; \quad \sigma_i \geq 0, \; \kappa_i \geq 0, \; \sigma_i\kappa_i = 0 \tag{10.30}$$

providing already the desired complementarity conditions. Note that the tangential relative acceleration $\dot{\gamma}_T$ is expressed by a non-negative linear combination of given directions $\mathbf{e}_i$. The scalars κ_i are accelerations that are complementary to the friction saturations σ_i: For a given direction $\mathbf{e}_i$, either the relative acceleration κ_i or the friction saturation σ_i vanishes.

We may now summarize the equations that describe the rolling contact problem: The Newton-Euler equations (10.16), the normal contact law for $g_N = 0$, $\gamma_N = 0$ from the third line in (10.18), the approximation of the tangent contact law consisting of the definition of the friction saturations σ_i in (10.26), the acceleration equation in (10.30), and the complementarities in (10.30). Finally, we need from kinematics both equations in the second line of (10.13) since both, $\dot{\gamma}_N$ and $\dot{\gamma}_T$ enter the contact laws. Altogether we have the following complete description of the rolling contact problem:

$$\begin{aligned}
&\mathbf{M}\dot{\mathbf{u}} - \mathbf{h} - \mathbf{w}_N\lambda_N - \mathbf{W}_T\boldsymbol{\lambda}_T = 0 \\
&\dot{\gamma}_N = \mathbf{w}_N^T\dot{\mathbf{u}} + \bar{w}_N \\
&\dot{\gamma}_T = \mathbf{W}_T^T\dot{\mathbf{u}} + \bar{\mathbf{w}}_T \\
&\lambda_N \geq 0, \; \dot{\gamma}_N \geq 0, \; \lambda_N\dot{\gamma}_N = 0 \\
&\sigma_i = \mu_0\lambda_N - \mathbf{e}_i^T\boldsymbol{\lambda}_T, \quad i = 1, \dots, n \\
&-\dot{\gamma}_T = \textstyle\sum_{i=1}^{n} \mathbf{e}_i\kappa_i \\
&\sigma_i \geq 0, \; \kappa_i \geq 0, \; \sigma_i\kappa_i = 0, \quad i = 1, \dots, n.
\end{aligned} \tag{10.31}$$

The remaining part of the section is devoted to the formulation of (10.31) as a linear complementarity problem in standard form with dimension $n+1$. The steps that have to be performed are similar to those presented in Chapter 8; the inversion of an expression $\mathbf{W}^T\mathbf{M}^{-1}\mathbf{W}$ can also be avoided. Following [126] we choose an indexing i in the last three equations in (10.31) in a manner that the directions $\mathbf{e}_i$ for $i = 1, 2$ are linearly independent. Introducing the abbreviations

$$\begin{aligned}
&\boldsymbol{\sigma}_P = \begin{pmatrix} \sigma_1 \\ \sigma_2 \end{pmatrix}, \quad \boldsymbol{\kappa}_P = \begin{pmatrix} \kappa_1 \\ \kappa_2 \end{pmatrix}, \quad \boldsymbol{\mu}_P = \begin{pmatrix} \mu_0 \\ \mu_0 \end{pmatrix}, \quad \mathbf{I}_P = (\mathbf{e}_1 \; \mathbf{e}_2) \\
&\boldsymbol{\sigma}_R = \begin{pmatrix} \sigma_3 \\ \vdots \\ \sigma_n \end{pmatrix}, \quad \boldsymbol{\kappa}_R = \begin{pmatrix} \kappa_3 \\ \vdots \\ \kappa_n \end{pmatrix}, \quad \boldsymbol{\mu}_R = \begin{pmatrix} \mu_0 \\ \vdots \\ \mu_0 \end{pmatrix}, \quad \mathbf{I}_R = (\mathbf{e}_3 \; \dots \; \mathbf{e}_n)
\end{aligned} \tag{10.32}$$

we may then rewrite the fifth line in (10.31) in matrix notation,

$$\begin{aligned} \boldsymbol{\sigma}_P &= \boldsymbol{\mu}_P\,\lambda_N - \mathbf{I}_P^T\,\boldsymbol{\lambda}_T, \\ \boldsymbol{\sigma}_R &= \boldsymbol{\mu}_R\,\lambda_N - \mathbf{I}_R^T\,\boldsymbol{\lambda}_T. \end{aligned} \tag{10.33}$$

Since $\mathbf{I}_P = (\mathbf{e}_1\;\mathbf{e}_2)$ is invertible we may solve the first equation in (10.33) for $\boldsymbol{\lambda}_T$,

$$\boldsymbol{\lambda}_T = \mathbf{I}_P^{-T}\,\boldsymbol{\mu}_P\,\lambda_N - \mathbf{I}_P^{-T}\boldsymbol{\sigma}_P, \tag{10.34}$$

and then substitute the result into the second equation,

$$\boldsymbol{\sigma}_R = (\boldsymbol{\mu}_R - \mathbf{I}_R^T\,\mathbf{I}_P^{-T}\,\boldsymbol{\mu}_P)\,\lambda_N + \mathbf{I}_R^T\,\mathbf{I}_P^{-T}\,\boldsymbol{\sigma}_P. \tag{10.35}$$

The accelerations from the sixth line in (10.31) are solved for $\boldsymbol{\kappa}_P$ which yields

$$\boldsymbol{\kappa}_P = -\mathbf{I}_P^{-1}\,\dot{\boldsymbol{\gamma}}_T - \mathbf{I}_P^{-1}\,\mathbf{I}_R\,\boldsymbol{\kappa}_R. \tag{10.36}$$

Finally, we put $\boldsymbol{\lambda}_T$ from (10.34) into the first equation in (10.31) and substitute $\dot{\boldsymbol{\gamma}}_T$ from the third line into (10.36). This gives, together with (10.35), the new system

$$\begin{aligned} &\mathbf{M}\dot{\mathbf{u}} - \mathbf{h} - (\mathbf{w}_N + \mathbf{W}_T\,\mathbf{I}_P^{-T}\,\boldsymbol{\mu}_P)\,\lambda_N + \mathbf{W}_T\,\mathbf{I}_P^{-T}\boldsymbol{\sigma}_P = 0 \\ &\dot{\gamma}_N = \mathbf{w}_N^T\dot{\mathbf{u}} + \bar{w}_N \\ &\boldsymbol{\kappa}_P = -\mathbf{I}_P^{-1}\,\mathbf{W}_T^T\,\dot{\mathbf{u}} - \mathbf{I}_P^{-1}\,\bar{\mathbf{w}}_T - \mathbf{I}_P^{-1}\,\mathbf{I}_R\,\boldsymbol{\kappa}_R \\ &\boldsymbol{\sigma}_R = (\boldsymbol{\mu}_R - \mathbf{I}_R^T\,\mathbf{I}_P^{-T}\,\boldsymbol{\mu}_P)\,\lambda_N + \mathbf{I}_R^T\,\mathbf{I}_P^{-T}\,\boldsymbol{\sigma}_P \\ &\dot{\gamma}_N \geq 0,\; \lambda_N \geq 0,\; \dot{\gamma}_N\lambda_N = 0 \\ &\boldsymbol{\kappa}_P \geq 0,\; \boldsymbol{\sigma}_P \geq 0,\; \boldsymbol{\kappa}_P^T\boldsymbol{\sigma}_P = 0 \\ &\boldsymbol{\sigma}_R \geq 0,\; \boldsymbol{\kappa}_R \geq 0,\; \boldsymbol{\sigma}_R^T\boldsymbol{\kappa}_R = 0 \end{aligned} \tag{10.37}$$

which describes the rolling contact problem. In a last step we solve the first equation in (10.37) for $\dot{\mathbf{u}}$ and substitute the result into the second and the third equation. With $\mathbf{W}_P := \mathbf{W}_T\,\mathbf{I}_P^{-T}$ and $\bar{\mathbf{w}}_P := \mathbf{I}_P^{-1}\,\bar{\mathbf{w}}_T$ we finally obtain

$$\begin{aligned} \begin{pmatrix} \dot{\gamma}_N \\ \boldsymbol{\kappa}_P \\ \boldsymbol{\sigma}_R \end{pmatrix} &= \begin{pmatrix} \mathbf{w}_N^T\,\mathbf{M}^{-1}(\mathbf{w}_N + \mathbf{W}_P\boldsymbol{\mu}_P) & -\mathbf{w}_N^T\,\mathbf{M}^{-1}\mathbf{W}_P & 0 \\ -\mathbf{W}_P^T\mathbf{M}^{-1}(\mathbf{w}_N + \mathbf{W}_P\boldsymbol{\mu}_P) & \mathbf{W}_P^T\mathbf{M}^{-1}\mathbf{W}_P & -\mathbf{I}_P^{-1}\,\mathbf{I}_R \\ \boldsymbol{\mu}_R - \mathbf{I}_R^T\,\mathbf{I}_P^{-T}\,\boldsymbol{\mu}_P & \mathbf{I}_R^T\,\mathbf{I}_P^{-T} & 0 \end{pmatrix} \begin{pmatrix} \lambda_N \\ \boldsymbol{\sigma}_P \\ \boldsymbol{\kappa}_R \end{pmatrix} \\ &\quad + \begin{pmatrix} \mathbf{w}_N^T\,\mathbf{M}^{-1}\mathbf{h} + \bar{w}_N \\ -\mathbf{W}_P^T\mathbf{M}^{-1}\mathbf{h} - \bar{\mathbf{w}}_P \\ 0 \end{pmatrix} \end{aligned} \tag{10.38}$$

which is, together with the inequalities and complementarities from the last three lines in (10.37), the desired LCP of dimension $n+1$ in standard form [126].

If the friction disc D_T is approximated, for example, by a square, the directions $\mathbf{e}_i$, $i = 1, \dots, 4$ are simply $\mathbf{e}_1^T = (1\ 0)$, $\mathbf{e}_2^T = (0\ 1)$, $\mathbf{e}_3^T = (-1\ 0)$, $\mathbf{e}_4^T = (0\ -1)$. In this case the matrices $\mathbf{I}_P$ and $\mathbf{I}_R$ as defined in (10.32) become $\mathbf{I}_P = -\mathbf{I}_R = \mathbf{E}$, where $\mathbf{E}$ is the 2×2 identity matrix, and the dimension of the LCP is 5. Thus, $\mathbf{W}_P \equiv \mathbf{W}_T$ and $\bar{\mathbf{w}}_P \equiv \bar{\mathbf{w}}_T$, and the entries in the last row and column of the LCP matrix in (10.38) become

$$\boldsymbol{\mu}_R - \mathbf{I}_R^T \mathbf{I}_P^{-T} \boldsymbol{\mu}_P = 2\boldsymbol{\mu}_T, \qquad \mathbf{I}_R^T \mathbf{I}_P^{-T} = -\mathbf{E}, \qquad -\mathbf{I}_P^{-1} \mathbf{I}_R = \mathbf{E} \tag{10.39}$$

with $\boldsymbol{\mu}_T := \boldsymbol{\mu}_P \equiv \boldsymbol{\mu}_R = (\mu_0\ \mu_0)^T$. Exactly this formulation but expanded to multi-contact configurations is used in [103].

In the case of one-dimensional friction, i.e. the planar friction problem, the directions $\mathbf{e}_1^T = (1\ 0)$ and $\mathbf{e}_3^T = (-1\ 0)$ reduce to $+1$ and -1, respectively, whereas $\mathbf{e}_2$ and $\mathbf{e}_4$ are no longer needed. The LCP (10.38) then writes

$$\begin{aligned}\begin{pmatrix} \dot{\gamma}_N \\ \kappa_1 \\ \sigma_3 \end{pmatrix} = &\begin{pmatrix} \mathbf{w}_N^T \mathbf{M}^{-1}(\mathbf{w}_N + \mathbf{w}_T \mu_0) & -\mathbf{w}_N^T \mathbf{M}^{-1} \mathbf{w}_T & 0 \\ -\mathbf{w}_T^T \mathbf{M}^{-1}(\mathbf{w}_N + \mathbf{w}_T \mu_0) & \mathbf{w}_T^T \mathbf{M}^{-1} \mathbf{w}_T & 1 \\ 2\mu_0 & -1 & 0 \end{pmatrix} \begin{pmatrix} \lambda_N \\ \sigma_1 \\ \kappa_3 \end{pmatrix} \\ &+ \begin{pmatrix} \mathbf{w}_N^T \mathbf{M}^{-1} \mathbf{h} + \bar{w}_N \\ -\mathbf{w}_T^T \mathbf{M}^{-1} \mathbf{h} - \bar{w}_T \\ 0 \end{pmatrix}\end{aligned} \tag{10.40}$$

and is of dimension 3. The index T denotes the first of the former two tangential directions. The tangential relative acceleration (10.30) becomes

$$\dot{\gamma}_T = \kappa_3 - \kappa_1, \tag{10.41}$$

and the friction saturations (10.26) are

$$\begin{aligned} \sigma_1 &= \mu_0 \lambda_N - \lambda_T, \\ \sigma_3 &= \mu_0 \lambda_N + \lambda_T \end{aligned} \tag{10.42}$$

with λ_T being the tangential contact force. Due to the complementarities $\kappa_i \geq 0$, $\sigma_i \geq 0$, $\kappa_i \sigma_i = 0$ we finally note that κ_3 and κ_1 correspond to the positive and negative parts of $\dot{\gamma}_T$ for $\mu_0 \lambda_N > 0$, respectively, i.e. $\dot{\gamma}_T = \dot{\gamma}_T^+ - \dot{\gamma}_T^-$.

The presented method of formulating linear complementarity problems for dynamic multibody systems is general. Every polytope obtained from the intersection of a finite number of half-spaces may be treated in this way. The dimension of the resulting LCP is always equal to the number of half-spaced used. For example, the most general force laws between two bodies in a multibody system are six-dimensional, corresponding to the number of degrees of freedom of an unconstrained body. In this case, a force vector $\boldsymbol{\lambda} \in \mathbb{R}^6$ would enter the Newton-Euler equations, and the elimination of $\boldsymbol{\lambda}$ would require six linearly independent directions $\mathbf{e}_i$ to be arranged in $\mathbf{I}_P$ from (10.32).

10.7 Friction Cones and NCP Formulations

With the results from the preceding section one is able to treat spatial friction at least by approximations. However, due to the pyramid model, the collinearity of the negative friction forces $\boldsymbol{\lambda}_T$ and the tangential relative accelerations $\dot{\boldsymbol{\gamma}}_T$ is lost. This lack of consistence with respect to the physical model of friction in mind (and also with respect to the sliding friction problem from section 10.2) may be reduced by increasing the number of vertices of the polytope which, however, leads to LCP's of higher dimensions. We therefore try to find a formulation which, on the one hand, incorporates the conic friction law (10.23), (10.24) and, on the other hand, requires as less equations as possible. Since this problem is obviously beyond linearity we try to use the next available standard formulation, i.e. the representation as a nonlinear complementarity problem (NCP).

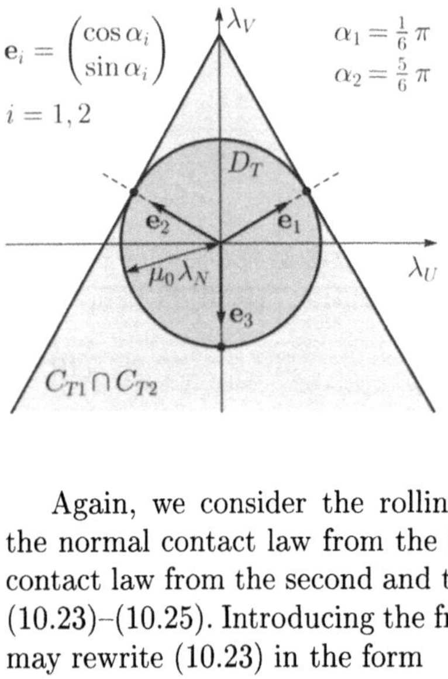

Figure 10.4. Friction disc and two additional constraints

Again, we consider the rolling contact problem which is described by the normal contact law from the third line in (10.18) and by the tangential contact law from the second and the third line in (10.20) or, equivalently, by (10.23)–(10.25). Introducing the friction saturation σ_D of the friction disk we may rewrite (10.23) in the form

$$D_T = \{\boldsymbol{\lambda}_T \mid \sigma_D(\boldsymbol{\lambda}_T) \geq 0\} \quad \text{with } \sigma_D(\boldsymbol{\lambda}_T) = \mu_0\lambda_N - |\boldsymbol{\lambda}_T|. \tag{10.43}$$

From now, the derivation of the NCP follows exactly the procedure presented in Section 10.6. A crucial step necessary to avoid inversions of terms $\mathbf{W}^T\mathbf{M}^{-1}\mathbf{W}$ had been performed in equations (10.34) and (10.37) where the tangential forces $\boldsymbol{\lambda}_T$ had been eliminated from the Newton-Euler equations. This, however, requires at least two equations which can be solved for $\boldsymbol{\lambda}_T$, preferably being linear. We therefore introduce two additional linear inequalities on the tangential forces, not changing D_T when intersection is performed: Let

$$C_{Ti} = \{\boldsymbol{\lambda}_T \mid \sigma_i(\boldsymbol{\lambda}_T) \geq 0\} \quad \text{with } \sigma_i(\boldsymbol{\lambda}_T) = \mu_0 \lambda_N - \mathbf{e}_i^T \boldsymbol{\lambda}_T, \quad \text{and } i = 1, 2, \tag{10.44}$$

where $\mathbf{e}_1$, $\mathbf{e}_2$ are two unit vectors as defined in Figure 10.4. The set of the admissible tangential forces may now be written as

$$C_T := \bigcap_{i=1}^{2} C_{Ti} \cap D_T \equiv D_T, \tag{10.45}$$

and the corresponding normal cone becomes

$$N_{C_T}(\boldsymbol{\lambda}_T) = \sum_{i=1}^{2} N_{C_{Ti}}(\boldsymbol{\lambda}_T) + N_{D_T}(\boldsymbol{\lambda}_T) \equiv N_{D_T}(\boldsymbol{\lambda}_T). \tag{10.46}$$

Since $C_T \equiv D_T$ we also have $N_{C_T}(\boldsymbol{\lambda}_T) \equiv N_{D_T}(\boldsymbol{\lambda}_T)$. Thus, the tangential contact law (10.24) may be equivalently expressed by

$$-\dot{\boldsymbol{\gamma}}_T \in N_{C_T}(\boldsymbol{\lambda}_T). \tag{10.47}$$

Writing down the normal cone conditions for (10.47) we obtain

$$-\dot{\boldsymbol{\gamma}}_T \in \sum_{i=1}^{2} \mathbf{e}_i \kappa_i - \partial\sigma_D(\boldsymbol{\lambda}_T)\,\kappa_D; \quad \kappa_j \geq 0,\ \sigma_j \geq 0,\ \kappa_j \sigma_j = 0, \quad j = 1, 2, D, \tag{10.48}$$

where, in contrast to (10.30), no further simplification is possible because σ_D is not differentiable at $\boldsymbol{\lambda}_T = 0$, i.e. $\partial\sigma_D(\boldsymbol{\lambda}_T = 0) = B(0,1)$ is set-valued.

Thus, in order to avoid a set-valued expression as in (10.48), we try to find an equivalent contact law which can be written as an equation. We set

$$-\dot{\boldsymbol{\gamma}}_T = \sum_{i=1}^{2} \mathbf{e}_i \kappa_i + \mathbf{e}_D(\boldsymbol{\lambda}_T)\,\kappa_D; \quad \kappa_j \geq 0,\ \sigma_j \geq 0,\ \kappa_j \sigma_j = 0, \quad j = 1, 2, D, \tag{10.49}$$

where the function $\boldsymbol{\lambda}_T \to \mathbf{e}_D(\boldsymbol{\lambda}_T)$ is defined as

$$\mathbf{e}_D(\boldsymbol{\lambda}_T) = \begin{cases} \boldsymbol{\lambda}_T / |\boldsymbol{\lambda}_T| & \text{if } \boldsymbol{\lambda}_T \neq 0 \\ \mathbf{e}_3 & \text{if } \boldsymbol{\lambda}_T = 0 \end{cases} \tag{10.50}$$

with $\mathbf{e}_3$ being the unit vector from Figure 10.4.

Proposition 10.7.1. *The contact laws (10.48) and (10.49) are equivalent.*

Proof. For $\boldsymbol{\lambda}_T \neq 0$ the friction saturation σ_D as defined in (10.43) is differentiable. Thus, $\partial\sigma_D(\boldsymbol{\lambda}_T) = \nabla\sigma_D(\boldsymbol{\lambda}_T) = -\boldsymbol{\lambda}_T/|\boldsymbol{\lambda}_T|$ in (10.48) which is as in (10.49), (10.50). For $\boldsymbol{\lambda}_T = 0$ equations (10.48) and (10.49) become

$$
\begin{aligned}
-\dot{\gamma}_T &\in \sum_{i=1}^{2} \mathbf{e}_i \kappa_i + B(0,1)\,\kappa_D, \\
-\dot{\gamma}_T &= \sum_{i=1}^{2} \mathbf{e}_i \kappa_i + \mathbf{e}_3 \kappa_D
\end{aligned}
\tag{10.51}
$$

with $\sigma_i = \mu_0 \lambda_N \geq 0$, $\kappa_i \geq 0$, $\sigma_i \kappa_i = 0$ for $i = 1, 2, D$. Let $\mu_0 \lambda_N > 0$. In this case we have $\sigma_i > 0$, $\kappa_i = 0$ for $i = 1, 2, D$, and both equations in (10.51) reduce to $\dot{\gamma}_T = 0$. Now let $\mu_0 \lambda_N = 0$. We obtain $\sigma_i = \mu_0 \lambda_N = 0$ and $\kappa_i \geq 0$ without any further restrictions. The left equation in (10.51) becomes $-\dot{\gamma}_T \in \mathbb{R}^2$ which is also true for the right equation because $(\mathbf{e}_1, \mathbf{e}_2, \mathbf{e}_3)$ have been chosen such that the union of the convex cones generated by $(\mathbf{e}_i, \mathbf{e}_j)$, $i, j = 1, 2, 3$ is $\mathbb{R}^2$. □

As in (10.31) we may now state the rolling contact problem. The first four expressions consisting of the Newton-Euler equations, the kinematic expressions, and the normal contact law may be adopted unchanged from (10.31), whereas the tangential contact law is now described by equation (10.49) together with the definitions of σ_D and σ_i from (10.43) and (10.44), respectively. Altogether we have

$$
\begin{aligned}
&\mathbf{M}\dot{\mathbf{u}} - \mathbf{h} - \mathbf{w}_N \lambda_N - \mathbf{W}_T \boldsymbol{\lambda}_T = 0 \\
&\dot{\gamma}_N = \mathbf{w}_N^T \dot{\mathbf{u}} + \bar{w}_N \\
&\dot{\boldsymbol{\gamma}}_T = \mathbf{W}_T^T \dot{\mathbf{u}} + \bar{\mathbf{w}}_T \\
&\lambda_N \geq 0,\ \dot{\gamma}_N \geq 0,\ \lambda_N \dot{\gamma}_N = 0 \\
&\sigma_i = \mu_0 \lambda_N - \mathbf{e}_i^T \boldsymbol{\lambda}_T, \quad i = 1, 2 \\
&\sigma_D = \mu_0 \lambda_N - |\boldsymbol{\lambda}_T| \\
&-\dot{\boldsymbol{\gamma}}_T = \textstyle\sum_{i=1}^{2} \mathbf{e}_i \kappa_i + \mathbf{e}_D(\boldsymbol{\lambda}_T)\,\kappa_D \\
&\sigma_j \geq 0,\ \kappa_j \geq 0,\ \sigma_j \kappa_j = 0, \quad j = 1, 2, D.
\end{aligned}
\tag{10.52}
$$

Since the directions $\mathbf{e}_1$ and $\mathbf{e}_2$ are linearly independent we may use the same notation as in (10.32). We set

$$
\boldsymbol{\sigma}_P = \begin{pmatrix} \sigma_1 \\ \sigma_2 \end{pmatrix}, \quad \boldsymbol{\kappa}_P = \begin{pmatrix} \kappa_1 \\ \kappa_2 \end{pmatrix}, \quad \boldsymbol{\mu}_P = \begin{pmatrix} \mu_0 \\ \mu_0 \end{pmatrix}, \quad \mathbf{I}_P = \begin{pmatrix} \mathbf{e}_1 & \mathbf{e}_2 \end{pmatrix} \tag{10.53}
$$

in order to solve the fifth equation in (10.52) for $\boldsymbol{\lambda}_T$,

$$
\boldsymbol{\lambda}_T = \mathbf{I}_P^{-T} \boldsymbol{\mu}_P \lambda_N - \mathbf{I}_P^{-T} \boldsymbol{\sigma}_P. \tag{10.54}
$$

Putting (10.54) into the sixth and seventh equation in (10.52) yields

$$
\sigma_D = \mu_0 \lambda_N - |\mathbf{I}_P^{-T} (\boldsymbol{\mu}_P \lambda_N - \boldsymbol{\sigma}_P)|, \tag{10.55}
$$

$$\boldsymbol{\kappa}_P = -\mathbf{I}_P^{-1}\dot{\boldsymbol{\gamma}}_T - \mathbf{I}_P^{-1}\mathbf{e}_D(\lambda_N, \boldsymbol{\sigma}_P)\,\kappa_D, \tag{10.56}$$

where the accelerations $\dot{\boldsymbol{\gamma}}_T$ have additionally been solved for $\boldsymbol{\kappa}_P$. Note that the direction $\mathbf{e}_D$ in (10.56) as defined in (10.49) now depends on λ_N and $\boldsymbol{\sigma}_P$. As in (10.37) we eliminate $\boldsymbol{\lambda}_T$ with the help of (10.54) from the Newton-Euler equations in (10.52), and we substitute $\dot{\boldsymbol{\gamma}}_T$ from the third line in (10.52) into (10.56). Altogether we have now

$$\begin{aligned}
&\mathbf{M}\dot{\mathbf{z}} - \mathbf{h} - (\mathbf{w}_N + \mathbf{W}_T\mathbf{I}_P^{-T}\boldsymbol{\mu}_P)\lambda_N + \mathbf{W}_T\mathbf{I}_P^{-T}\boldsymbol{\sigma}_P = 0\\
&\dot{\gamma}_N = \mathbf{w}_N^T\dot{\mathbf{u}} + \bar{w}_N\\
&\boldsymbol{\kappa}_P = -\mathbf{I}_P^{-1}\mathbf{W}_T^T\dot{\mathbf{u}} - \mathbf{I}_P^{-1}\bar{\mathbf{w}}_T - \mathbf{I}_P^{-1}\mathbf{e}_D(\lambda_N, \boldsymbol{\sigma}_P)\,\kappa_D\\
&\sigma_D = \mu_0\lambda_N - |\mathbf{I}_P^{-T}(\boldsymbol{\mu}_P\lambda_N - \boldsymbol{\sigma}_P)|\\
&\dot{\gamma}_N \geq 0,\ \lambda_N \geq 0,\ \dot{\gamma}_N\lambda_N = 0\\
&\boldsymbol{\kappa}_P \geq 0,\ \boldsymbol{\sigma}_P \geq 0,\ \boldsymbol{\kappa}_P^T\boldsymbol{\sigma}_P = 0\\
&\sigma_D \geq 0,\ \kappa_D \geq 0,\ \sigma_D\kappa_D = 0,
\end{aligned} \tag{10.57}$$

where in a last step the accelerations $\dot{\mathbf{u}}$ have to be eliminated from the second and the third line by using the first equation. Similarly to (10.38) the remaining three equations may be written in the form

$$\begin{aligned}
\begin{pmatrix}\dot{\gamma}_N\\ \boldsymbol{\kappa}_P\\ \sigma_D\end{pmatrix} &= \begin{pmatrix}\mathbf{w}_N^T\mathbf{M}^{-1}(\mathbf{w}_N + \mathbf{W}_P\boldsymbol{\mu}_P) & -\mathbf{w}_N^T\mathbf{M}^{-1}\mathbf{W}_P & 0\\ -\mathbf{W}_P^T\mathbf{M}^{-1}(\mathbf{w}_N + \mathbf{W}_P\boldsymbol{\mu}_P) & \mathbf{W}_P^T\mathbf{M}^{-1}\mathbf{W}_P & 0\\ \mu_0 & 0 & 0\end{pmatrix}\begin{pmatrix}\lambda_N\\ \boldsymbol{\sigma}_P\\ \kappa_D\end{pmatrix}\\
&+ \begin{pmatrix}0\\ -\mathbf{I}_P^{-1}\mathbf{e}_D(\lambda_N, \boldsymbol{\sigma}_P)\,\kappa_D\\ -|\mathbf{I}_P^{-T}(\boldsymbol{\mu}_P\lambda_N - \boldsymbol{\sigma}_P)|\end{pmatrix} + \begin{pmatrix}\mathbf{w}_N^T\mathbf{M}^{-1}\mathbf{h} + \bar{w}_N\\ -\mathbf{W}_P^T\mathbf{M}^{-1}\mathbf{h} - \bar{\mathbf{w}}_P\\ 0\end{pmatrix},
\end{aligned} \tag{10.58}$$

where the abbreviations $\mathbf{W}_P = \mathbf{W}_T\mathbf{I}_P^{-T}$ and $\bar{\mathbf{w}}_P = \mathbf{I}_P^{-1}\bar{\mathbf{w}}_T$ have been used. Equation (10.58) together with the inequalities and complementarities from the last three lines in (10.57) constitutes a *nonlinear complementarity problem* in standard form, i.e. a problem of the form $\mathbf{y} = \mathbf{f}(\mathbf{x})$, $\mathbf{y} \geq 0$, $\mathbf{x} \geq 0$, $\mathbf{y}^T\mathbf{x} = 0$, with $\mathbf{f}(\mathbf{x})$ being here $\mathbf{f}(\mathbf{x}) = \mathbf{A}\mathbf{x} + \mathbf{g}(\mathbf{x}) + \mathbf{c}$.

The dimension of the NCP in (10.58) is 4. We expect that this is also the minimal number of equations necessary to describe spatial friction problems in NCP-standard form: In the case $\mu_0\lambda_N = 0$ the friction law (10.47) reduces to $\dot{\boldsymbol{\gamma}}_T \in \mathbb{R}^2$. In order to express *any* $\dot{\boldsymbol{\gamma}}_T \in \mathbb{R}^2$ by a *positive* linear combination $\sum \mathbf{e}_i\kappa_i$ at least three directions $\mathbf{e}_i$ are necessary. The scalars $\kappa_i \geq 0$ enter the NCP as independent variables which makes, together with the normal contact law, an overall dimension of at least 4.

10.8 A Differentiable NCP for Rolling Contacts

There is a serious lack in the NCP-formulation of the last section: The non-linear function $\mathbf{g}(\mathbf{x})$ contained in the right-hand side $\mathbf{f}(\mathbf{x})$ of (10.58) is neither differentiable nor continuous. The direction $\mathbf{e}_D$ defined in (10.49) may jump at points $\boldsymbol{\lambda}_T = 0$, i.e. if $\boldsymbol{\mu}_P \lambda_N = \boldsymbol{\sigma}_P$. In order to overcome this problem we will finally present a NCP formulation with a continuous and differentiable right-hand side but with an overall dimension increased to 5.

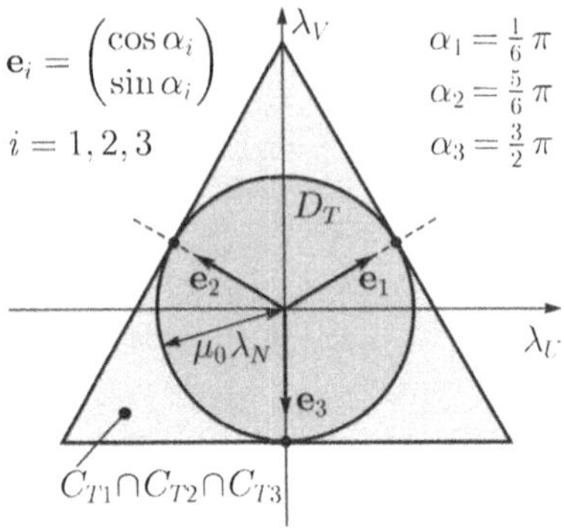

Figure 10.5. Friction disc and three additional constraints

For this purpose we characterize the friction disk D_T by a parabolic function σ_D which is differentiable everywhere

$$D_T = \{\boldsymbol{\lambda}_T \mid \sigma_D(\boldsymbol{\lambda}_T) \geq 0\} \quad \text{with } \sigma_D(\boldsymbol{\lambda}_T) = \mu_0^2 \lambda_N^2 - |\boldsymbol{\lambda}_T|^2, \tag{10.59}$$

but we note that the gradient of σ_D vanishes for $\boldsymbol{\lambda}_T = 0$. The subdifferential $\partial\sigma_D(\boldsymbol{\lambda}_T) = \nabla\sigma_D(\boldsymbol{\lambda}_T)$ is therefore not capable of describing the normal cone to D_T at $\boldsymbol{\lambda}_T = 0$. We therefore introduce additional inequalities as in (10.44) but now representing the whole triangle as it is depicted in Figure 10.5, i.e. the index i is now running from 1 to 3. Obviously the normal cones to D_T and $\bigcap_{i=1}^{3} C_{Ti}$ at $\boldsymbol{\lambda}_T = 0$ are identical and may be represented by using the three directions $\mathbf{e}_i$, $i = 1, 2, 3$ that are defined in Figure 10.5. The set of the admissible tangential forces C_T and the corresponding normal cone $N_{C_T}(\boldsymbol{\lambda}_T)$ may be expressed in the same manner as in (10.45) and (10.46), and the tangential contact law (10.24) becomes

$$-\dot{\boldsymbol{\gamma}}_T \in N_{C_T}(\boldsymbol{\lambda}_T), \tag{10.60}$$

since $C_T \equiv D_T$ and thus $N_{C_T}(\boldsymbol{\lambda}_T) \equiv N_{D_T}(\boldsymbol{\lambda}_T)$. Since the gradient of σ_D vanishes for $\boldsymbol{\lambda}_T = 0$, the normal cone conditions as in (10.48) are no longer valid in the general case. However, we formally write

$$-\dot{\boldsymbol{\gamma}}_T = \sum_{i=1}^{3} \mathbf{e}_i \kappa_i + 2\boldsymbol{\lambda}_T \kappa_D; \quad \kappa_j \geq 0,\ \sigma_j \geq 0,\ \kappa_j \sigma_j = 0, \quad j = 1, 2, 3, D \tag{10.61}$$

and prove, again, equivalence:

Proposition 10.8.1. *The friction laws (10.61) and (10.24) are equivalent.*

Proof. Let $0 < |\boldsymbol{\lambda}_T| \le \mu_0 \lambda_N$. In this case the (relative) interior of C_T is not empty, and the gradients of σ_j, $j = 1,2,3,D$ do not vanish. Therefore, (10.61) are the normal cone conditions for $N_{C_T}(\boldsymbol{\lambda}_T)$. Since $N_{C_T}(\boldsymbol{\lambda}_T) \equiv N_{D_T}(\boldsymbol{\lambda}_T)$ equivalence of (10.61) and (10.25) holds. Now let $0 = |\boldsymbol{\lambda}_T| \le \mu_0 \lambda_N$. In this case (10.61) reduces to $-\dot{\boldsymbol{\gamma}}_T = \sum_{i=1}^{3} \mathbf{e}_i \kappa_i$ with $\sigma_i = \mu_0 \lambda_N \ge 0$, $\kappa_i \ge 0$, $\sigma_i \kappa_i = 0$ and $i = 1,2,3$. Assume that $\mu_0 \lambda_N > 0$. We obtain $\sigma_i > 0$, $\kappa_i = 0$ for $i = 1,2,3$. Thus $\dot{\boldsymbol{\gamma}}_T = 0$, which is the second condition in (10.25) for $\boldsymbol{\lambda}_T = 0$. Assume that $\mu_0 \lambda_N = 0$. We obtain $\sigma_i = 0$, $\kappa_i \ge 0$ for $i = 1,2,3$. Since the convex cone generated by $(\mathbf{e}_1, \mathbf{e}_2, \mathbf{e}_3)$ is $\mathbb{R}^2$ we have $\dot{\boldsymbol{\gamma}}_T \in \mathbb{R}^2$ which is the third condition in (10.25). □

Using the tangential contact law from (10.61) together with the definitions of the "friction saturations" σ_D and σ_i from (10.59) and (10.44) we may now state the rolling contact problem in the form

$$\begin{aligned}
&\mathbf{M}\dot{\mathbf{u}} - \mathbf{h} - \mathbf{w}_N \lambda_N - \mathbf{W}_T \boldsymbol{\lambda}_T = 0 \\
&\dot{\gamma}_N = \mathbf{w}_N^T \dot{\mathbf{u}} + \bar{w}_N \\
&\dot{\boldsymbol{\gamma}}_T = \mathbf{W}_T^T \dot{\mathbf{u}} + \bar{\mathbf{w}}_T \\
&\lambda_N \ge 0,\ \dot{\gamma}_N \ge 0,\ \lambda_N \dot{\gamma}_N = 0 \\
&\sigma_i = \mu_0 \lambda_N - \mathbf{e}_i^T \boldsymbol{\lambda}_T, \quad i = 1,2,3 \\
&\sigma_D = \mu_0^2 \lambda_N^2 - |\boldsymbol{\lambda}_T|^2 \\
&-\dot{\boldsymbol{\gamma}}_T = \textstyle\sum_{i=1}^{3} \mathbf{e}_i \kappa_i + 2 \boldsymbol{\lambda}_T \kappa_D \\
&\sigma_j \ge 0,\ \kappa_j \ge 0,\ \sigma_j \kappa_j = 0, \quad j = 1,2,3,D.
\end{aligned} \tag{10.62}$$

The derivation of the nonlinear complementarity problem is now straightforward and will be performed in the same manner as in Section 10.7. By using the abbreviations from (10.53), the fifth equation leads to the tangential forces (10.54) which are then substituted in the Newton-Euler equations as well as in the remaining two friction saturations σ_3 and σ_D. For the latter, this yields

$$\begin{aligned}
\sigma_3 &= (\mu_0 - \mathbf{e}_3^T \mathbf{I}_P^{-T} \boldsymbol{\mu}_P)\, \lambda_N + \mathbf{e}_3^T \mathbf{I}_P^{-T} \boldsymbol{\sigma}_P, \\
\sigma_D &= \mu_0^2 \lambda_N^2 - |\mathbf{I}_P^{-T} (\boldsymbol{\mu}_P \lambda_N - \boldsymbol{\sigma}_P)|^2.
\end{aligned} \tag{10.63}$$

Solving the relative accelerations $\dot{\boldsymbol{\gamma}}_T$ in the seventh line of (10.62) for $\boldsymbol{\kappa}_P$,

$$\boldsymbol{\kappa}_P = -\mathbf{I}_P^{-1} \dot{\boldsymbol{\gamma}}_T - \mathbf{I}_P^{-1} \mathbf{e}_3 \kappa_3 - 2\, \mathbf{I}_P^{-1} \mathbf{I}_P^{-T} (\boldsymbol{\mu}_P \lambda_N - \boldsymbol{\sigma}_P)\, \kappa_D, \tag{10.64}$$

we may rewrite the system of equations (10.62) to be

$$
\begin{aligned}
&\mathbf{M}\dot{\mathbf{u}} - \mathbf{h} - (\mathbf{w}_N + \mathbf{W}_T\,\mathbf{I}_P^{-T}\boldsymbol{\mu}_P)\,\lambda_N + \mathbf{W}_T\,\mathbf{I}_P^{-T}\boldsymbol{\sigma}_P = 0 \\
&\dot{\gamma}_N = \mathbf{w}_N^T\dot{\mathbf{u}} + \bar{w}_N \\
&\boldsymbol{\kappa}_P = -\mathbf{I}_P^{-1}\,\mathbf{W}_T^T\,\dot{\mathbf{u}} - \mathbf{I}_P^{-1}\,\bar{\mathbf{w}}_T - \mathbf{I}_P^{-1}\,\mathbf{e}_3\kappa_3 - 2\,\mathbf{I}_P^{-1}\,\mathbf{I}_P^{-T}(\boldsymbol{\mu}_P\lambda_N - \boldsymbol{\sigma}_P)\,\kappa_D \\
&\sigma_3 = (\mu_0 - \mathbf{e}_3^T\,\mathbf{I}_P^{-T}\boldsymbol{\mu}_P)\lambda_N + \mathbf{e}_3^T\,\mathbf{I}_P^{-T}\boldsymbol{\sigma}_P \\
&\sigma_D = \mu_0^2\lambda_N^2 - |\mathbf{I}_P^{-T}(\boldsymbol{\mu}_P\lambda_N - \boldsymbol{\sigma}_P)|^2 \qquad (10.65) \\
&\dot{\gamma}_N \ge 0,\ \lambda_N \ge 0,\ \dot{\gamma}_N\lambda_N = 0 \\
&\boldsymbol{\kappa}_P \ge 0,\ \boldsymbol{\sigma}_P \ge 0,\ \boldsymbol{\kappa}_P^T\boldsymbol{\sigma}_P = 0 \\
&\sigma_3 \ge 0,\ \kappa_3 \ge 0,\ \sigma_3\kappa_3 = 0 \\
&\sigma_D \ge 0,\ \kappa_D \ge 0,\ \sigma_D\kappa_D = 0.
\end{aligned}
$$

Finally, the accelerations $\dot{\mathbf{u}}$ are eliminated which gives, together with the inequalities and complementarities from the last four lines in (10.65), a nonlinear complementarity problem in standard form of dimension 5,

$$
\begin{pmatrix}\dot{\gamma}_N\\ \boldsymbol{\kappa}_P\\ \sigma_3\\ \sigma_D\end{pmatrix}
=
\begin{pmatrix}
\mathbf{w}_N^T\mathbf{M}^{-1}(\mathbf{w}_N + \mathbf{W}_P\boldsymbol{\mu}_P) & -\mathbf{w}_N^T\mathbf{M}^{-1}\mathbf{W}_P & 0 & 0\\
-\mathbf{W}_P^T\mathbf{M}^{-1}(\mathbf{w}_N + \mathbf{W}_P\boldsymbol{\mu}_P) & \mathbf{W}_P^T\mathbf{M}^{-1}\mathbf{W}_P & -\mathbf{I}_P^{-1}\,\mathbf{e}_3 & 0\\
\mu_0 - \mathbf{e}_3^T\,\mathbf{I}_P^{-T}\boldsymbol{\mu}_P & \mathbf{e}_3^T\mathbf{I}_P^{-T} & 0 & 0\\
0 & 0 & 0 & 0
\end{pmatrix}
\begin{pmatrix}\lambda_N\\ \boldsymbol{\sigma}_P\\ \kappa_3\\ \kappa_D\end{pmatrix}
+
\begin{pmatrix}0\\ -2\,\mathbf{I}\,\boldsymbol{\lambda}_P\,\kappa_D\\ 0\\ \mu_0^2\lambda_N^2 - \boldsymbol{\lambda}_P^T\,\mathbf{I}\,\boldsymbol{\lambda}_P\end{pmatrix}
+
\begin{pmatrix}\mathbf{w}_N^T\mathbf{M}^{-1}\mathbf{h} + \bar{w}_N\\ -\mathbf{W}_P^T\mathbf{M}^{-1}\mathbf{h} - \bar{\mathbf{w}}_P\\ 0\\ 0\end{pmatrix}.
\qquad (10.66)
$$

Hereby, the abbreviations $\mathbf{W}_P := \mathbf{W}_T\mathbf{I}_P^{-T}$, $\bar{\mathbf{w}}_P := \mathbf{I}_P^{-1}\bar{\mathbf{w}}_T$, and $\mathbf{I} := \mathbf{I}_P^{-1}\mathbf{I}_P^{-T}$ have been used, where $\mathbf{I}$ is a symmetric and positive definite 2×2 matrix, and

$$
\boldsymbol{\lambda}_P(\lambda_N, \boldsymbol{\sigma}_P) := \boldsymbol{\mu}_P\lambda_N - \boldsymbol{\sigma}_P. \qquad (10.67)
$$

The NCP (10.66) is of the form

$$
\mathbf{y} = \mathbf{f}(\mathbf{x}), \quad \mathbf{y} \ge 0,\ \mathbf{x} \ge 0,\ \mathbf{y}^T\mathbf{x} = 0 \qquad (10.68)
$$

where $\mathbf{f}(\mathbf{x})$ in (10.66) is represented by a linear part, a nonlinear part, and a constant,

$$
\mathbf{f}(\mathbf{x}) = \mathbf{A}\mathbf{x} + \mathbf{g}(\mathbf{x}) + \mathbf{c}. \qquad (10.69)
$$

Since $\mathbf{f}(\mathbf{x})$ is continuous and differentiable we may finally derive its Jacobian. With

$$
D_x\mathbf{f} = \mathbf{A} + D_x\mathbf{g} \qquad (10.70)
$$

and $D_x\mathbf{g}$ explicitly depending on λ_N, $\boldsymbol{\sigma}_P$ and $\boldsymbol{\kappa}_D$, we obtain

$$D_x \mathbf{g} = \begin{pmatrix} 0 & 0 & 0 & 0 \\ 0 - 2\,\mathbf{I}\,\boldsymbol{\mu}_P \kappa_D & 2\,\mathbf{I}\,\kappa_D & 0 & -2\,\mathbf{I}\,\boldsymbol{\lambda}_P \\ 0 & 0 & 0 & 0 \\ 2\mu_0^2 \lambda_N - 2\,\boldsymbol{\lambda}_P^T \mathbf{I}^T \boldsymbol{\mu}_P & 2\,\boldsymbol{\lambda}_P^T \mathbf{I}^T & 0 & 0 \end{pmatrix} \tag{10.71}$$

with symmetric $\mathbf{I}$. Finally we draw attention to the similar structure of the LCP matrix from (10.38) and the Jacobian (10.70). In both cases the second column of the corresponding matrix, multiplied by the friction coefficients $\boldsymbol{\mu}_P$, appears in the first column. This indicates that the tangential forces $\boldsymbol{\lambda}_T$ have been eliminated from the Newton-Euler equations by the use of the friction saturations $\boldsymbol{\sigma}_P$.

10.9 Example and Remarks

The friction pyramid model as presented in Section 10.3 has been used for about four years and has been applied, for example, to the friction problem at the legs and in the gears of a tube crawling robot [125] as well as to the investigation of the wheel/rail contact of the undercarriage of a roller coaster. The NCP formulation, however, is new and has not yet been applied to systems with complicated contact structure. Thus, we will give at least an idea about the treatment of the NCP (10.68), and we will present some calculations obtained for a test example, i.e. a point mass moving on a horizontal surface under the influence of friction and force excitation.

The numerical integration of this example has been performed by using a time-stepping algorithm with the unilateral contact constraint and Coulomb's frictional law being directly evaluated on the displacement and velocity level, respectively. The occurring NCP's (10.66) have been solved by applying a semi-smooth Newton algorithm on the associated Fischer-Burmeister NCP-functions as reported in [32]. We shortly review the main idea of this approach:

A function $\phi : \mathbb{R}^2 \to \mathbb{R}$ is called an *NCP-function* if

$$\phi(x, y) = 0 \quad \Leftrightarrow \quad x \geq 0,\ y \geq 0,\ xy = 0. \tag{10.72}$$

Solving the NCP $y = f(x)$, $x \geq 0$, $y \geq 0$, $xy = 0$ is thus equivalent to finding the zeros of $\phi(x, f(x))$ when ϕ is an NCP-function. There are many possible choices for such a ϕ, one of them being the *Fischer-Burmeister function*

$$\phi(x, y) = \sqrt{x^2 + y^2} - x - y \tag{10.73}$$

from [33] which is depicted in Figure 10.6. This function is in addition strictly convex and locally Lipschitzian, hence almost everywhere differentiable. The only point at which differentiability fails is at $(x, y) = (0, 0)$, forming a corner in the graph of ϕ. In [32] a modified Newton algorithm is presented, especially designed for finding the zeros of (10.73). This algorithm can also handle the discontinuity in the gradient of ϕ at the point $(0,0)$. The (non-defined) gradient is here replaced by any member of the B-subdifferential, denoted by ∂_B; this is, at any point (x,y), the set consisting of the limits of all sequences $\nabla\phi(x_k,y_k)$ with (x_k,y_k) converging to (x,y) such that ϕ is differentiable at (x_k,y_k). For the NCP-function (10.73) the B-subdifferential at $(0,0)$ is a compact set not containing zero, whereas it corresponds to the gradient of ϕ for any other point in $\mathbb{R}^2$. Furthermore, note that the gradient of ϕ does not vanish at any point in $\mathbb{R}^2$, in particular not at $(x=0, y>0)$ or $(x>0, y=0)$.

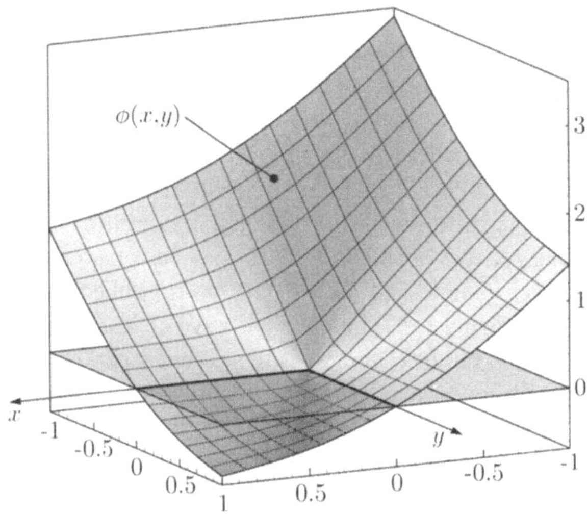

Figure 10.6. The Fischer-Burmeister NCP-function

In order to solve the NCP (10.68) by this method, one has to introduce one separate NCP-function (10.73) for each complementary pair (x_i, y_i). Let therefore $\mathbf{\Phi} : \mathbb{R}^n \times \mathbb{R}^n \to \mathbb{R}^n$ be the vector-valued function associated with the NCP of dimension n ($n=5$ for the friction problem (10.68)) defined by

$$\mathbf{\Phi}(\mathbf{x},\mathbf{y}) := \begin{pmatrix} \phi(x_1,y_1) \\ \vdots \\ \phi(x_n,y_n) \end{pmatrix} \tag{10.74}$$

and $\mathbf{\Theta} : \mathbb{R}^n \to \mathbb{R}^n$ the function composed as

$$\mathbf{\Theta}(\mathbf{x}) := \mathbf{\Phi}(\mathbf{x}, \mathbf{f}(\mathbf{x})) \tag{10.75}$$

with $\mathbf{y} = \mathbf{f}(\mathbf{x})$, i.e. $y_i = f_i(\mathbf{x})$. A solution of the NCP (10.68) is thus a zero of the function $\mathbf{x} \to \mathbf{\Theta}(\mathbf{x})$ and vice versa. For the calculation of the search direction, the algorithm in [32] also requires the (generalized) Jacobian of (10.75). By applying the chain rule one obtains

$$\partial_B \mathbf{\Theta}(\mathbf{x}) = \partial_B \mathbf{\Phi}(\mathbf{x}, \mathbf{f}(\mathbf{x})) \begin{pmatrix} \mathbf{E} \\ D_x \mathbf{f} \end{pmatrix} \tag{10.76}$$

with $\mathbf{E}$ being the $n \times n$ identity matrix and $D_x\mathbf{f}$ the Jacobian from (10.70).

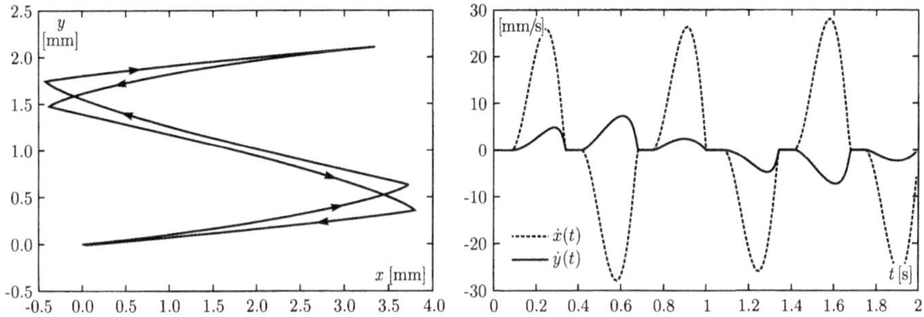

Figure 10.7. A point mass under spatial friction and force excitation

Figure 10.7 shows some results for a point mass (mass $m = 1\,\text{kg}$, gravity $g = 9.81\,\text{m/s}^2$) moving on a horizontal surface under the influence of friction (friction coefficient $\mu = 0.06$). The mass is excited by sinusoidal forces $F_x = A_x \sin\omega_x t$ and $F_y = A_y \sin\omega_y t$ in x and y direction, respectively, with the parameters $A_x = 1.4\,\mu m g$, $A_y = 0.3\,\mu m g$, $\omega_x = 3\pi$, $\omega_y = \pi$ chosen for the numerical evaluation. Note that the force in y-direction alone is too small to make the mass moving, since always $|F_y| < \mu m g$. This can be seen in the right diagram where the velocities $\dot{x}(t)$ and $\dot{y}(t)$ are depicted as functions of time t: For $0 < t < 1$ the force F_y acts in positive y-direction ($F_y > 0$), leading to positive values of $\dot{y}$ within the sliding regime. There are three periods of sliding in this interval, each of them starting when the overall force $\sqrt{F_x^2 + F_y^2}$ reaches the friction bound $\mu m g$. Here, F_x changes sign twice, whereas F_y does not. Note also that the velocities are differentiable at the transition points to sliding, but not at the transitions to sticking. The latter are accompanied with discontinuous changes in the friction force that jumps from the boundary of the friction disc to some point in the interior, causing an acceleration jump. In the left diagram the x-y positions of the mass are depicted for one cycle of excitation. The six corners in the displacement curve correspond to the six periods of sticking that are recognized from the right diagram when $\dot{x}(t) = \dot{y}(t) = 0$.

In this chapter, we have presented different formulations of the unilateral contact problem with Coulomb friction in dynamic multibody systems. By using the same methodical approach we have formulated pyramidic and conic friction laws in terms of linear and nonlinear complementarity problems, respectively. As an important result, these formulations have been obtained in standard form and, moreover, in a way that also linearly dependent contact constraints can be treated. We have observed similar structures arising in the

Jacobians of the linear and the nonlinear complementarity problem, due to the fact that the formulation processes follow the same rules.

As we have already pointed out, the treatment of the Coulomb friction law as presented in this chapter should be understood as an example. The way in which standard LCP formulations may be obtained when approximating arbitrary set-valued force interactions by polyhedral convex sets is general. The NCP formulation was simply based on the fact that the intersection of convex sets remains convex, and that additional inequalities may be added if convenient. This method works if the normal cone can be generated by a finite number of given directions.

11. Velocity Jumps due to C^0-Constraints

In all of the preceding chapters we have discussed the impact-free non-smooth motion of rigid body systems with the pre- and post-impact velocities fully integrated in the directional Newton-Euler equations. The impact itself, however, has been excluded. Since this book is not intended to be a treatise on impacts, we continue to stay away from any general approach. In this chapter, we just want to present some special types of velocity jumps by an example, somehow different from what is intuitively understood by an impact.

Impacts in rigid body dynamics are characterized by the occurrence of impulsive forces leading to instantaneous changes in the velocities of the system. Such velocity jumps are usually associated with collisions of the bodies between each other or in interaction with the environment.

Besides the collisions, there are other sources for velocity jumps, such as the restriction of a dynamic system to C^0-submanifolds. In the following, this case will be discussed by the example of a roller coaster. After a short review of impacts in mechanical systems in Section 11.1, we will show in Section 11.2 that a velocity jump occurs when some simplified mechanical model of the roller coaster is used. In section 11.3 the impact-free bilaterally constrained motion of the system is analyzed with special emphasis on the physical meaning of the Lagrange multipliers and the geometric structure of the constraint. The remaining part of the chapter is then devoted to the evaluation of the velocity jump. We propose two different methods on how to choose the underlying impact law and how to evaluate the velocity jump for scleronomic multibody systems with one non-differentiable bilateral constraint, when conservation of kinetic energy is desired:

The first approach, presented in Section 11.4, is based on a time scaling procedure, which leads to first order differential algebraic equations, similar to those of classical dynamics, with the displacements being fixed. Constraint forces are introduced in the classical way, but now representing the densities of the overall impact impulse vector with respect to some real measure. We show that conservation of energy is guaranteed by this force law, and that the values of the post-impact velocities and also of the impact impulse are invariant under re-parameterization.

In our second approach in Section 11.5 we will show that every velocity jump which preserves kinetic energy, may be interpreted as a reflection of the velocities with respect to some hyperplane when the kinetic metric is used. This provides an alternative method to compute the post-impact velocities by a direct evaluation of the reflection law, in contrast to the more longwinded time-scaling procedure.

Finally, as a main result, we show in Section 11.6 that any reflection of the velocities may be interpreted as an ordinary non-dissipative collision at a priori unknown (generalized) locations of the rigid body system. As a consequence, this confirms the approach chosen by [35], where the constitutive laws describing the impact behavior have to be formulated in a much more general form when simultaneity of the impact events is presupposed.

11.1 On Impacts in Mechanical Systems

In order to treat velocity jumps we return first to the measure equality (5.2) from [91] which provides a general framework for non-smooth rigid body dynamics, and which replaces the generalized Newton-Euler equations from classical dynamics, i.e.

$$\mathbf{M}(\mathbf{q},t)\,\mathrm{d}\mathbf{u} - \mathbf{h}(\mathbf{q},\mathbf{u},t)\,\mathrm{d}t = \mathrm{d}\mathbf{R}. \tag{11.1}$$

We recall that t denotes time, $\mathbf{u}(t)$ are functions of bounded variations that represent the generalized velocities of the system, $\mathrm{d}\mathbf{u}$ is the differential measure of $\mathbf{u}$, and $\mathbf{q}(t)$ are the absolutely continuous displacements obtained from integration of $\mathbf{u}(t)$. For a detailed discussion of the differential measures we refer, again, to [90]. Furthermore, $\mathbf{M}(\mathbf{q},t)$ is the symmetric and positive definite mass matrix, and $\mathbf{h}(\mathbf{q},\mathbf{u},t)$ the gyroscopic acceleration vector consisting of the Christoffel symbols. All other forces acting on the system are expressed as measures and are collected in the right-hand side $\mathrm{d}\mathbf{R}$.

For the evaluation of a velocity jump at time $t^\star$ only the atomic portion of this equation has to be regarded, denoted by lower index A as already done in (4.13), (5.3) with $\mathrm{d}\mathbf{R}_A = \mathbf{F}\,\mathrm{d}\eta$. This is

$$\mathbf{M}(\mathbf{q}^\star,t^\star)\,\mathrm{d}\mathbf{u}_A = \mathrm{d}\mathbf{R}_A \tag{11.2}$$

with $\mathbf{q}^\star := \mathbf{q}(t^\star)$. Integration of (11.2) over the singleton $\{t^\star\}$ yields

$$\begin{aligned}\int_{\{t^\star\}} \mathbf{M}(\mathbf{q}^\star,t^\star)\,\mathrm{d}\mathbf{u}_A &= \mathbf{M}(\mathbf{q}^\star,t^\star)\,(\mathbf{u}_+(t^\star) - \mathbf{u}_-(t^\star))\\ \int_{\{t^\star\}} \mathrm{d}\mathbf{R}_A &= \mathbf{R}_A(t^\star)\end{aligned} \tag{11.3}$$

where $\mathbf{u}_+(t^\star)$ and $\mathbf{u}_-(t^\star)$ denote the right and the left velocity at $t^\star$, and $\mathbf{R}_A(t^\star)$ is the mass of the atom of $\mathrm{d}\mathbf{R}_A$ at $t^\star$, a force impulsion that causes the velocity jump.

Equations (11.2), (11.3) are general. They have, with respect to velocity jumps, the same axiomatic character as the classical Newton-Euler equations for smooth motion. However, since nothing has been said so far about a functional relation between $\mathbf{R}_A$ and $(\mathbf{u}_+, \mathbf{u}_-)$, there is not enough information contained in (11.2), (11.3) to calculate the right velocity $\mathbf{u}_+(t^\star)$ from a given $\mathbf{u}_-(t^\star)$ for some special impact configuration. This missing information has to be provided by the *impact law*. Nearly all publications on velocity jumps treat exclusively the collisions. We refer especially to the book [21] with some hundreds of references given there. The impact laws for collisions are developed in various manners. For example, one finds different definitions of the restitution coefficient in [19], [54] or [139]. Single collisions with friction are treated in [144], [15], and [61] by a time scaling method which we will also use, for different reasons, in Section 11.4. Multiple impacts may be found in [108], [37], [44], [144] and, of course, in [91]. All these different approaches, however, have one thing in common: Applied to a single frictionless collision, the direction of the contact force is a priori assumed to be known and is *not* allowed to change during the impact. For this situation the measure $\mathrm{d}\mathbf{R}_A$ in (11.2) therefore takes the form

$$\mathrm{d}\mathbf{R}_A = \mathbf{w}(\mathbf{q}^\star, t^\star)\,\mathrm{d}\Lambda \tag{11.4}$$

with $\mathrm{d}\Lambda$ being the scalar value of the contact impulsion measure $\mathrm{d}\mathbf{R}_A$ and $\mathbf{w}(\mathbf{q}, t)$ the generalized force direction, in the frictionless case corresponding to the gradient $\nabla_q g(\mathbf{q}, t)$ of some inequality constraint $g(\mathbf{q}, t) \geq 0$.

In contrast to the situation (11.4) there are problems in which jumps in the velocity coordinates $\mathbf{u}$ are caused by *instantaneous changes* in the gradient of the constraints. We will discuss this phenomenon by the example of a car moving on a rail.

11.2 Mechanical Model and Problem

The problem of velocity jumps due to C^0-constraints turned up during the development of a simulation software for the roller coaster "Wild Mouse," see e.g. [138] for further references. The track of this relatively small roller coaster is depicted in Figure 11.1 and has an overall length of about 400 m, arranged in a 36 m × 20 m × 15m box. In rough outline, it consists of the station, the elevator, a section which is formed like a meander, and a total of four bumps. Maximum speed is reached in the first bump and is at about 45 km/h. As a special feature, the entire track is composed of straight and circular rails only, best seen at the meandric part in Figure 11.1. Figure 11.2 shows a close-up of the original coach. Since this design was intended to be substantially changed, a simulation software had to be developed in order to get first ideas about the dynamic behavior of the modified vehicle as well as first estimates of occurring forces and accelerations. Calculations with extensive variations of the design parameters were expected, thus a very fast computer code was required, based on a simple mechanical model as a

Figure 11.1. The track of the "Wild Mouse"

necessary pre-condition. Figure 11.3 shows a print-out of the implemented track with which numerical simulations have been performed on both, the modified and the original vehicle, the latter in order to verify the mechanical model chosen.

We are now going to present the mechanical model of the roller coaster. In view of the obligation to maintain the company's secrecies only the original version of the vehicle is discussed. Furthermore, we investigate only the dynamics at the connection points of two succeeding circular sections of the

Figure 11.2. The coach of the "Wild Mouse"

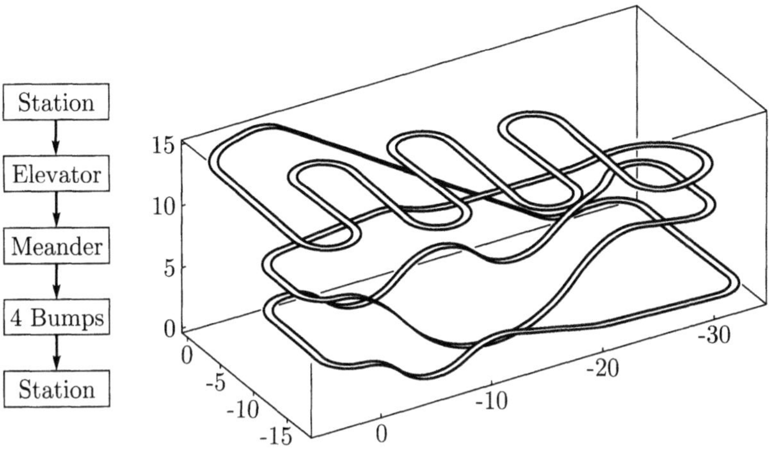

Figure 11.3. Implemented track

track when they have different curvatures, such as it appears in the bumps or in the meander. We neglect the inclination of the track as well as any other forces acting, except of the constraint forces between the rail and the car. Of course, all these simplifications are not made in the simulation software. We have introduced them to fully concentrate on the dynamics at the transition point which will generate a velocity jump. Since all the classical forces like from damping or from gravity are non-atomic, they would not contribute to the velocity jump anyway. The transition point is also the only part that does not fit in with classical mechanics. Everything else may be treated by standard methods.

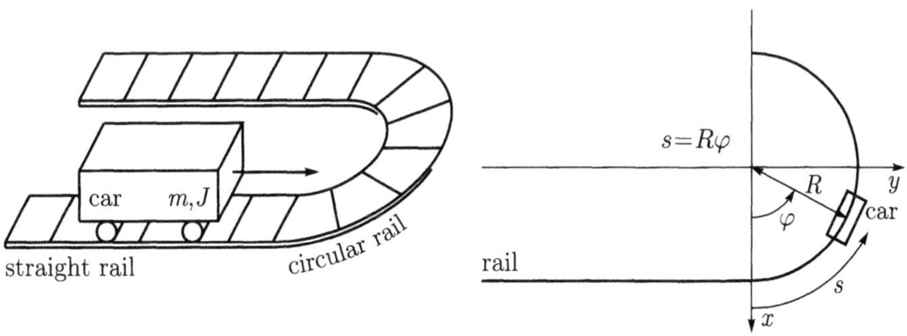

Figure 11.4. Simplified system and mechanical model

Henceforth we assume the track to be composed of a straight and a circular rail as in the meander, shown in the left diagram of Figure 11.4. In order to avoid tedious kinematics resulting from a vehicle with two axles and to keep the mathematical formulation as short as any possible we have chosen a model with one axle only. This is depicted in the right diagram of Figure 11.4, and was also used as a basis for the modified vehicle with additional degrees of freedom. For the latter, the computation time on a Pentium 200 was one second per ride. The planar mechanical model in Figure 11.4 has to be understood as follows: The rail is composed of a straight line ($s < 0$) and a circular section ($s > 0$) with radius R, where the natural parameter s (arc length) is used to determine the position of the car (mass m, moment of inertia J) on the rail. The orientation of the car is described by φ, and the velocities corresponding to (s, φ) are denoted by (v, ω).

Suppose that the car is running along the straight line ($s < 0, \varphi = 0$) with a given velocity ($v > 0, \omega = 0$). The choice of the model with only one axle causes a *velocity jump* in (v, ω) when the car passes $s = 0$, i.e. at the entrance of the circular region. To see this we state the kinetic energy for $s < 0$ and $s > 0$,

$$
\begin{aligned}
T_- &= \frac{1}{2} m v_-^2 + \frac{1}{2} J \omega_-^2 \quad \text{with } \omega_- = 0, \\
T_+ &= \frac{1}{2} m v_+^2 + \frac{1}{2} J \omega_+^2 \quad \text{with } v_+ = R\omega_+.
\end{aligned}
\tag{11.5}
$$

At the transition from $s < 0$ to $s > 0$ we may have conservation of the kinetic energy at best, which corresponds to the behavior of classical perfect bilateral constraints,

$$
T_- = T_+ \quad \Rightarrow \quad \frac{1}{2} m v_-^2 = \frac{1}{2} m v_+^2 + \frac{1}{2} J \frac{v_+^2}{R^2}. \tag{11.6}
$$

From (11.6) we obtain

$$
v_+ = \frac{1}{\mu} v_-, \quad \omega_+ = \frac{1}{R} v_+ \quad \text{with } \mu = \sqrt{1 + \frac{J}{mR^2}} > 1, \tag{11.7}
$$

hence $v_+ < v_-$ and $\omega_+ > \omega_- = 0$, i.e. a discontinuity in $v(t)$ and $\omega(t)$ at $t^\star$ with $s(t^\star) = 0$ and $\varphi(t^\star) = 0$. In the following we analyze the reason for the velocity jump, and we propose a solution method, not based on energy assumptions as in (11.6). This is absolutely necessary for the modified system with additional degrees of freedom, since there is *not* enough information contained in (11.6) on how to distribute the kinetic energy among the participating coordinates.

11.3 Bilaterally Constrained Motion

In this section, we derive the equations of motion of the car for both cases, the straight and the circular rail, by Lagrange's equations of first kind. The velocity jump itself is not yet treated; the formulation of the constraints, however, will already give some insight into the structure behind the problem.

According to Figure 11.4, the equations of motion without regarding the constraint between s and φ are simply $m\dot{v} = 0$, $J\dot{\omega} = 0$ with $v = \dot{s}$ and $\omega = \dot{\varphi}$ or, in matrix notation, $\mathbf{M}\dot{\mathbf{u}} = 0$ with $\mathbf{u} = \dot{\mathbf{q}}$.

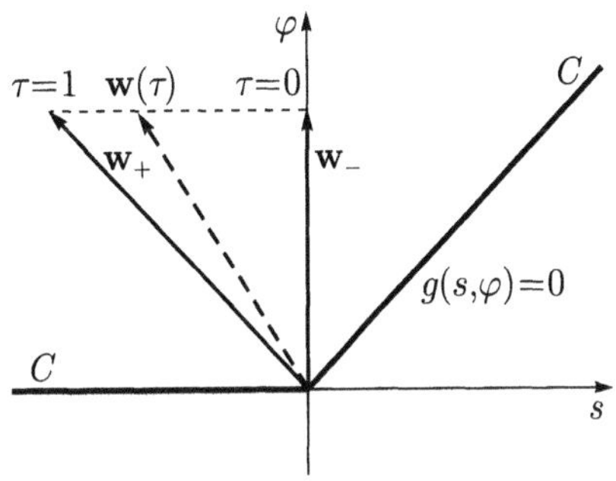

Figure 11.5. Bilateral constraint between s and φ

There are obviously two different bilateral constraints, corresponding to the two regions of the rail, that have to be used to align the car's orientation φ with the tangent of the rail at some position s. We will denote them by $g_-(\mathbf{q})$ and $g_+(\mathbf{q})$ with associated gradients $\mathbf{w}_-(\mathbf{q}) := \nabla g_-(\mathbf{q})$ and $\mathbf{w}_+(\mathbf{q}) := \nabla g_+(\mathbf{q})$, where the lower index "−" refers to the straight line and "+" to the circle, respectively. The equations for the constrained motion thus read

$$\mathbf{M}\dot{\mathbf{u}} - \mathbf{w}_-(\mathbf{q})\,\lambda = 0, \quad g_-(\mathbf{q}) = 0 \quad \text{for the straight rail,} \tag{11.8}$$

$$\mathbf{M}\dot{\mathbf{u}} - \mathbf{w}_+(\mathbf{q})\,\lambda = 0, \quad g_+(\mathbf{q}) = 0 \quad \text{for the circular rail.} \tag{11.9}$$

For our example, the constraints are simply

$$g_-(s,\varphi) = \varphi \quad \text{and} \quad g_+(s,\varphi) = \varphi - \tfrac{s}{R}, \tag{11.10}$$

where the first constraint is defined for the straight rail, i.e. $s \leq 0$, and the second for the circle ($s \geq 0$). Putting both cases together, the constraint equation is

$$g(s,\varphi) = 0 \quad \text{with } g(s,\varphi) = \begin{cases} \varphi & \text{for } s \leq 0 \\ \varphi - \frac{s}{R} & \text{for } s \geq 0 \end{cases} \tag{11.11}$$

which restricts the displacements (s,φ) to the set $C = \{(s,\varphi)\,|\,g(s,\varphi) = 0\}$, depicted in Figure 11.5. Obviously, this set is tangentially *non-regular* at the point $(s,\varphi) = (0,0)$ with contingent cone $K_C(0,0) \sim C$, tangent cone $T_C(0,0) = (0,0)$, normal cone $N_C(0,0) = \mathbb{R}^2$, and these three cones as defined in [9]. The gradients of (11.11) at $(s,\varphi) \in C$ are

$$\begin{aligned} \mathbf{w}_-^T &= \nabla g(s,\varphi) = (0,1) && \text{for } (s < 0, \varphi = 0) \\ \mathbf{w}_+^T &= \nabla g(s,\varphi) = (\tfrac{-1}{R},1) && \text{for } (s > 0, \varphi = \tfrac{s}{R}) \end{aligned} \tag{11.12}$$

with a discontinuity in the gradient mapping $(s, \varphi) \to \nabla g(s, \varphi)$ at $(s, \varphi) = (0, 0)$, which is the *source* for the velocity jump. Note that configurations with different tangent and contingent cone do usually *not* occur when the overall constraint is obtained by the *intersection* of smooth sets of constraints, even of inequality constraints. Here, the situation is different: The two bilateral constraints for the straight and the circular rail have to be regarded as two submanifolds with boundaries, *pasted together* along their boundaries such that a corner is formed. Such cases can not be treated within the framework of classical mechanics.

Let us now state the equations of the impact-free motion for our example. With the constraints (11.11) and their gradients (11.12), Lagrange's equations of first kind (11.8), (11.9) become

$$\begin{pmatrix} m & 0 \\ 0 & J \end{pmatrix} \begin{pmatrix} \dot{v} \\ \dot{\omega} \end{pmatrix} = \begin{pmatrix} 0 \\ 1 \end{pmatrix} \lambda, \quad g(s, \varphi) = \varphi = 0 \quad \text{for } s < 0 \tag{11.13}$$

for the straight rail and, equivalently, for the circular rail

$$\begin{pmatrix} m & 0 \\ 0 & J \end{pmatrix} \begin{pmatrix} \dot{v} \\ \dot{\omega} \end{pmatrix} = \begin{pmatrix} \frac{-1}{R} \\ 1 \end{pmatrix} \lambda, \quad g(s, \varphi) = \varphi - \tfrac{s}{R} = 0 \quad \text{for } s > 0. \tag{11.14}$$

Within this setting, the gradients $\mathbf{w}$ from (11.12) and the Lagrange multipliers λ have to be regarded as the directions and the scalar values of some generalized forces $\mathbf{f} = \mathbf{w}\lambda$, respectively, with both, $\mathbf{w}$ and λ, having a pure physical origin: The interpretation of the generalized force $(0, 1)^T\lambda$ in (11.13) is obvious. The multiplier λ corresponds to a constraint moment that acts only on the angular coordinate φ in order to prevent the car from turning ($\varphi = 0$). In contrast, the generalized constraint force $(-\frac{1}{R}, 1)^T\lambda$ in (11.14) influences both coordinates, acting in s-direction as a force and in φ-direction as a moment, which has also mechanical reasons.

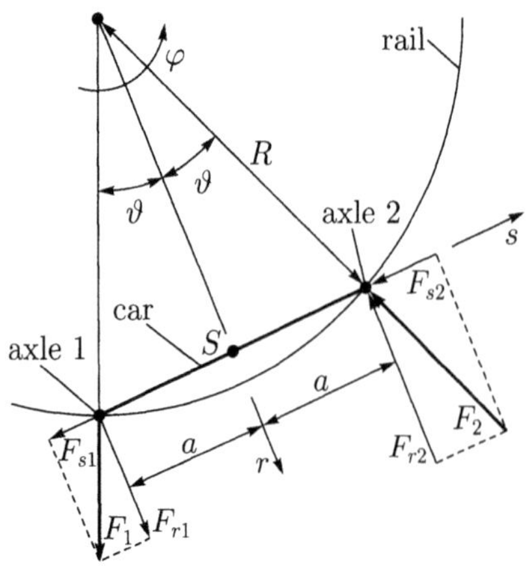

Figure 11.6. Constraint forces

In order to see this, we consider a car with two axles on a circular rail as depicted in Figure 11.6. The contacts between the two axles and the rail are modeled as two bilateral perfect constraints with contact forces F_1 and F_2 acting on the car in radial direction as a consequence. The resultant force F_s in s-direction and the resultant moment M_φ^S in φ-direction with respect to point S on the car become

$$\begin{aligned} F_s &= -F_{s1} - F_{s2} = -(F_1 + F_2)\sin\vartheta \\ M_\varphi^S &= aF_{r1} + aF_{r2} = (F_1 + F_2)R\sin\vartheta\cos\vartheta \end{aligned} \tag{11.15}$$

and constitute the generalized force

$$\mathbf{f} = \begin{pmatrix} F_s \\ M_\varphi^S \end{pmatrix} = FR\sin\vartheta \begin{pmatrix} \frac{-1}{R} \\ \cos\vartheta \end{pmatrix} \tag{11.16}$$

with $F := F_1 + F_2$. In order to obtain the model in Figure 11.4 with one axle only, we let the distance $2R\sin\vartheta$ of the axles tend to zero. By doing so, one must allow infinite values of the force F to realize some finite moment M_φ^S, due to the vanishing moment arm. Let therefore $\vartheta \to 0$ and $F \to \infty$ in (11.16) such that $FR\sin\vartheta =: \lambda$ remains finite. With this limit, (11.16) becomes

$$\mathbf{f} = \begin{pmatrix} \frac{-1}{R} \\ 1 \end{pmatrix} \lambda \tag{11.17}$$

which is the constraint force in equation (11.14). If, in addition, $R \to \infty$, we obtain again the constraint force in (11.13).

11.4 Velocity Jump by Time-Scaling

In this section we treat the velocity jump, that is the dynamics of the system when passing from the straight rail to the circle. The evaluation of the velocity jump is here done by a time-scaling method which does not use any energy considerations and which may be applied for systems with even more degrees of freedom. The advantage of this approach is that the impact laws to be used for the description of the velocity jump transform to ordinary force laws, which might be chosen, for example, in analogy to those from classical perfect constraints. We will show for this case that conservation of kinetic energy holds, and that the post-impact velocities are not affected when using different parameterizations for the impact.

For the parameterization of the impact we introduce an artificial time $\tau \in [0, 1]$, and we set $d\mathbf{u}_A = \mathbf{u}_A' \, d\tau$, $d\mathbf{R}_A = \mathbf{R}_A' \, d\tau$ in (11.2) with $d\tau$ being the Lebesgue measure on $[0, 1]$. Hence, $\mathbf{u}_A$ and $\mathbf{R}_A$ become functions of τ, whereas the displacements $\mathbf{q}^\star$ stay fixed. We denote $\mathbf{u}_A(\tau = 0) := \mathbf{u}_-$ and $\mathbf{u}_A(\tau = 1) := \mathbf{u}_+$. Within this time interval, we may now change the gradient of g (11.12) at $(0,0)$ continuously from $\mathbf{w}_- =: \mathbf{w}(\tau = 0)$ to $\mathbf{w}_+ =: \mathbf{w}(\tau = 1)$ as depicted in Figure 11.5 by setting

$$\mathbf{w}(\tau) := (1-\tau)\mathbf{w}_- + \tau\mathbf{w}_+, \quad \tau \in [0,1]. \tag{11.18}$$

With this $\mathbf{w}(\tau)$, the function

$$\gamma(\mathbf{u}_A(\tau),\tau) := \mathbf{w}^T(\tau)\,\mathbf{u}_A(\tau) = 0 \tag{11.19}$$

defines a smooth velocity constraint for $\tau \in [0,1]$, agreeing with the pre- and post-impact velocity constraints $R\omega_- = 0$ and $R\omega_+ - v_+ = 0$ from (11.5).

We have still to choose some "force law" that characterizes the impulsion density $\mathbf{R}'_A$, i.e. a physical condition, not yet introduced. This force law will complete the mathematical description of the problem and has to be regarded as the impact law. In order to obtain conservation of kinetic energy as a result, we choose, for every $\tau \in [0,1]$, the direction of $\mathbf{R}'_A$ as $\mathbf{w}$ in (11.18). This choice turns $\mathbf{R}'_A$ into a constraint force, fulfilling the classical orthogonality condition $\mathbf{R}'^T_A\mathbf{u}_A = 0$, cp. (11.19). With

$$\mathbf{R}'_A(\tau) := \mathbf{w}(\tau)\,\lambda(\tau), \tag{11.20}$$

$$\begin{aligned} \mathrm{d}\mathbf{u}_A &= \mathbf{u}'_A(\tau)\,\mathrm{d}\tau, \\ \mathrm{d}\mathbf{R}_A &= \mathbf{R}'_A(\tau)\,\mathrm{d}\tau \end{aligned} \tag{11.21}$$

substituted in (11.2) we obtain together with (11.19) the index-two-problem

$$\begin{aligned} \mathbf{M}\,\mathbf{u}'_A(\tau) &= \mathbf{w}(\tau)\,\lambda(\tau), \\ \gamma(\mathbf{u}_A(\tau),\tau) &= \mathbf{w}^T(\tau)\,\mathbf{u}_A(\tau) = 0 \end{aligned} \tag{11.22}$$

and, after differentiation of the velocity constraint $\gamma(\mathbf{u}_A(\tau),\tau)$ with respect to τ, the index-one-problem:

Problem 11.4.1. With the initial condition $\mathbf{u}_- = \mathbf{u}_A(\tau = 0)$ fulfilling the velocity constraint $\gamma(\mathbf{u}_A(0),0) = \mathbf{w}_-^T\mathbf{u}_- = 0$, find $\mathbf{u}_+ = \mathbf{u}_A(\tau = 1)$ such that

$$\begin{aligned} \mathbf{M}\mathbf{u}'_A &= \mathbf{w}\lambda, \\ \gamma' &= \mathbf{w}^T\mathbf{u}'_A + \mathbf{w}'^T\mathbf{u}_A = 0 \end{aligned} \tag{11.23}$$

with $\mathbf{w}(\tau)$ as defined in (11.18).

Equation (11.23) already describes the complete evolution of the impact and will later be evaluated for our example. Integration of (11.21) over $\{t^\star\}$ yields, together with (11.20),

$$\begin{aligned} \int_{\{t^\star\}} \mathbf{M}\,\mathrm{d}\mathbf{u}_A &= \int_0^1 \mathbf{M}\,\mathbf{u}'_A(\tau)\,\mathrm{d}\tau, \\ \int_{\{t^\star\}} \mathrm{d}\mathbf{R}_A &= \int_0^1 \mathbf{R}'_A(\tau)\,\mathrm{d}\tau = \int_0^1 \mathbf{w}(\tau)\,\lambda(\tau)\,\mathrm{d}\tau \end{aligned} \tag{11.24}$$

which gives the connection to (11.2) and enables us to calculate the impact impulsion $\mathbf{R}_A$ also from $\mathbf{R}_A = \mathbf{M}\,(\mathbf{u}_+ - \mathbf{u}_-)$.

Equations (11.22) may be formulated in a more compact form: For $\tau \in [0,1]$ we denote by $L(\tau)$ the hyperplane in $\mathbb{R}^f$ defining the admissible velocities, $L(\tau) = \{\mathbf{u}_A(\tau)\,|\,\mathbf{w}^T(\tau)\,\mathbf{u}_A(\tau) = 0\}$, and by $N(\tau) = \{\mathbf{w}(\tau)\,\lambda\}$ the straight line orthogonal to $L(\tau)$. The impulsion density $\mathbf{R}'_A = \mathbf{w}\lambda$ from (11.20) is then an element of $N(\tau)$, such that we may rewrite equations (11.22) as

$$-\mathbf{M}\,\mathbf{u}'_A(\tau) \in N(\tau), \quad \mathbf{u}_A(\tau) \in L(\tau), \quad \tau \in [0,1]. \tag{11.25}$$

Proposition 11.4.1. *The system (11.25) is non-dissipative: Kinetic energy T is preserved during the velocity jump.*

Proof. With $T(\tau) = \frac{1}{2}\mathbf{u}_A^T(\tau)\,\mathbf{M}\,\mathbf{u}_A(\tau)$ and $\mathbf{M} = \mathbf{M}^T$ positive definite we obtain $\mathrm{d}T(\tau) = \mathbf{u}_A^T(\tau)\,\mathbf{M}\,\mathrm{d}\mathbf{u}_A(\tau) = \mathbf{u}_A^T(\tau)\,\mathrm{d}\mathbf{R}_A(\tau)$. Therefore,

$$T_+ - T_- = \int_0^1 \mathrm{d}T(\tau) = \int_0^1 \mathbf{u}_A^T(\tau)\,\mathbf{R}'_A(\tau)\,\mathrm{d}\tau \equiv 0 \tag{11.26}$$

since $\mathbf{R}'^T_A\mathbf{u}_A = 0$ for every $\tau \in [0,1]$ by construction. □

After the values of $\mathbf{u}_+$ have been computed, the difference in the kinetic energy may also be expressed as

$$T_+ - T_- = \tfrac{1}{2}\,(\mathbf{u}_+ + \mathbf{u}_-)^T\,\mathbf{M}\,(\mathbf{u}_+ - \mathbf{u}_-), \tag{11.27}$$

which leads in the case of energy conservation to the well-known relation

$$(\mathbf{u}_+ + \mathbf{u}_-)^T\,\mathbf{R}_A = 0, \tag{11.28}$$

since $\mathbf{R}_A = \mathbf{M}\,(\mathbf{u}_+ - \mathbf{u}_-)$ by (11.2).

For the evaluation of the velocity jump via time scaling we have chosen some parameterization τ of the impact interval. One may ask about the results when another parameterization is chosen.

Proposition 11.4.2. *The values $\mathbf{R}_A$ and $(\mathbf{u}_+ - \mathbf{u}_-)$ obtained from (11.25) are invariant under re-parameterizations $\tau \to \sigma(\tau)$, as long as the associated real measure $\mathrm{d}\sigma$ is $\mathrm{d}\tau$-continuous.*

Proof. We set $\tilde{L}(\sigma(\tau)) := L(\tau)$, $\tilde{N}(\sigma(\tau)) := N(\tau)$ and $\tilde{\mathbf{u}}_A(\sigma(\tau)) := \mathbf{u}_A(\tau)$ with $\sigma(0) =: \sigma_0$ and $\sigma(1) =: \sigma_1$. The existence of a density $\sigma'(\tau)$ of $\mathrm{d}\sigma$ with respect to $\mathrm{d}\tau$, uniquely determined $\mathrm{d}\tau$-almost everywhere, is a priori secured by the Radon-Nikodym theorem, see e.g. [127]. The points at which $\sigma'(\tau)$ vanishes make a $\mathrm{d}\sigma$-negligible subset and do therefore not need to be considered when integration with respect to σ is performed. We therefore obtain

$$\mathbf{u}_+ - \mathbf{u}_- = \int_{\{t^*\}} \mathrm{d}\mathbf{u}_A = \int_{\sigma_0}^{\sigma_1} \tilde{\mathbf{u}}'_A(\sigma)\,\mathrm{d}\sigma, \tag{11.29}$$

where $\tilde{\mathbf{u}}'_A(\sigma)$ has to be determined from (11.25). With

$$\mathbf{u}'_A(\tau) = \tilde{\mathbf{u}}'_A(\sigma)\,\sigma'(\tau), \tag{11.30}$$

the first inclusion in (11.25) in terms of σ becomes

$$-\mathbf{M}\,\tilde{\mathbf{u}}'_A(\sigma) \in \tfrac{1}{\sigma'(\tau)}\,\tilde{N}(\sigma) \equiv \tilde{N}(\sigma) \tag{11.31}$$

and may be solved for $\tilde{\mathbf{u}}'_A(\sigma)$ as long as $\sigma'(\tau) \neq 0$, this means for almost every $\sigma \in [\sigma_0, \sigma_1]$. We also note that $\frac{1}{\sigma'(\tau)}\,\tilde{N}(\sigma) \equiv \tilde{N}(\sigma)$ since $\tilde{N}$ is a linear space and hence not changed by multiplication with scalars. The new parameterization therefore leads to

$$-\mathbf{M}\,\tilde{\mathbf{u}}'_A(\sigma) \in \tilde{N}(\sigma), \quad \tilde{\mathbf{u}}_A(\sigma) \in \tilde{L}(\sigma), \quad \sigma \in [\sigma_0, \sigma_1] \tag{11.32}$$

which, together with (11.29), completes the proof. □

Let us now return to our example. In order to obtain the post-impact velocities via time-scaling we proceed with equation (11.23) which becomes

$$\begin{pmatrix} m & 0 \\ 0 & J \end{pmatrix} \begin{pmatrix} v' \\ \omega' \end{pmatrix} = \begin{pmatrix} -\frac{\tau}{R} \\ 1 \end{pmatrix} \lambda, \quad \gamma' = -\tfrac{\tau}{R} v' + \omega' - \tfrac{1}{R} v = 0 \tag{11.33}$$

with initial conditions $v(\tau = 0) = v_- > 0$ and $\omega(\tau = 0) = \omega_- = 0$. Elimination of λ and ω' from (11.33) yields the differential equation

$$(mR^2 + J\tau^2)\,v' = -J\tau v, \quad v(\tau = 0) = v_- \tag{11.34}$$

which may be integrated analytically. After separation of the variables we obtain

$$\begin{aligned} \int_{v_-}^{v_+} \frac{\mathrm{d}v}{v} &= -\int_0^1 \frac{J\tau}{mR^2 + J\tau^2}\,\mathrm{d}\tau = -\frac{1}{2}\ln\left(mR^2 + J\tau^2\right)\Big|_0^1 \\ &= \ln\left(1 + \frac{J}{mR^2}\right)^{-\frac{1}{2}} = \ln\frac{1}{\mu} \end{aligned} \tag{11.35}$$

with μ as defined in (11.7). We therefore have $(v, \omega)(\tau = 1) = (v_+, \omega_+)$ with the same values as in (11.7). Finally, the impact impulsion $\mathbf{R}_A = \mathbf{M}\,(\mathbf{u}_A(\tau = 1) - \mathbf{u}_A(\tau = 0))$ reads

$$\mathbf{R}_A = (1 - \tfrac{1}{\mu})\,m v_- \begin{pmatrix} -1 \\ R(1 + \mu) \end{pmatrix} \tag{11.36}$$

lying in this case in the linear space spanned by $(\mathbf{w}_-, \mathbf{w}_+)$, cp. Figure 11.5.

11.5 Velocity Jump by Reflection

An alternative method to handle the velocity jump is provided by reflections, as long as the impact is assumed to be non-dissipative. Since this pre-supposition is in agreement with the physical behavior that we want to implement, we can put our problem into this framework. As it will turn out later, the main advantage of this method is that the post-impact velocities may be calculated by just evaluating algebraic equations, in contrast to the time scaling method of Section 11.4 where differential equations had to be solved. In addition, much more insight into the underlying geometrical meaning of impacts is gained. The treatment of the impact that we are going to present is based on the following observation:

Proposition 11.5.1. *Every velocity jump* $\mathbf{u}_+ \neq \mathbf{u}_-$ *in a scleronomic system at which kinetic energy is preserved may be interpreted as a* reflection *with* $\mathbf{u}_+$ *being the mirror image of* $\mathbf{u}_-$ *with respect to a hyperplane with normal in the direction of* $\mathbf{M}^{-1}\mathbf{R}_A$ *(see e.g. [91]).*

Proof. We will use the following notation: We denote by E the Euclidean space $\mathbb{R}^f$ with inner product $(\mathbf{y},\mathbf{x})_E := \mathbf{y}^T\mathbf{M}\mathbf{x}$ and by $E^\star$ the dual of E with inner product $(\mathbf{g},\mathbf{f})_{E^\star} = \mathbf{g}^T\mathbf{M}^{-1}\mathbf{f}$, induced by the duality mapping $M : E \to E^\star$, $\mathbf{v} \to \mathbf{M}\mathbf{v} = (\mathbf{v},\cdot)_E$. Let now $\mathbf{u}_+$ and $\mathbf{u}_-$ be given as elements of E. We have

$$\mathbf{M}(\mathbf{u}_+ - \mathbf{u}_-) = \mathbf{R}_A \quad \text{and} \quad \tfrac{1}{2}\|\mathbf{u}_+\|_E^2 = \tfrac{1}{2}\|\mathbf{u}_-\|_E^2, \tag{11.37}$$

where the first equation, resulting from (11.2), defines the impact impulsion $\mathbf{R}_A$ as a member of $E^\star$, and the second equation reflects conservation of kinetic energy, where $\|\mathbf{x}\|_E := \sqrt{(\mathbf{x},\mathbf{x})_E}$. With respect to this norm, the velocities $\mathbf{u}_+$ and $\mathbf{u}_-$ have the same length, which means for their *parallelogram decomposition*

$$\mathbf{v} := \tfrac{1}{2}(\mathbf{u}_- + \mathbf{u}_+), \quad \mathbf{v}^\perp := \tfrac{1}{2}(\mathbf{u}_- - \mathbf{u}_+) \tag{11.38}$$

that $\mathbf{v} \perp \mathbf{v}^\perp$, since $(\mathbf{v},\mathbf{v}^\perp)_E = 0$ as it is obvious from (11.37), (11.38). Furthermore, note that $\mathbf{v}^\perp \neq 0$, since $\mathbf{u}_+ \neq \mathbf{u}_-$ by assumption. Solving (11.38) for $\mathbf{u}_+$ and $\mathbf{u}_-$,

$$\mathbf{u}_- = \mathbf{v} + \mathbf{v}^\perp, \quad \mathbf{u}_+ = \mathbf{v} - \mathbf{v}^\perp, \tag{11.39}$$

and substituting the result in (11.37) yields

$$2\mathbf{M}\mathbf{v}^\perp = -\mathbf{R}_A \quad \text{and} \quad (\mathbf{v},\mathbf{v}^\perp)_E = 0, \quad \mathbf{v}^\perp \neq 0. \tag{11.40}$$

In order to interpret (11.37) as a reflection, we just have to take the direction of $\mathbf{v}^\perp$, i.e. the component of $\mathbf{u}$ in (11.39) which is inverted, as the normal defining a hyperplane in E which then contains $\mathbf{v}$, i.e. the component of

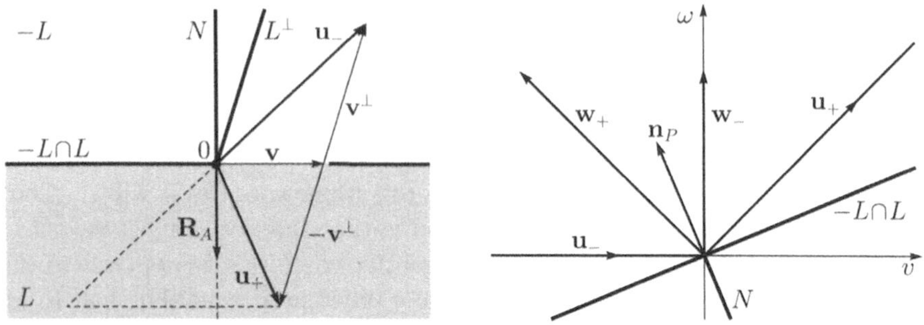

Figure 11.7. General reflection and reflection for the "Wild Mouse"

u which is not changed. We will do this step a bit more specific by using cones instead of linear spaces to gain additional information about directions. We thus define in $E^\star$ the ray $N := \{-\mathbf{R}_A \lambda \,|\, \lambda \geq 0\}$ generated by $-\mathbf{R}_A$, and by $L := \{\mathbf{y} \,|\, \mathbf{f}^T\mathbf{y} \leq 0, \quad \forall \mathbf{f} \in N\}$ the polar of N, in this case being a closed half-space in E. Let further $L^\perp \subset E$ be the dual of N which is also a ray, and $N^\perp \subset E^\star$ be the dual of L which is another closed half-space as depicted in the left part of Figure 11.7. In terms of L, we may then express $L^\perp$ as $L^\perp = \{\mathbf{x} \,|\, (\mathbf{y}, \mathbf{x})_E \leq 0, \quad \forall \mathbf{y} \in L\}$, and analogously $N^\perp$ as $N^\perp = \{\mathbf{g} \,|\, (\mathbf{g}, \mathbf{f})_{E^\star} \leq 0, \quad \forall \mathbf{f} \in N\}$. Since $0 \neq -\mathbf{R}_A \in N$ by construction, we obtain from the first equation in (11.40) that $0 \neq \mathbf{v}^\perp \in L^\perp$. On the other hand, we have by the second equation in (11.40) that $\mathbf{v} \perp \mathbf{v}^\perp$. Therefore, $\mathbf{v}$ is an element of the boundary of L and hence contained in L. Summarizing these results we may write

$$\mathbf{v} \in L, \quad \mathbf{v}^\perp \in L^\perp, \quad (\mathbf{v}, \mathbf{v}^\perp)_E = 0, \tag{11.41}$$

which constitutes, in the sense of proximations, the unique decomposition of $\mathbf{u}_- = \mathbf{v} + \mathbf{v}^\perp$ with respect to the cones L and $L^\perp$, see e.g. [121]. Equation (11.39) shows that $\mathbf{u}_+$ is the *mirror image* of $\mathbf{u}_-$ with respect to the hyperplane $(-L \cap L)$, because the component $\mathbf{v} \in (-L \cap L)$ of $\mathbf{u}_-$ remains unchanged for $\mathbf{u}_+$, whereas the component $\mathbf{v}^\perp$ orthogonal to $(-L \cap L)$ occurs with different signs. Finally, one observes that

$$\mathbf{u}_- \in -L, \quad \mathbf{u}_+ \in L, \tag{11.42}$$

since $\mathbf{u}_- = \mathbf{v} + \mathbf{v}^\perp$ with $\mathbf{v} \in (-L \cap L) \subset -L$ and $0 \neq \mathbf{v}^\perp \in L^\perp \subset -L$ for the first inclusion, and $\mathbf{u}_+ = \mathbf{v} - \mathbf{v}^\perp$ with $\mathbf{v} \in (-L \cap L) \subset L$ and $0 \neq -\mathbf{v}^\perp \in -L^\perp \subset L$ for the second inclusion. □

With the help of the reflections one may now avoid the time-scaling method of equation (11.23) but directly compute the post-impact velocities for the mouse. Hereby, the choice of the impact law corresponds to the *orientation* of the reflecting hyperplane with normal $\mathbf{n}_P \in E^\star$ and is done by the two conditions

$$\begin{aligned} &\text{(i)} \quad \cos\angle(\mathbf{n}_-, \mathbf{n}_P) = \cos\angle(\mathbf{n}_+, \mathbf{n}_P) \\ &\text{(ii)} \quad \mathbf{n}_P \in \text{span}\,\{\mathbf{n}_-, \mathbf{n}_+\}. \end{aligned} \tag{11.43}$$

Here, $\mathbf{n}_-$ and $\mathbf{n}_+$ are the unit vectors associated with $\mathbf{w}_-$ and $\mathbf{w}_+$, respectively. The first condition is necessary to ensure kinematic compatibility of the post-impact velocities with the constraint, whereas the second condition ensures that the impact impulsion $\mathbf{R}_A$ acts only between the coordinates for which the constraint is defined. Thus, other coordinates are not directly affected by $\mathbf{R}_A$, but they may be indirectly influenced due to the couplings in the mass matrix.

In order to construct now the reflection we have to determine the normal $\mathbf{n}_P$ of the reflecting hyperplane as the *angle bisector* of $\mathbf{w}_-$ and $\mathbf{w}_+$, defining the direction of N. We set

$$\mathbf{n}_- := \frac{\mathbf{w}_-}{||\mathbf{w}_-||_{E^\star}}, \quad \mathbf{n}_+ := \frac{\mathbf{w}_+}{||\mathbf{w}_+||_{E^\star}}, \tag{11.44}$$

and we look for a unit vector $\mathbf{n}_P$ such that (11.43) holds. One solution of this problem is

$$\mathbf{n}_P = \frac{\mathbf{n}_- + \mathbf{n}_+}{||\mathbf{n}_- + \mathbf{n}_+||_{E^\star}}, \tag{11.45}$$

where the second condition in (11.43) is obviously satisfied, and the first condition holds since $(\mathbf{n}_-, \mathbf{n}_- + \mathbf{n}_+)_{E^\star} = (\mathbf{n}_+, \mathbf{n}_- + \mathbf{n}_+)_{E^\star} = (\mathbf{n}_-, \mathbf{n}_+)_{E^\star} + 1$. Note that this $\mathbf{n}_P$ defines already the direction of $\mathbf{R}_A$, i.e. $\mathbf{R}_A = ||\mathbf{R}_A||_{E^\star}\, \mathbf{n}_P\, \alpha$ for some $\alpha \in \{+1, -1\}$.

Once the direction $\mathbf{n}_P$ of the impact impulsion $\mathbf{R}_A$ has been determined, the post-impact velocity $\mathbf{u}_+$ may be directly computed via $\mathbf{v}^\perp$. From the right picture in Figure 11.7 we obtain $\mathbf{v}^\perp$ by orthogonal projection of $\mathbf{u}_-$ on $L^\perp$

$$\mathbf{v}^\perp = (\mathbf{u}_-, \mathbf{M}^{-1}\mathbf{n}_P)_E\, \mathbf{M}^{-1}\mathbf{n}_P = \mathbf{M}^{-1}\mathbf{n}_P\, \mathbf{n}_P^T\, \mathbf{u}_- \tag{11.46}$$

which is then used to determine $\mathbf{u}_+$ in (11.38),

$$\mathbf{u}_+ = \mathbf{u}_- - 2\mathbf{v}^\perp = (\mathbf{E} - 2\mathbf{M}^{-1}\mathbf{n}_P\, \mathbf{n}_P^T)\, \mathbf{u}_-. \tag{11.47}$$

Finally, the impact impulsion $\mathbf{R}_A$ becomes by (11.40)

$$\mathbf{R}_A = -2\mathbf{M}\, \mathbf{v}^\perp = -2\mathbf{n}_P\, \mathbf{n}_P^T\, \mathbf{u}_-. \tag{11.48}$$

Note again that conservation of kinetic energy holds since we have constructed $\mathbf{u}_+$ to be, relatively to some hyperplane, the mirror image of $\mathbf{u}_-$ with respect to the kinetic metric. Conservation of kinetic energy may be equivalently expressed by either the norm condition $||\mathbf{u}_+||_E = ||\mathbf{u}_-||_E$ or the orthogonality $(\mathbf{u}_+ + \mathbf{u}_-, \mathbf{u}_+ - \mathbf{u}_-)_E = 0$, see (11.37), (11.40).

For our example, we determine first the unit vectors $\mathbf{n}_-$ and $\mathbf{n}_+$ in (11.44). With $\mathbf{w}_-$, $\mathbf{w}_+$ from (11.12) and μ as defined in (11.7) they read

$$\mathbf{n}_- = \sqrt{J}\begin{pmatrix}0\\1\end{pmatrix}, \quad \mathbf{n}_+ = \frac{\sqrt{J}}{\mu}\begin{pmatrix}-\frac{1}{R}\\1\end{pmatrix}. \tag{11.49}$$

The angle bisector $\mathbf{n}_P$ is then obtained from (11.45),

$$\mathbf{n}_P = \frac{\sqrt{J}}{R}\frac{1}{\sqrt{2\mu(\mu+1)}}\begin{pmatrix}-1\\R(1+\mu)\end{pmatrix}, \tag{11.50}$$

and determines the direction of $\mathbf{R}_A$, cp. (11.36). With the help of the velocity component $\mathbf{v}^\perp$ from (11.46),

$$\mathbf{v}^\perp = \frac{v_-}{2}\begin{pmatrix}1-\frac{1}{\mu}\\\frac{-1}{R\mu}\end{pmatrix}, \tag{11.51}$$

one may finally compute the post-impact velocities as $\mathbf{u}_+ = \mathbf{u}_- - 2\mathbf{v}^\perp$ and the impact impulsion as $\mathbf{R}_A = -2\mathbf{M}\mathbf{v}^\perp$, see (11.47) and (11.48), respectively. Again, their values are in accordance with (11.7) and (11.36).

11.6 Reflections and Collisions - Remarks

In rigid body dynamics one usually associates a velocity jump with some kind of collision, where some relative velocity is inverted for the non-dissipative case. This does apparently not apply for our example. There is, however, a connection between reflections and elastic collisions which we are going to elaborate now.

By the parallelogram decomposition of the last section we have obtained the velocities in the form

$$\begin{aligned}&\mathbf{u}_- = \mathbf{v} + \mathbf{v}^\perp, \quad \mathbf{u}_+ = \mathbf{v} - \mathbf{v}^\perp\\ &\text{with } \mathbf{v} \in -L \cap L \text{ and } 0 \neq \mathbf{v}^\perp \in L^\perp,\end{aligned} \tag{11.52}$$

where L is a half-space, $-L \cap L$ the reflecting hyperplane, and $L^\perp$ the ray normal to L, all as subsets of E. We choose a basis $\mathbf{e}_i$, $i = 1, \ldots, f$ in E such that $\mathbf{e}_i$, $i = 1, \ldots, f-1$ spans $-L \cap L$ and $\mathbf{e}_f \in L^\perp$. Let $\mathbf{b}^j$ denote the associated dual basis, i.e. the basis in $E^\star$ with $\mathbf{b}^{jT}\mathbf{e}_i = \delta_i^j$, where $\mathbf{b}^j$, $j = 1, \ldots, f-1$ spans $-N^\perp \cap N^\perp$ and $\mathbf{b}^f \in N$ due to the orthogonality of $-L \cap L$ and $L^\perp$. Writing down the velocities (11.52) in this basis,

$$\mathbf{u}_- = \mathbf{e}_i\, u^i + \mathbf{e}_f\, u^f, \quad \mathbf{u}_+ = \mathbf{e}_i\, u^i - \mathbf{e}_f\, u^f, \tag{11.53}$$

we may identify $\mathbf{v}$ and $\mathbf{v}^\perp$ as

$$\mathbf{v} = \mathbf{e}_i\, u^i, \quad \mathbf{v}^\perp = \mathbf{e}_f\, u^f \tag{11.54}$$

with i running from 1 to $f-1$, and $u^f > 0$ since $L^\perp$ is a ray. We may now solve (11.53) for u^f by multiplying with $\mathbf{b}^f$,

$$\mathbf{b}^{fT}\mathbf{u}_- = u^f, \quad \mathbf{b}^{fT}\mathbf{u}_+ = -u^f \quad (u^f > 0), \tag{11.55}$$

and we may regard these expressions as a linear function of $\mathbf{u}$,

$$\gamma_P(\mathbf{u}) := -\mathbf{b}^{fT}\mathbf{u}, \quad \mathbf{b}^f \in N, \tag{11.56}$$

with values being inverted by the impact: We have $\gamma_{P+} = -\gamma_{P-}$ with $\gamma_{P+} := \gamma_P(\mathbf{u}_+) = u^f > 0$ and $\gamma_{P-} := \gamma_P(\mathbf{u}_-) = -u^f < 0$. In rigid body dynamics the inversion of such an expression is usually associated with an *elastic impact*, which gives now the connection to the collisions:

As soon as equation (11.56) has been stated, one might ask for a configuration of the rigid body system for which this expression can be *interpreted* as the relative velocity between certain colliding points. If this succeeds, one has immediately obtained the corresponding collision. This question might be called the *inverse problem of elastic collisions*: To determine the points at which the rigid body system collides from the only knowledge of the pre- and post-impact velocities of an energy-preserving velocity jump.

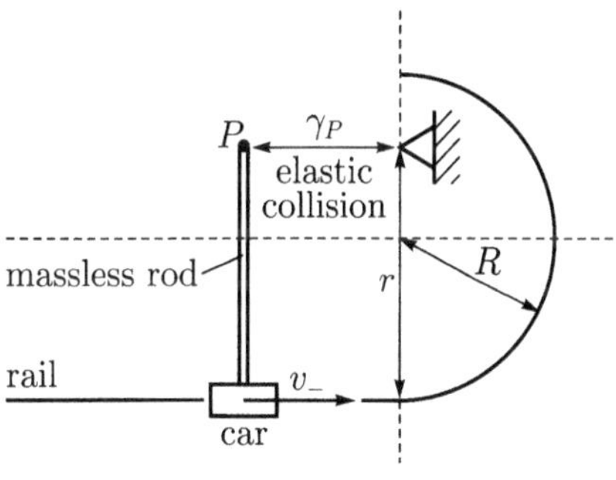

Figure 11.8. Equivalent elastic collision for the "Wild Mouse"

Let us now determine the *collision point* for the "Wild Mouse". Since $\mathbf{n}_P$ in (11.50) has been constructed to lie in $-N$, see Figure 11.7, we can choose $\mathbf{b}^f$ as

$$N \ni \mathbf{b}^f = \begin{pmatrix} 1 \\ -R(1+\mu) \end{pmatrix} \tag{11.57}$$

With $\mathbf{u} = (v, \omega)^T$ the relative velocity (11.56) becomes

$$\gamma_P(v,\omega) = -v + r\omega \quad \text{with } r = R(1+\mu) > R \tag{11.58}$$

and defines a collision point P as depicted in Figure 11.8: The rigid car is extended by a massless rod that impacts against an obstacle placed at the distance r from the straight rail. The velocities before the impact in (11.5) have been ($v_- > 0, \omega_- = 0$) leading to $\gamma_{P-} = -v_- < 0$. The post-impact velocities (11.7) are ($v_+ = \frac{1}{\mu}v_-, \omega_+ = \frac{1}{R\mu}v_-$), hence $\gamma_{P+} = \frac{v_-}{\mu}(-1 + \frac{r}{R}) = +v_- > 0$. Physically, the impact impulsion acting on point P takes the car in a state of rotation ω_+ by a *simultaneous* reduction of its velocity to v_+. By a proper choice of r as derived in equation (11.58), these two magnitudes may be adjusted such that they obey the post-impact velocity constraint $v_+ = R\omega_+$. Note again that problems of this kind can *not* be solved by the common impact theory, since the points of collision and hence the direction of the impact impulsion are a priori *unknown*. Note also that a *completely inelastic collision* of the car's rod against an obstacle placed at the *midpoint* of the circular rail would yield compatible post-impact velocities, but would *not* conserve kinetic energy.

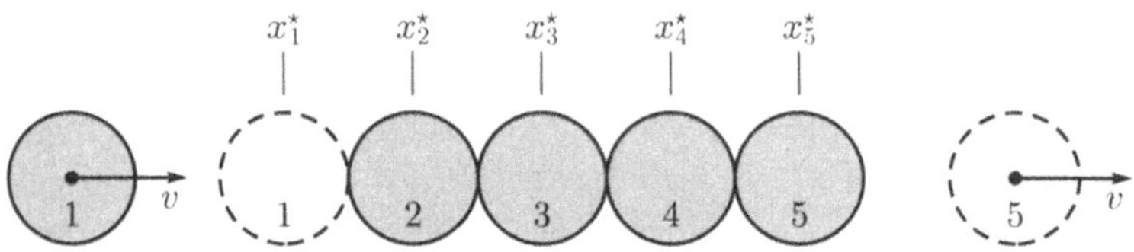

Figure 11.9. Impact at Newton's cradle

It is, however, *not* always possible to detect the collision points in a multibody system as points which come *physically* in contact at the impact. We demonstrate this by Newton's cradle which is depicted in Figure 11.9: Denote by $\mathbf{q}^\star := (x_1^\star, \dots, x_5^\star)^T$ the displacements of the five identical balls at the time of impact and by $\mathbf{u} := (u_1, \dots, u_5)^T$ their velocities. Suppose that ball 1 is approaching from $x_1 < x_1^\star$ with a constant positive velocity $u_{1-} := v > 0$, and that all the other balls are not moving, $u_{2-} = \dots = u_{5-} = 0$. We know by experiment that, with good approximation, the first ball stops after the impact, and that only the last ball leaves with a velocity $u_{5+} = v$. The pre- and post-impact velocities thus are $\mathbf{u}_- = (v\ 0\ 0\ 0\ 0)^T$ and $\mathbf{u}_+ = (0\ 0\ 0\ 0\ v)^T$, respectively, from which we calculate the impact impulsion $\mathbf{R}_A = \mathbf{M}(\mathbf{u}_+ - \mathbf{u}_-) = m(-v\ 0\ 0\ 0\ v)^T \in -N$. The relative velocity γ_P from (11.56), which is inverted by the impact, becomes $\gamma_P(\mathbf{u}) = -u_1 + u_5$, and is thus *measured between ball 1 and ball 5*, although these two balls may *never* contact each other directly.

Direct consequences on these observations are: The points that we recognize in the three-dimensional space as the contacts between two succeeding bodies, may *not* define the relative velocity which is inverted by an energy-preserving impact. If, on the other hand, one regards the impact laws as constitutive laws being expressed, for example, by a coefficient of restitution equal to one for the case of reflections, and one wants to evaluate the velocity jump without any supplementary knowledge on the chronological course of the impact events, then one *must* take into account constitutive laws between *any* of the contacting bodies. This means, in last consequence, that impacts have *directly to be modeled in the configuration space.* A theoretical framework of this approach can be found in [35], where constitutive laws are rigorously introduced to describe impacts.

12. Electropneumatic Drilling Machine

The design of a professional drilling machine has to meet two guidelines. First, the removal rate in hard material like concrete should be optimal; second, the vibration loads at the handle should be kept below the acceleration loads as defined by the usual work regulations. Such considerations are the starting point for analyzing the vibrations of drilling machines more carefully while providing some means to study parameter influences in a faster and broader way [132].

The operating principle of an electropneumatic drilling machine can be seen from Figure 12.1. The drilling machine consists of five bodies: gearbox (1) including handle, motor and driving piston; primary piston (2); secondary piston (3); tool (4); and pressure cylinder (5), including the tool guide. The motion of each body is described by one translational degree of freedom. The constant rotation of the motor is transformed via a crank mechanism to a given sinusoidal translation of the driving piston that excites the primary piston through compression and decompression of the enclosed air cushion (6). Thus, the primary piston is supplied by the cycle of crank rotation with kinetic energy, which is then partially transferred through a sequence of impacts to the secondary piston and the tool. The motion of the secondary

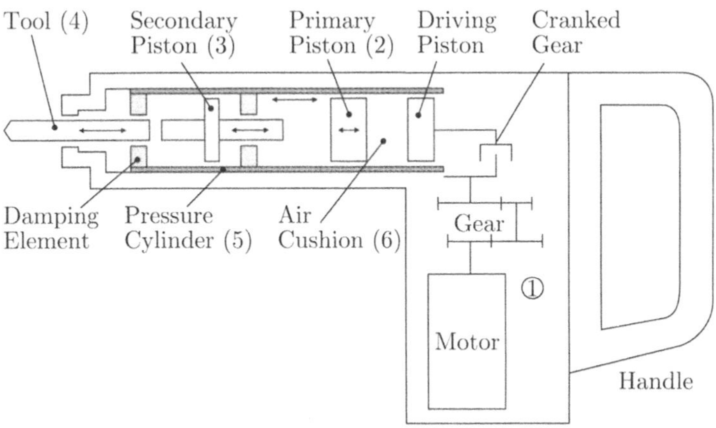

Figure 12.1. Components of an electropneumatic drilling machine

piston is restrained in one direction by a backstop. For assembly reasons, there is also some backlash in the coupling connecting bodies (5) and (1). In all axial guides dry friction resulting from sealing rings is to be considered. In addition to the hysteresis characteristic for the tool-environment model, we have three single-valued force elements with vertices describing elastic contacts, two unilateral constraints, and five frictional guides that are modeled by dry friction characteristics.

12.1 Mechanical Model

The first step for modeling machines or mechanisms as multibody systems consists of the definition of their degrees of freedom and existing force characteristics. To this extent we classify the system to consist of perfect rigid joints, rigid joints with backlash, rigid joints with friction, and force couplings with given force-time, force-distance or force-velocity characteristics which may be set-valued. Generally, the representation (5.19) of a mechanical system is developed by starting with a set of generalized coordinates $\mathbf{q} \in \mathbb{R}^f$ that describe uniquely the kinematics when the system is only subjected to the subset of perfect bilateral holonomic constraints. In this stage, the mechanical model consists often of several individual subsystems.

The forces acting between the bodies need now to be specified. The single-valued and set-valued force characteristics related to the contact problems occurring in the drilling machine may be organized into four classes, see Figure 12.2:

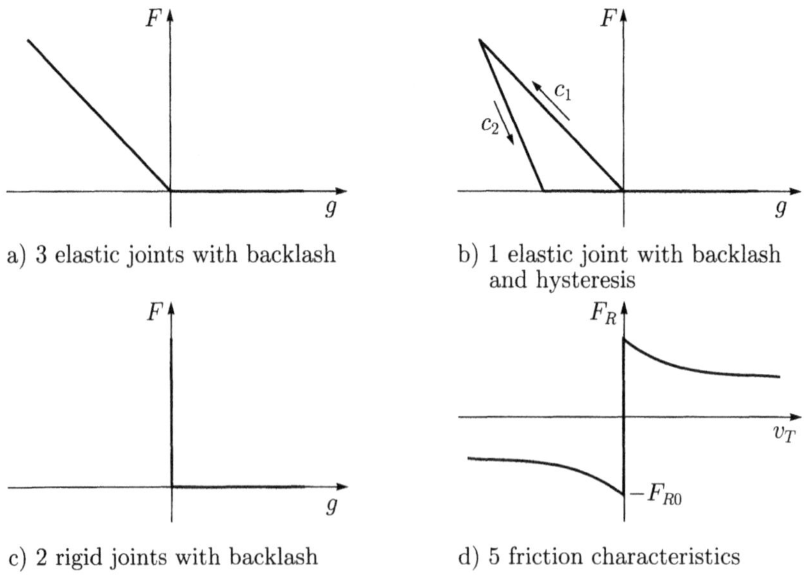

Figure 12.2. Force characteristics related to contact problems

1. Elastic joints with backlash (Figure 12.2a): For elastic joints with backlash the transferred forces are determined by the local deformations $g(\mathbf{q})$. As long as there is no contact ($g > 0$) the transferred force must vanish, whereas there is a linear or nonlinear dependency on the deformation ($g < 0$) for two gearing bodies.
2. Elastic behavior with hysteresis (Figure 12.2b): This is a modified nonlinear, non-smooth force-displacement characteristic where the force also depends on the phases of compression and decompression, respectively.
3. Rigid joints with backlash (Figure 12.2c): If contact deformations can be neglected, there are unilaterally acting constraints ($g(\mathbf{q}) \geq 0$). The joint forces for contacting bodies can be calculated by formulating the unilateral constraint on the acceleration level which yields algebraic inequality constraints for the solution of the equations of motion. Thus, the mechanical model gets structure variant as the number of constraints

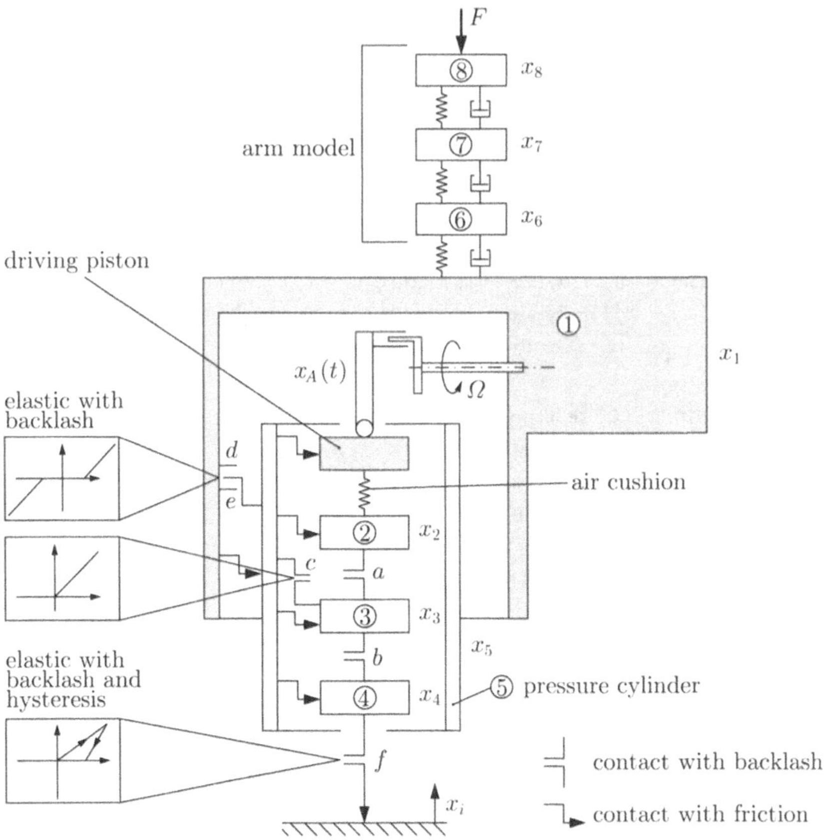

Figure 12.3. Mechanical model of the drilling machine

and, equivalently, the number of degrees of freedom may change with time.

4. Frictional guides (Figure 12.2d): Due to the linear elastic behavior of the pre-stressed sealing rings in the radial direction, the friction in the guides does not depend on radial displacements. Dry friction characteristics, i.e. Coulomb friction with given normal force, can therefore be used for modeling the tangential contact behavior.

For the mechanical model under consideration the equations presented in the preceding chapters apply, particularly equation (5.19). The relevant vectors and matrices are derived from Figure 12.3 in a straightforward way:

Generalized coordinates:

$$\mathbf{q} = (x_1\ x_2\ x_3\ x_4\ x_5\ x_6\ x_7\ x_8)^T \tag{12.1}$$

Mass matrix:

$$\mathbf{M} = \operatorname{diag}\{m_i\}, \quad i = 1, \ldots, 8 \tag{12.2}$$

Smooth part $\mathbf{h}_1$ of external forces $\mathbf{h} = \mathbf{h}_1 + \mathbf{h}_2$:

$$\mathbf{h}_1 = \begin{pmatrix} \begin{pmatrix} c_1(x_6 - x_1) + d_1(\dot{x}_6 - \dot{x}_1) - d_5(\dot{x}_1 + \dot{x}_A - \dot{x}_5) \\ -m_A\ddot{x}_A + F_A(x_A, x_2) \end{pmatrix} \\ -d_5(\dot{x}_2 - \dot{x}_5) - F_A(x_A, x_2) \\ 0 \\ 0 \\ d_5(\dot{x}_1 + \dot{x}_A - \dot{x}_5) + d_5(\dot{x}_2 - \dot{x}_5) \\ c_2(x_7 - x_6) + d_2(\dot{x}_7 - \dot{x}_6) - c_1(x_6 - x_1) - d_1(\dot{x}_6 - \dot{x}_1) \\ c_3(x_8 - x_7) + d_3(\dot{x}_8 - \dot{x}_7) - c_2(x_7 - x_6) - d_2(\dot{x}_7 - \dot{x}_6) \\ -F - c_3(x_8 - x_7) - d_3(\dot{x}_8 - \dot{x}_7) \end{pmatrix} \tag{12.3}$$

c_i	Stiffness coefficients
d_i	Damping coefficients
F	Pressing force
$x_A(t)$	Kinematic excitation from driving unit
m_A	Condensed mass of driving unit
F_A	Force from adiabatic air cushion

Non-smooth part $\mathbf{h}_2$ of external forces $\mathbf{h} = \mathbf{h}_1 + \mathbf{h}_2$:

$$\mathbf{h}_2 = \begin{pmatrix} 0 & 1 & -1 & 0 \\ 0 & 0 & 0 & 0 \\ -1 & 0 & 0 & 0 \\ 0 & 0 & 0 & 1 \\ 1 & -1 & 1 & 0 \\ 0 & 0 & 0 & 0 \\ 0 & 0 & 0 & 0 \\ 0 & 0 & 0 & 0 \end{pmatrix} \begin{pmatrix} F_c \\ F_d \\ F_e \\ F_f \end{pmatrix}, \quad \begin{pmatrix} g_c \\ g_d \\ g_e \\ g_f \end{pmatrix} = \begin{pmatrix} x_5 - x_3 \\ x_1 - x_5 \\ x_5 - x_1 + \varepsilon \\ x_4 \end{pmatrix} \tag{12.4}$$

$$\text{with} \quad F_i(g_i, \dot{g}_i) = \begin{cases} -c_i g_i - d_i \dot{g}_i & \text{if } g_i < 0,\ c_i g_i + d_i \dot{g}_i < 0 \\ 0 & \text{else} \qquad (\text{for } i = c, d, e) \end{cases}$$

$$\text{and} \quad F_f(g_f, \dot{g}_f) = \begin{cases} -c_{f1} g_f - d_{f1} \dot{g}_f & \text{if } g_f < 0,\ \dot{g}_f < 0 \\ -c_{f2}(g_f + a) - d_{f2} \dot{g}_f & \text{if } g_f + a < 0,\ \dot{g}_f > 0, \\ & \quad c_{f2}(g_f + a) + d_{f2} \dot{g}_f < 0 \\ 0 & \text{else} \end{cases}$$

c_i, d_i	Stiffness coefficients and damping coefficients
Subscript c	Damping element between pressure cylinder (5) and secondary piston (3)
Subscript d, e	Force element with backlash ε between casing (1) and pressure cylinder (5)
Subscript f	Hysteresis characteristic between tool (4) and environment

Unilateral frictionless contact-impact constraints:

$$\begin{aligned} & g_{Na} = x_2 - x_3 \geq 0, \quad \tilde{w}_{Na} = \bar{w}_{Na} = 0 \\ & g_{Nb} = x_3 - x_4 \geq 0, \quad \tilde{w}_{Nb} = \bar{w}_{Nb} = 0 \\ & \mathbf{w}_{Na} = (0\ 1\ {-1}\ 0\ 0\ 0\ 0\ 0)^T \\ & \mathbf{w}_{Nb} = (0\ 0\ 1\ {-1}\ 0\ 0\ 0\ 0)^T \end{aligned} \tag{12.5}$$

Subscript a	Contact constraint between primary piston (2) and secondary piston (3)
Subscript b	Contact constraint between secondary piston (3) and tool (4)

Dry friction elements with given normal forces:

$$\begin{aligned} & \dot{g}_{T1} = \dot{x}_4 - \dot{x}_5, && \tilde{w}_{T1} = \bar{w}_{T1} = 0 \\ & \dot{g}_{T2} = \dot{x}_3 - \dot{x}_5, && \tilde{w}_{T2} = \bar{w}_{T2} = 0 \\ & \dot{g}_{T3} = \dot{x}_2 - \dot{x}_5, && \tilde{w}_{T3} = \bar{w}_{T3} = 0 \\ & \dot{g}_{T4} = \dot{x}_1 + \dot{x}_A(t) - \dot{x}_5, && \tilde{w}_{T4} = \dot{x}_A(t),\ \bar{w}_{T4} = \ddot{x}_A(t) \\ & \dot{g}_{T5} = \dot{x}_1 - \dot{x}_5, && \tilde{w}_{T5} = \bar{w}_{T5} = 0 \\ & \mathbf{w}_{T1} = (0\ \ 0\ \ 0\ \ 1\ \ {-1}\ \ 0\ \ 0\ \ 0)^T \\ & \mathbf{w}_{T2} = (0\ \ 0\ \ 1\ \ 0\ \ {-1}\ \ 0\ \ 0\ \ 0)^T \\ & \mathbf{w}_{T3} = (0\ \ 1\ \ 0\ \ 0\ \ {-1}\ \ 0\ \ 0\ \ 0)^T \\ & \mathbf{w}_{T4} = (1\ \ 0\ \ 0\ \ 0\ \ {-1}\ \ 0\ \ 0\ \ 0)^T \\ & \mathbf{w}_{T5} = (1\ \ 0\ \ 0\ \ 0\ \ {-1}\ \ 0\ \ 0\ \ 0)^T \end{aligned} \tag{12.6}$$

Subscript 1	Dry friction between tool (4) and pressure cylinder (5)
Subscript 2	Dry friction between secondary piston (3) and pressure cylinder (5)
Subscript 3	Dry friction between primary piston (2) and pressure cylinder (5)
Subscript 4	Dry friction between driving piston and pressure cylinder (5)
Subscript 5	Dry friction between pressure cylinder (5) and casing (1)

We have a total of five masses of the machine and three masses of the arm-hand model, each mass with one degree of freedom. In addition, the machine possesses one adiabatic air cushion, three classical spring-damper elements, three elastic contacts with backlash, one elastic contact with backlash and hysteresis, five contacts with dry friction leading to stick-slip behavior, and two unilateral constraints that lead to impulsive motion. The numerical solution has been obtained by piece-wise integration of the equations of motion, which includes on the one side the evaluation of the linear complementarity problem in order to obtain the correct accelerations for each integration step, and on the other side the evaluation of the impulsive behavior which has been done by applying a unilateral Poisson impact law such as it is presented in [37], [116].

12.2 Simulations

The simulations were performed for an electropneumatic drilling machine of a southern European manufacturer. Two goals could be realized: a reduction of the vibration load on the handle and an improvement of the machine's operation, particularly with respect to the removal rate. In the following we present some typical results.

Figure 12.4 portrays characteristic motions of the most important machine components. The curves describe the absolute changes in position, and the penetration of secondary piston and damping element indicate compression of the springs. The sequence of impacts is as follows. The primary piston pushes the secondary piston, which hits the tool and bounces back. It then pushes the primary piston into the reverse direction and transmits an upward velocity to it. On the other side, the tool pushes into the material and rebounds with about 50% of its energy. It again hits the secondary mass, which transmits its energy partially to the damping element.

Within the free-motion range of the damping element several short impacts between tool and secondary piston will take place which cannot be seen in Figure 12.4 due to a rough time resolution. By the jarring blow of tool and secondary piston the machine housing performs an upward motion before

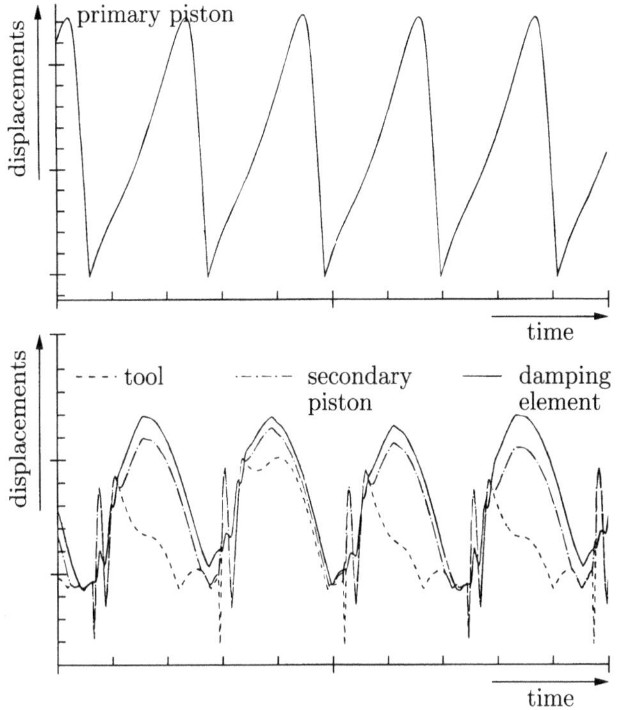

Figure 12.4. Motion of the machine components

it will be pressed down again by the arm force. This sequence of events is confirmed by experience.

In order to justify the mechanical model, simulations and measurements for the machine housing have been compared. The upper diagram of Figure 12.5 portrays the acceleration time history of the handle, which is proportional to the sum of forces applied to the gearbox. Five cycles of crank rotations are shown. The thin graph corresponds to the model shown in Figure 12.3, with positive values representing an acceleration toward the hand-arm system. In the time history the amplitudes are dominated by the occurrence of maximum air pressure and the kickback reaction of the tool and secondary piston in the backstop (c). This causes body (5) to bounce between the stops (d) and (e), which leads to acceleration peaks of the gearbox in both directions.

Figure 12.5 also depicts results derived from a model without backlash in the coupling between bodies (1) and (5). The comparison shows that dynamic loads in the presence of backlash produce excessive joint forces which, on the one hand, result in high-frequency vibrations of the handle and, on the other hand, cause higher rates of wear in the stops. In the lower diagram of Figure 12.5 we have contrasted experimental results with our numerical data. Both curves are in good agreement. This correlation proves the validity of our

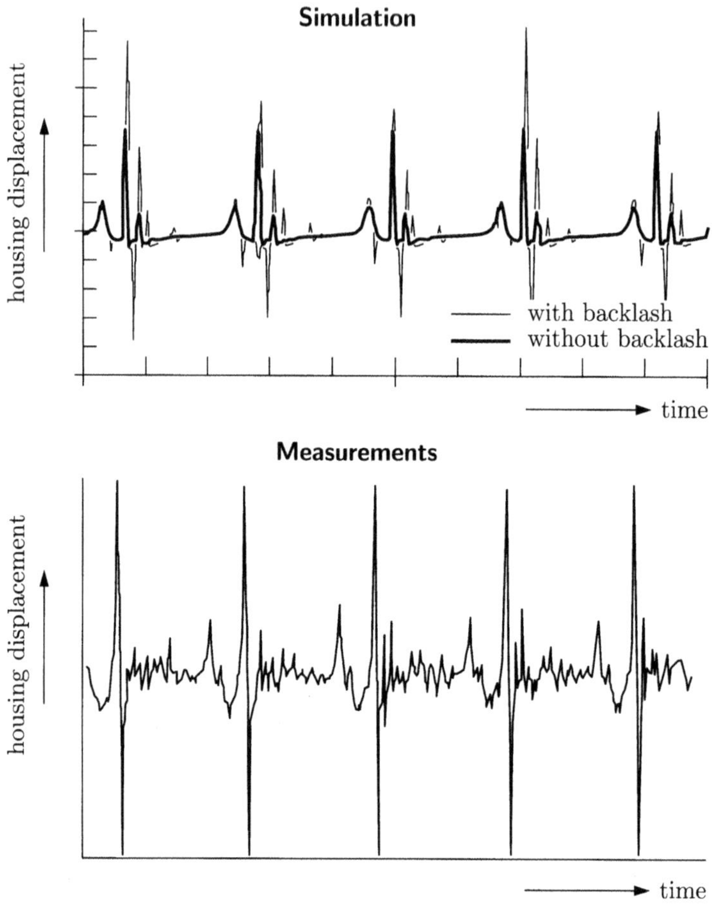

Figure 12.5. Comparison of simulations and measurements

mechanical and mathematical model with respect to the machine's handle vibrations and confirms the usefulness and physical correctness of set-valued force laws.

13. Percussion Drilling Machine

Commercial percussion drilling machines possess for their operation in the percussion drill mode two cam disks with some teeth. By moving the upper disk at the motor side with constant angular speed, for example, and by pressing the drill into the material, a hammering process is started that consists of a combined drill/impact motion. Machines of that type must be designed in such a way that the removal rate in hard materials is optimized while the acceleration load at the handle is minimized. These requirements can be achieved by proper design of the cams especially with respect to the number of cams and their geometric form. The goal of the subsequent investigation is therefore to find the optimum number and shape of the cams [42].

13.1 Mechanical Model of the Drilling Machine

The mechanical model shown in Figure 13.1 is basically composed of two subsystems, the operator's hand-arm model and the percussion drilling machine. The hand-arm model, consisting of the three masses (m_3, m_4, m_5), takes into account the dynamic interaction between the operator and the machine, and is necessary to obtain correct information about the acceleration load on the handle of the drill. It has been already used in Chapter 12 and was taken from [82]. The drilling machine is modeled by three rigid bodies: Body 0 (mass m_0, moment of inertia J_0) contains the drill and the first percussion disk (disk 0); body 1 (m_1, J_1) is composed of the second disk (disk 1) together with the gearbox and the motor; and body 2 corresponds to the flexible mounted handle of the machine. The linear displacements of the three bodies are described by the absolute coordinates x_0, x_1 and x_2. Furthermore, two rotational degrees of freedom (φ_1, φ_0) are used in the model. The inertial angular displacement of body 1 is denoted by φ_1, whereas $\varphi_0 + \Omega t$ describes the angular displacement of body 0 relative to body 1. Here, the motor of the drilling machine, rotating with constant angular velocity Ω, contributes with Ωt, and the elasticity of the gear box, being taken into account by a spring-damper element (c_φ, d_φ), with φ_0. The drill on body 0 is activated by the motor on body 1 via the elastic gear box.

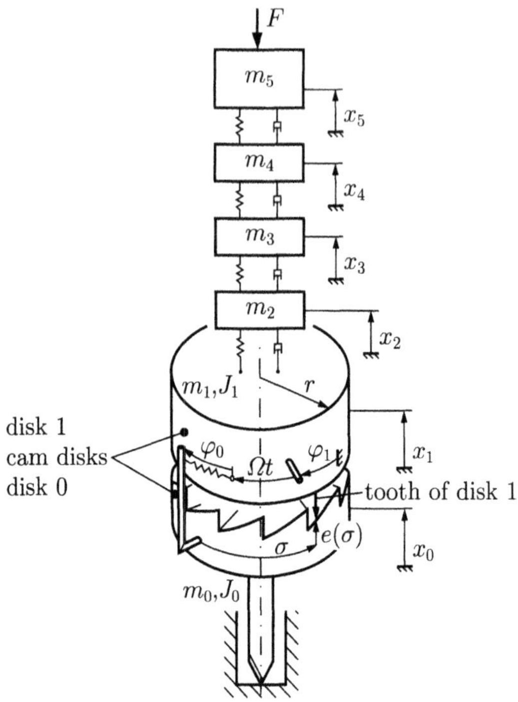

Figure 13.1. The mechanical model of the percussion drill

The two percussion disks with radius r sit on the bodies 0 and 1. Both disks hold cams with a similar geometrical forms, described by a function $e(\sigma)$ on its outer surface. During operation the cams slip on each other until they lose contact on their tops. The following free-flight state of the drilling machine continues until the two disks hit each other again because of the pressing force F. The resulting impact then causes the drilling feed. For the mechanical model of the contact process we have thus to regard free-flight phases, impacts, and friction between the two cam disks. We will, however, not present the impact theory used for this model. It can be found, for example, in [116].

13.2 Mathematical Model for Non-Contact

During the free-flight state the two cam disks are separated and the drill does not touch the environment. In that state the motion of the system is described by a linear differential equation system,

$$\mathbf{M}\dot{\mathbf{u}} + \mathbf{D}\mathbf{u} + \mathbf{C}\mathbf{q} - \mathbf{h} = 0 \quad \in \mathbb{R}^8, \tag{13.1}$$

where $\mathbf{q}$ is the set of generalized coordinates specified in Figure 13.1,

$$\mathbf{q} = (\, x_0 \; x_1 \; x_2 \; x_3 \; x_4 \; x_5 \; \varphi_0 \; \varphi_1 \,)^T, \tag{13.2}$$

and $\mathbf{u} = \dot{\mathbf{q}}$. The matrices $\mathbf{M}$, $\mathbf{D}$ and $\mathbf{C}$ represent the masses within the system, the damping and the stiffness coefficients, and the vector $\mathbf{h}$ contains the pressing force. All terms may be obtained in a straightforward manner from Figure 13.1. In particular, we have

$$\mathbf{M} = \begin{pmatrix} m_0 & 0 & & & & & & \\ 0 & m_1 & 0 & & & & & \\ & 0 & m_2 & 0 & & & & \\ & & 0 & m_3 & 0 & & & \\ & & & 0 & m_4 & 0 & & \\ & & & & 0 & m_5 & 0 & \\ & & & & & 0 & J_0 & J_0 \\ & & & & & & J_0 & J_0+J_1 \end{pmatrix},$$

$$\mathbf{D} = \begin{pmatrix} 0 & 0 & & & & & & \\ 0 & d_1 & -d_1 & & & & & \\ & -d_1 & d_1+d_2 & -d_2 & & & & \\ & & -d_2 & d_2+d_3 & -d_3 & & & \\ & & & -d_3 & d_3+d_4 & -d_4 & & \\ & & & & -d_4 & d_4 & 0 & \\ & & & & & 0 & d_\varphi & 0 \\ & & & & & & 0 & 0 \end{pmatrix}, \tag{13.3}$$

$$\mathbf{C} = \begin{pmatrix} 0 & 0 & & & & & & \\ 0 & c_1 & -c_1 & & & & & \\ & -c_1 & c_1+c_2 & -c_2 & & & & \\ & & -c_2 & c_2+c_3 & -c_3 & & & \\ & & & -c_3 & c_3+c_4 & -c_4 & & \\ & & & & -c_4 & c_4 & 0 & \\ & & & & & 0 & c_\varphi & 0 \\ & & & & & & 0 & 0 \end{pmatrix}, \quad \mathbf{h} = \begin{pmatrix} 0 \\ 0 \\ 0 \\ 0 \\ 0 \\ -F \\ 0 \\ 0 \end{pmatrix}.$$

13.3 The Contact Model

Two possible contact points are modeled in order to investigate the impulse transfer from the upper cam disk via the lower cam disk and the tool to the environment. The first contact point is between the cams of the disks; the second one is between the tool and the bottom of the drill hole.

To be able to state the constraint magnitudes of the contact between the cams, we have to describe their contours by a function $e(\sigma)$ which is continuous in its second derivative. This continuity is necessary in order to avoid jumps in the contact forces during the sliding process. In Figure 13.2 the function

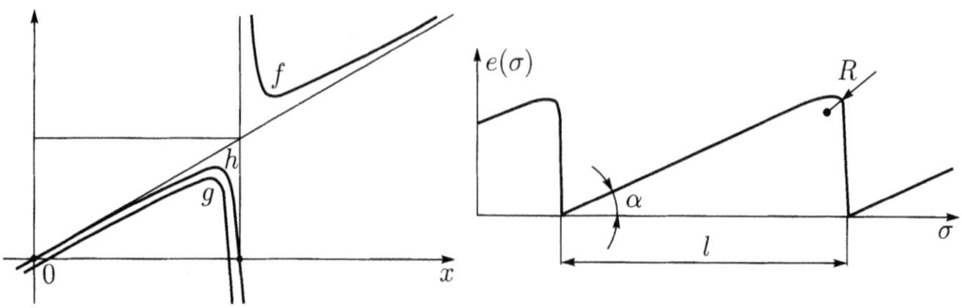

Figure 13.2. Contour functions

$$f(x) = cx\left(\frac{x}{x-b}\right)^{a} \tag{13.4}$$

is plotted, which fulfills this requirement. Using the linear transformations

$$f(x) \to g(x) \to h(x) \to e(\sigma) \tag{13.5}$$

we fit $f(x)$ to the searched outline $e(\sigma)$. The parameters a, b and c in (13.4) can be used to determine the length of the cams l, the angle of inclination α and the radius of curvature R at the top of the cams.

The variable σ depends on the generalized coordinate φ_0 and explicitly on the time t due to the rotation of the motor. This can be seen from Figure 13.1 or, more clearly, from the winding-off of the contact kinematics in Figure 13.3. With it, the contact parameter σ is easily determined to be

$$\sigma = r\,(\Omega t + \varphi_0 - k), \tag{13.6}$$

where k takes into account the numbers of cams which have already passed point D during operation. By applying the equations of spatial contact kine-

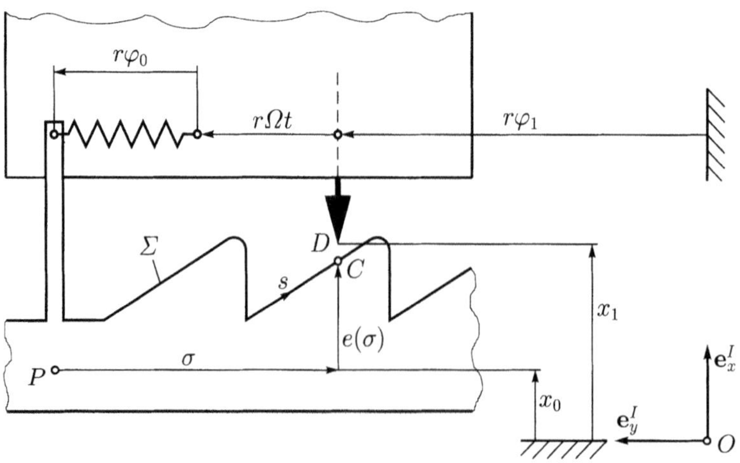

Figure 13.3. Contact kinematics of the cams

matics from sections 10.1 and 10.2 to the planar case, we obtain the contour vector as

$$ {}_I\mathbf{r}_{P\Sigma} = \begin{pmatrix} e(\sigma) \\ -\sigma \\ 0 \end{pmatrix}, \tag{13.7} $$

and the tangent and normal unit vectors become

$$ {}_I\mathbf{t} = \frac{1}{\sqrt{1+e'^2}} \begin{pmatrix} e'(\sigma) \\ -1 \\ 0 \end{pmatrix}, \quad {}_I\mathbf{n} = \frac{1}{\sqrt{1+e'^2}} \begin{pmatrix} -1 \\ -e'(\sigma) \\ 0 \end{pmatrix}. \tag{13.8} $$

With (13.7), (13.8) the angle of inclination $\bar{\alpha}$ at the contact point C becomes

$$ \tan\bar{\alpha} = e'(\sigma) \Rightarrow \frac{1}{\sqrt{1+e'^2}} = \cos\bar{\alpha}. \tag{13.9} $$

Finally, we will need the absolute velocities of the contact points C and D. They are straightforwardly derived from Figure 13.3 and are

$$ {}_I\mathbf{v}_C = \begin{pmatrix} \dot{x}_0 \\ r(\dot{\varphi}_0 + \dot{\varphi}_1 + \Omega) \\ 0 \end{pmatrix}, \quad {}_I\mathbf{v}_D = \begin{pmatrix} \dot{x}_1 \\ r\dot{\varphi}_1 \\ 0 \end{pmatrix}. \tag{13.10} $$

Now we have to state the distance g_{N1} between the contact points. This can be done according to equation (10.9) or, much easier by the relation

$$ g_{N1}^\star = x_1 - x_0 - e(\sigma), \tag{13.11} $$

compare Figure 13.3. Note that $g_{N1}^\star$ does not correspond to the natural distance g_{N1} between the contact points but shows the same properties: If the cams are separated then $g_{N1}^\star > 0$ holds, while $g_{N1}^\star = 0$ indicates the closed contact. The relative velocities in the normal and tangential directions are derived by means of equations (10.10),

$$ \begin{aligned} \gamma_{N1} &= \mathbf{n}^T(\mathbf{v}_C - \mathbf{v}_D), \\ \gamma_{T1} &= \mathbf{t}^T(\mathbf{v}_C - \mathbf{v}_D), \end{aligned} \tag{13.12} $$

which leads, after insertion of (13.8), (13.9), (13.10), to the expressions

$$ \begin{aligned} \gamma_{N1} &= (-\dot{x}_0 + \dot{x}_1 - e'r\dot{\varphi}_0 - e'r\Omega)\cos\bar{\alpha}, \\ \gamma_{T1} &= (e'\dot{x}_0 - e'\dot{x}_1 - r\dot{\varphi}_0 - r\Omega)\cos\bar{\alpha}. \end{aligned} \tag{13.13} $$

The changes in the relative velocities $\dot{\gamma}_{N1}$ and $\dot{\gamma}_{T1}$, called for short the relative accelerations, follow from a time differentiation of (13.13). For this purpose we set

$$\begin{aligned} L_N &:= (-\dot{x}_0 + \dot{x}_1 - e' r \dot{\varphi}_0 - e' r \Omega) \\ L_T &:= (e' \dot{x}_0 - e' \dot{x}_1 - r \dot{\varphi}_0 - r\Omega) \end{aligned} \tag{13.14}$$

in order to get

$$\begin{aligned} \dot{\gamma}_{N1} &= \dot{L}_N \cos\bar{\alpha} + L_N (\cos\bar{\alpha})^{\bullet}, \\ \dot{\gamma}_{T1} &= \dot{L}_T \cos\bar{\alpha} + L_T (\cos\bar{\alpha})^{\bullet}. \end{aligned} \tag{13.15}$$

Since the relative accelerations are needed only for detecting the transitions from sticking to sliding and for contact to separation, i.e. for contact states with vanishing relative velocities, equation (13.15) simplifies under this restriction to

$$\begin{aligned} \dot{\gamma}_{N1}(\gamma_{N1} = 0) &= \dot{L}_N \cos\bar{\alpha}, \\ \dot{\gamma}_{T1}(\gamma_{T1} = 0) &= \dot{L}_T \cos\bar{\alpha}, \end{aligned} \tag{13.16}$$

where the relations ($\gamma_{N1} = 0 \Leftrightarrow L_N = 0$), ($\gamma_{T1} = 0 \Leftrightarrow L_T = 0$) have been used. After re-substitution and differentiation of L_N and L_T from (13.14) the accelerations (13.16) become

$$\begin{aligned} \dot{\gamma}_{N1} &= (-\ddot{x}_0 + \ddot{x}_1 - e' r \ddot{\varphi}_0 - e'' \dot{\sigma} r \dot{\varphi}_0 - e'' \dot{\sigma} r \Omega) \cos\bar{\alpha} \\ \dot{\gamma}_{T1} &= (e' \ddot{x}_0 - e' \ddot{x}_1 - r \ddot{\varphi}_0 + e'' \dot{\sigma} \dot{x}_0 - e'' \dot{\sigma} \dot{x}_1) \cos\bar{\alpha} \end{aligned} \tag{13.17}$$

with $\dot{\sigma}$ being the time derivative of the contact parameter (13.6),

$$\dot{\sigma} = r(\Omega + \dot{\varphi}_0). \tag{13.18}$$

According to (10.12) the relative velocities and accelerations are stated as

$$\gamma_{N1} = \mathbf{w}_{N1}^T \mathbf{u} + \tilde{w}_{N1}, \quad \gamma_{T1} = \mathbf{w}_{T1}^T \mathbf{u} + \tilde{w}_{T1}, \tag{13.19}$$

$$\dot{\gamma}_{N1} = \mathbf{w}_{N1}^T \dot{\mathbf{u}} + \bar{w}_{N1}, \quad \dot{\gamma}_{T1} = \mathbf{w}_{T1}^T \dot{\mathbf{u}} + \bar{w}_{T1}. \tag{13.20}$$

Comparing (13.13) with (13.19) and (13.17) with (13.20) and concerning the structure of the coordinate vector $\mathbf{q}$ in (13.2), all the constraint terms are identified to be

$$\begin{aligned} \mathbf{w}_{N1}^T &= (\,-1\; 1\; 0\; 0\; 0\; 0\; -e'r\; 0\,) \cos\bar{\alpha}, \\ \tilde{w}_{N1} &= -e'' r \Omega \cos\bar{\alpha}, \\ \bar{w}_{N1} &= -e'' r^2 (\Omega + \dot{\varphi}_0)^2 \cos\bar{\alpha}, \\ \mathbf{w}_{T1}^T &= (\,e'\; -e'\; 0\; 0\; 0\; 0\; -r\; 0\,) \cos\bar{\alpha}, \\ \tilde{w}_{T1} &= -r\Omega \cos\bar{\alpha}, \\ \bar{w}_{T1} &= e'' r (\Omega + \dot{\varphi}_0)(\dot{x}_0 - \dot{x}_1) \cos\bar{\alpha}. \end{aligned} \tag{13.21}$$

The frictionless contact constraint between the drill and the environment is trivial. From Figure 13.1 we get immediately

$$\begin{aligned} g_{N0} &= x_0 \geq 0, \\ \mathbf{w}_{N0}^T &= (\,1\; 0\; 0\; 0\; 0\; 0\; 0\; 0\,), \\ \tilde{w}_{N0} &= \bar{w}_{N0} = 0. \end{aligned} \tag{13.22}$$

13.4 State Transitions

Equations (13.1)–(13.3) completely describe the free-flight state of the system where the cam disks are separated and the drill is not in contact with the environment. Due to the pressing force the cam disks move closer and closer during operation until they hit each other. This event is modeled by a cascade of two completely inelastic impacts in succession. During the first collision cam disk 1 hits disk 0, which stands idle in the axial direction, see Figure 13.4. Only constraint 1 is taken into account in the impact equations, which are solved by using Newton's impact law with a coefficient of restitution equal to zero. After this impact both disks are in contact and are moving in the axial direction toward the bottom of the drill hole. There the second impact takes place. The velocities shortly after the first impact and just before the second are the same. Again, the second impact is modeled inelastically, but now both constraints, 0 and 1, are considered in the evaluation process. As a result we get the relative velocities after that impact which are equal to zero, and the associated normal impulses $\boldsymbol{\Lambda}_N$.

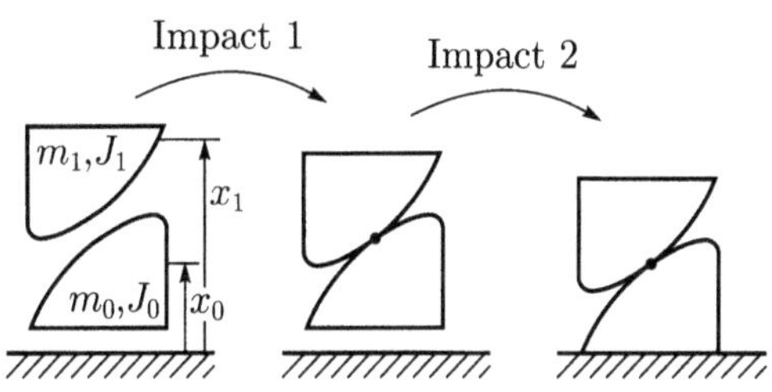

Figure 13.4. Contact dynamics: Impact process

The removal rate V_S per impact which is achieved by percussion drilling is obtained proportionally to the second power of the impulse transferred to the drill hole,

$$\boldsymbol{\Lambda}_N = \begin{pmatrix} \Lambda_{N1} \\ \Lambda_{N0} \end{pmatrix}, \quad V_S = k\Lambda_{N0}^2, \quad k = \text{const.} \tag{13.23}$$

With that we attain the removal rate per time unit by averaging the impacts which take place during the simulation time.

When the impact process has ended, the frictional contact between the cams may open due to the force regime or may remain closed. The second case turned out to show only sliding without sticking. Thus the friction force λ_{T1} can be expressed by their normal force λ_{N1} with Coulomb's law for sliding, which results together with the frictionless unilateral contact constraint of the drill (index 0) in the expressions presented in Section 10.5. Solving the underlying system of equations (10.21) enables us to determine the unknown accelerations $\dot{\mathbf{u}}$ as well as the contact forces λ_{N1}, λ_{T1}, λ_{N0} which are depicted for the cam's contact in Figure 13.5. The transitions between the contact states of the cams are evaluated by supervising certain kinematic and kinetic indicator magnitudes which are the distance indicator $g^{\star}_{N1}$ form (13.11), the normal contact force λ_{N1} in Figure 13.5, and the relative normal acceleration $\dot{\gamma}_{N1}$ from (13.20).

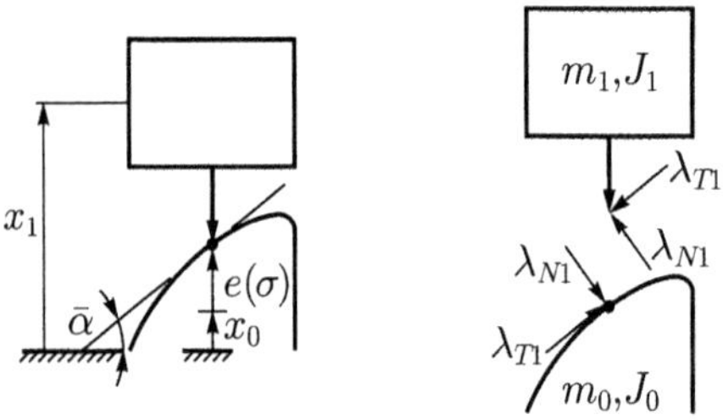

Figure 13.5. Contact dynamics: Contact forces

Figure 13.6 shows the transitions of the cam's contact that occurred during numerical simulation. The free-flight state Z_1 is valid as long as the distance between the disks, indicated by $g^{\star}_{N1}$, is greater than zero. When this distance disappears, $g^{\star}_{N1} = 0$, the two inelastic impacts Z_2 take place. Impacts lead to jumps in the velocities; therefore the normal force λ_{N1} may contain a discontinuity at the moment of impact, too. Thus, two transitions after the impact are possible: For a positive normal relative acceleration ($\dot{\gamma}_{N1} > 0$) the system returns to the free-flight state Z_1, and for a positive normal force ($\lambda_{N1} > 0$) sliding Z_3 takes place. During sliding the normal force decreases continuously to zero. Then the two disks lose contact and again the state free-flight Z_1 is valid.

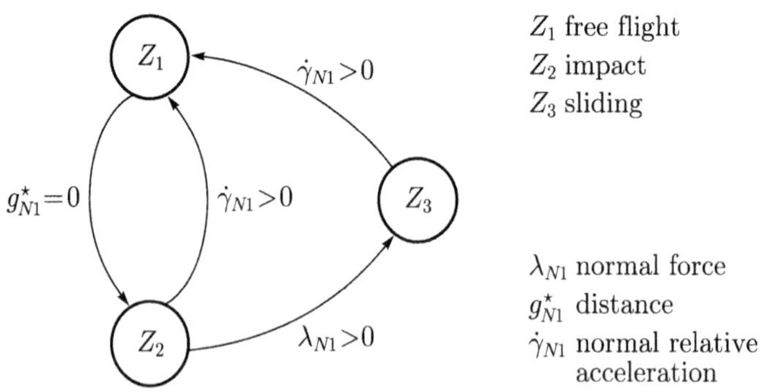

Figure 13.6. Transitions in the cam's contacts

13.5 Results

The left diagram in Figure 13.7 shows the behavior of the machine for percussion drilling in concrete. The removal rate is plotted as a function of the rotational speed of the motor for different pressing forces. We can see that the removal rate grows with increasing speed of rotation and increasing pressing forces. This can be explained by the fact that, on the one hand, the percussion frequency and, on the other hand, the impulse transferred by the impact grow. The lines result from calculation and are confirmed by measurements, represented by the dots.

In the right picture of Figure 13.7 the shape of disk 0 is plotted for different rotational speeds, together with the time behavior of the coordinate x_1 which corresponds to the axial displacement of disk 1. The corners and the different smoothing out on the top of the cams result from the arising impacts and effects of dry friction. With an increasing number of revolutions the contact point of the cams moves closer to the top, until a state is reached where the cams jump from one top to the next. With a further increase of the rotational speed only every second cam is hit.

This transition is shown in the phase space portraits, Figure 13.8. It should be kept in mind that mechanical systems with impulsive and stick-slip processes are highly nonlinear due to unsteady transitions and changing of degrees of freedom. In Figure 13.8 the first two pictures correspond to the case that every cam of the disks is hit. It is easy to make out the sliding and the free-flight state, followed by an impact. By increasing the rotational speed, picture (b), the hitting point of the cams moves upward and the limit cycle

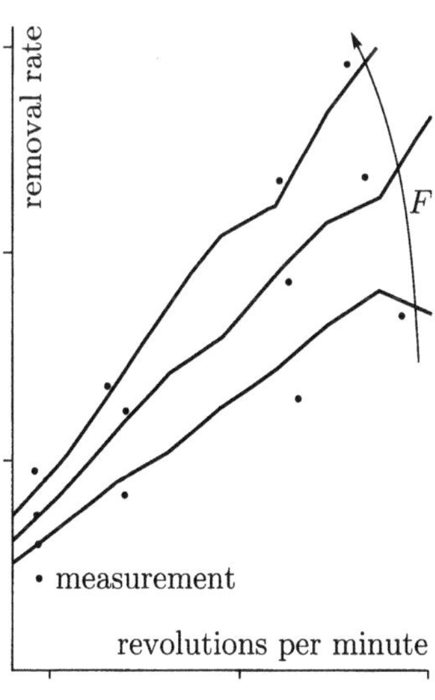

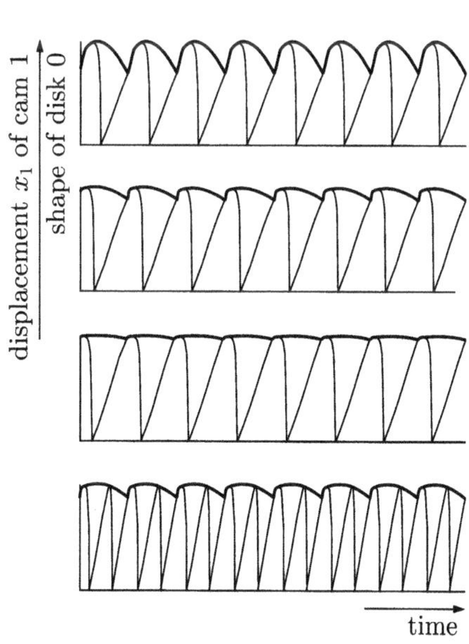

Figure 13.7. Drilling in concrete and axial displacements of disk 1

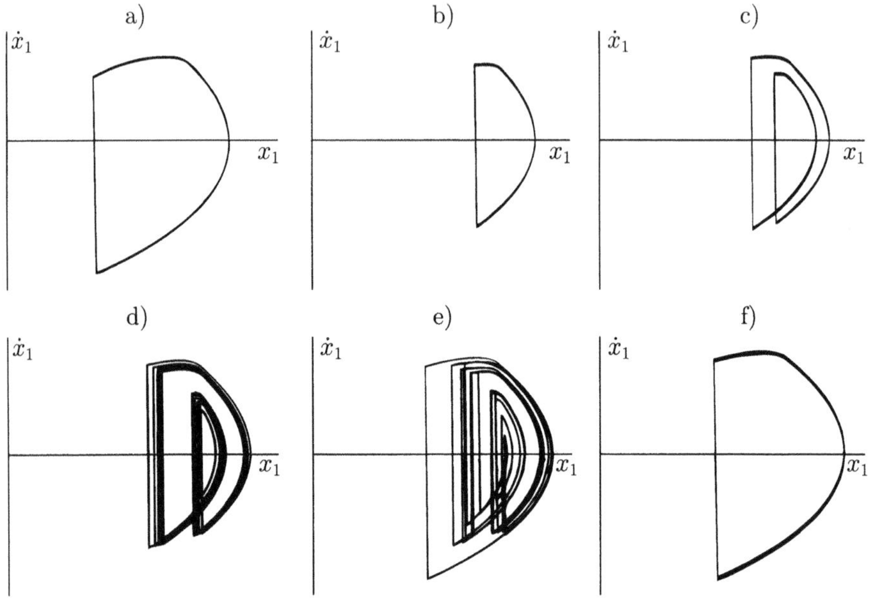

Figure 13.8. Phase space portraits

gets smaller. The maximal altitude of x_1 in the free-flight state, however, remains nearly constant. Further increasing of the rotational speed leads to a bifurcation of the solution. Picture (c) shows a two-periodic solution. The cams are hit at two different points which are alternately closer to the top and closer to the ground of the cams. If the rotational speed rises even more, the solution bifurcates again by period doubling. Picture (d) shows a periodic

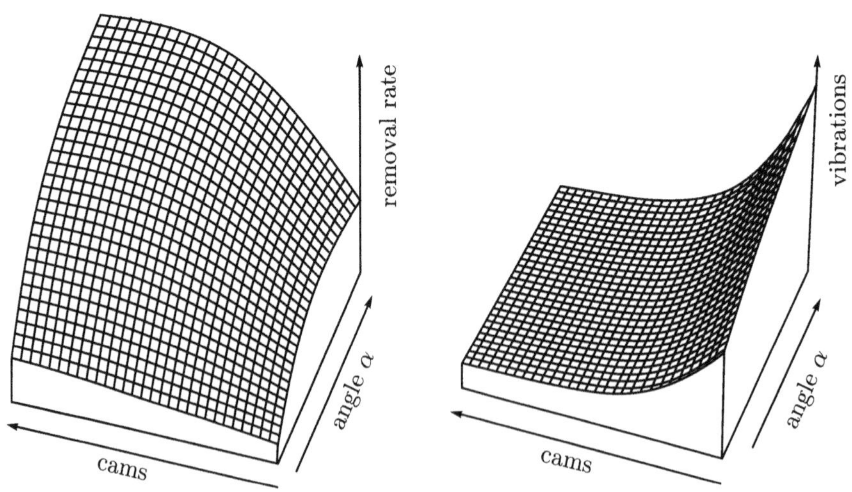

Figure 13.9. Removal rate and handle vibrations

solution of higher order which becomes irregular in picture (e) and finally changes in picture (f) to the solution where only every second cam is hit. The same behavior is observed at transitions from the second cam hit to the third one.

The aim of this examination is to analyze the influence of the geometric design of the disks on the removal rate and handle vibrations, where the removal rate was calculated as a function of the impulses (13.23), and the handle vibrations were determined proportional to the root-mean-square value of the Fourier coefficients of the handle accelerations. The angle of inclination α and the number of cams of one disk were chosen as design parameters. Figure 13.9 shows the result of this variation for parameter intervals where only that solution exists at which every cam is hit. With an increasing number of cams the hitting point of the cam disks moves toward the top of the cams. Thereby the frequency of impact increases and the handle amplitude decreases, which is shown by an increasing removal rate and decreasing handle vibrations. If the angle of inclination is enlarged, a more powerful impulse in axial direction takes place. The result is a better removal rate, but also slightly higher handle vibrations.

14. Turbine Blade Damper

For a reduction of blade vibration amplitudes in airborne gas turbines special damper devices are used which usually are arranged between the blade platforms of neigboring blades. These dampers possess a cross section like a parabola or a circle and are formed out of sheet steel. The dampers are arranged between the platforms of the blades in such a way that they are pressed by centrifugal forces against the oblique plane, which will vibrate together with the blades. This motion causes some relative movement between damper and oblique plane thus generating dissipation by Coulomb friction. The design goal for dampers of that kind consists of maximizing dissipation which requires relative motion or, to put it another way, to minimize stiction time [47], [117], [116], [43].

14.1 The Damper Model and the Non-Contact Case

The mechanical model in Figure 14.1 consists of two platforms (masses m_1, m_3) with turbine blades (masses m_2, m_4) connected by a friction damper

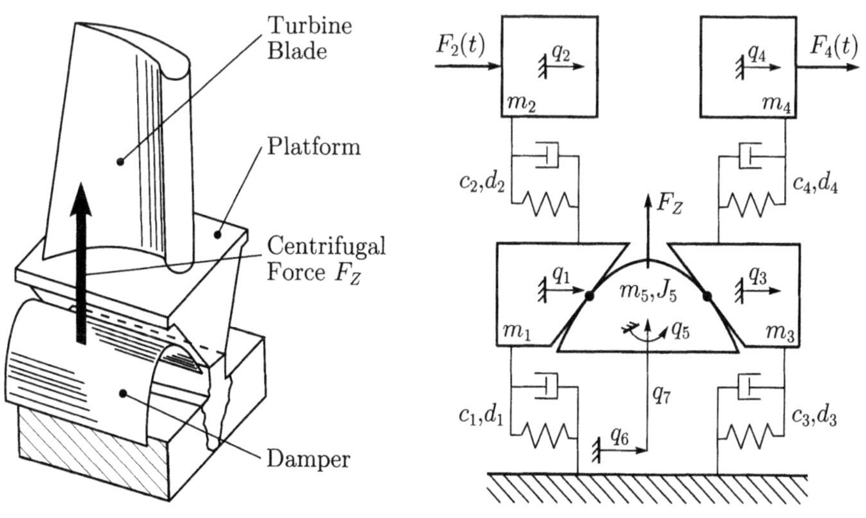

Figure 14.1. Turbine blade damper and mechanical model

(mass m_5, moment of inertia J_5). The displacements of the platforms and of the turbine blades are described by the translational coordinates (q_1, q_2, q_3, q_4). In the unconstrained case, the damper possesses two translational degrees of freedom q_6, q_7 and one rotational degree of freedom q_5. The vector of the generalized coordinates is

$$\mathbf{q} = (\, q_1 \; q_2 \; q_3 \; q_4 \; q_5 \; q_6 \; q_7 \,)^T \in \mathbb{R}^7. \tag{14.1}$$

A constant radial force F_Z, representing the centrifugal force, affects the damper and presses it against the oblique planes of the platforms. The excitation of the turbine blades is taken into account by the forces $F_i(t) = A\sin(\omega t + \varphi_i)$, $i = 2, 4$, which result from an excitation by periodically changing gas forces. The equations of motion for the non-contact case, here being of the form

$$\mathbf{M}\dot{\mathbf{u}} - \mathbf{h}(\mathbf{q}, \mathbf{u}, t) = 0 \quad \text{with } \dot{\mathbf{q}} = \mathbf{u} \tag{14.2}$$

follow immediately from Figure 14.1. With the above-mentioned agreements, we are able to state the mass matrix $\mathbf{M}$ and the vector $\mathbf{h}$ of the external forces as

$$\mathbf{M} = \begin{pmatrix} m_1 & 0 & 0 & 0 & 0 & 0 & 0 \\ 0 & m_2 & 0 & 0 & 0 & 0 & 0 \\ 0 & 0 & m_3 & 0 & 0 & 0 & 0 \\ 0 & 0 & 0 & m_4 & 0 & 0 & 0 \\ 0 & 0 & 0 & 0 & J_S & 0 & 0 \\ 0 & 0 & 0 & 0 & 0 & m_5 & 0 \\ 0 & 0 & 0 & 0 & 0 & 0 & m_5 \end{pmatrix}$$

$$\mathbf{h} = \begin{pmatrix} -(c_1 + c_2)q_1 + c_2 q_2 - (d_1 + d_2)u_1 + d_2 u_2 \\ c_2 q_1 - c_2 q_2 + d_2 u_1 - d_2 u_2 + F_2(t) \\ -(c_3 + c_4)q_3 + c_4 q_4 - (d_3 + d_4)u_3 + d_4 u_4 \\ c_4 q_3 - c_4 q_4 + d_4 u_3 - d_4 u_4 + F_4(t) \\ 0 \\ 0 \\ F_Z \end{pmatrix}. \tag{14.3}$$

14.2 Contact Kinematics of the Damping Device

Before stating the equations of motion for contact, some kinematic terms have to be derived which will be done separately for the left contact as shown in Figure 14.2. In order to describe the displacements of the platform and the damper, we use as coordinates the tuple (z, φ, x, y). Here, z denotes the horizontal location of the platform's body-fixed frame P, and (x, y, φ) determines the translational and rotational displacements of the damper. The contours under investigation are Σ_1 and Σ_2. They are described by the functions $y = ax$ and $y = -p_1x^2 + p_2$ related to the body-fixed frames P and D, respectively. In the following the contact problem will be solved analytically by several steps, particulary the calculation of the contact parameters σ_i, the distance g_N, and the terms related to the normal and tangential relative velocities $\gamma_{N,T}$ and accelerations $\dot{\gamma}_{N,T}$. Notation is adapted to sections 10.1, 10.2, where the spatial contact problem was treated.

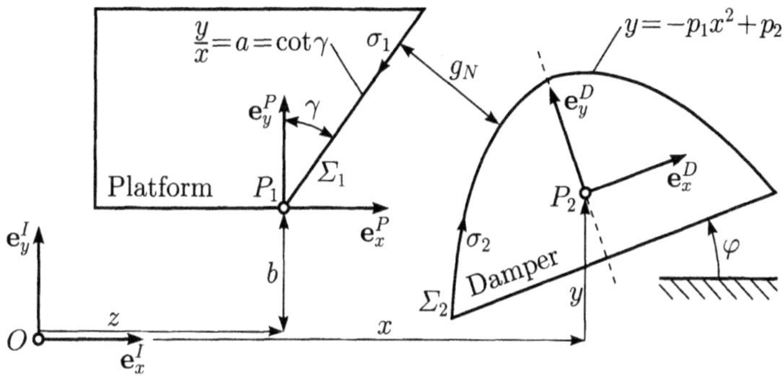

Figure 14.2. Contact problem of a parabola and a straight line

Let us first derive the kinematics of the platform. We choose a platform-fixed frame $(P_1, \mathbf{e}_x^P, \mathbf{e}_y^P, \mathbf{e}_z^P)$ as depicted in Figure 14.2. The displacement and orientation of the platform relative to the inertial frame $(O, \mathbf{e}_x^I, \mathbf{e}_y^I, \mathbf{e}_z^I)$ as well as the vector from P_1 to any point of the contour Σ_1 may be expressed as

$$
{}_I\mathbf{r}_{OP1} = \begin{pmatrix} z \\ b \\ 0 \end{pmatrix}, \quad {}_P\mathbf{r}_{P\Sigma 1} = \begin{pmatrix} -\sigma_1 \\ -a\sigma_1 \\ 0 \end{pmatrix}, \quad \mathbf{A}_{PI} = \mathbf{E} \tag{14.4}
$$

where $a = \cot\gamma > 0$ and $\mathbf{A}_{FK}$ denotes the transformation matrix from frame K to frame F, i.e. if ${}_F\mathbf{x}$ are the coordinates of $\mathbf{x}$ with respect to frame F and ${}_K\mathbf{x}$ the coordinates with respect to K, then ${}_F\mathbf{x} = \mathbf{A}_{FK}\,{}_K\mathbf{x}$. Since the platform can only move in $\mathbf{e}_x^I$-direction but cannot rotate we have $\mathbf{A}_{PI} = \mathbf{E}$ (identity) in (14.4). Thus, all platform-fixed points have the same velocity, unequal to zero only in the horizontal direction. This holds in particular for all points C_1 on Σ_1 which are not moving relative to Σ_1, i.e. all points C_1 on Σ_1 which are characterized by some constant σ_1 in (14.4). The velocities of the points P_1, C_1 with respect to O thus are

$$
{}_I\mathbf{v}_{P1} = \begin{pmatrix} \dot{z} \\ 0 \\ 0 \end{pmatrix}, \quad {}_P\mathbf{v}_{C1} = \begin{pmatrix} \dot{z} \\ 0 \\ 0 \end{pmatrix}, \quad {}_P\mathbf{\Omega}_1 = 0, \tag{14.5}
$$

with $\boldsymbol{\Omega}_1$ the angular velocity of the platform. Finally, take ${}_P\mathbf{r}'_{P\Sigma 1}(\sigma_1)$ to get the direction of the tangent $\mathbf{t}_1$ of Σ_1, pointing towards increasing σ_1, and determine from $\mathbf{t}_1$ the normal $\mathbf{n}_1 \perp \mathbf{t}_1$. This yields the unit vectors

$$ {}_P\mathbf{t}_1 = \frac{1}{\sqrt{1+a^2}} \begin{pmatrix} -1 \\ -a \\ 0 \end{pmatrix}, \quad {}_P\mathbf{n}_1 = \frac{1}{\sqrt{1+a^2}} \begin{pmatrix} a \\ -1 \\ 0 \end{pmatrix}. \tag{14.6} $$

with $\mathbf{n}_1$ as the outward normal vector. Equations (14.4)–(14.6) already describe the platform as far as needed for the further evaluation of the contact problem.

For the parabolic damper (damper-fixed frame $(P_2, \mathbf{e}_x^D, \mathbf{e}_y^D, \mathbf{e}_z^D)$) we proceed in exactly the same manner. With the parabola parameters $p_1, p_2 > 0$, the displacements are given by

$$ {}_I\mathbf{r}_{OP2} = \begin{pmatrix} x \\ y \\ 0 \end{pmatrix}, \quad {}_D\mathbf{r}_{P\Sigma 2} = \begin{pmatrix} \sigma_2 \\ -p_1\sigma_2^2 + p_2 \\ 0 \end{pmatrix}, \quad \mathbf{A}_{DI} = \begin{pmatrix} \mathrm{c}\varphi & \mathrm{s}\varphi & 0 \\ -\mathrm{s}\varphi & \mathrm{c}\varphi & 0 \\ 0 & 0 & 1 \end{pmatrix}, \tag{14.7} $$

where we will use from now the abbreviations c := cos, s := sin. The calculation of the velocities is straightforward. In particular, the velocities of the rigid body points C_2 on Σ_2 follow immediately from $\mathbf{v}_{C2} = \mathbf{v}_{P2} + \boldsymbol{\Omega}_2 \times \mathbf{r}_{P\Sigma 2}$ for any fixed σ_2. They read

$$ {}_I\mathbf{v}_{P2} = \begin{pmatrix} \dot{x} \\ \dot{y} \\ 0 \end{pmatrix}, \quad {}_D\mathbf{v}_{C2} = \begin{pmatrix} \dot{x}\,\mathrm{c}\varphi + \dot{y}\,\mathrm{s}\varphi - \dot{\varphi}(-p_1\sigma_2^2 + p_2) \\ -\dot{x}\,\mathrm{s}\varphi + \dot{y}\,\mathrm{c}\varphi + \dot{\varphi}\sigma_2 \\ 0 \end{pmatrix}, \quad {}_D\boldsymbol{\Omega}_2 = \begin{pmatrix} 0 \\ 0 \\ \dot{\varphi} \end{pmatrix}. \tag{14.8} $$

By means of ${}_D\mathbf{r}'_{P\Sigma 2}(\sigma_2)$, the tangent and the normal unit vector to the parabola at a point σ_2 become

$$ {}_D\mathbf{t}_2 = \frac{1}{\sqrt{1+4p_1^2\sigma_2^2}} \begin{pmatrix} 1 \\ -2p_1\sigma_2 \\ 0 \end{pmatrix}, \quad {}_D\mathbf{n}_2 = \frac{1}{\sqrt{1+4p_1^2\sigma_2^2}} \begin{pmatrix} 2p_1\sigma_2 \\ 1 \\ 0 \end{pmatrix}. \tag{14.9} $$

In contrast to the platform's trihedral (14.6), they explicitly depend on the parabola parameter σ_2.

Define now the distance vector $\mathbf{r}_D(\sigma_1, \sigma_2) := (\mathbf{r}_{OP2} + \mathbf{r}_{P\Sigma 2}) - (\mathbf{r}_{OP1} + \mathbf{r}_{P\Sigma 1})$ according to (10.6),

$$ {}_I\mathbf{r}_D = \begin{pmatrix} x - z + \sigma_1 + \sigma_2\,\mathrm{c}\varphi - (-p_1\sigma_2^2 + p_2)\,\mathrm{s}\varphi \\ y - b + a\sigma_1 + \sigma_2\,\mathrm{s}\varphi + (-p_1\sigma_2^2 + p_2)\,\mathrm{c}\varphi \\ 0 \end{pmatrix}, \tag{14.10} $$

that points from the platform's contour Σ_1 to the damper's contour Σ_2. In order to derive later the distance of the two bodies from this $\mathbf{r}_D$, we have to align it with both normals, which can be done as described in (10.7): First, we determine from the condition $\mathbf{n}_1^T \mathbf{t}_2 = 0$ the contact parameter σ_2,

$$\sigma_2 = \frac{-1}{2p_1} \frac{a\cos\varphi - \sin\varphi}{a\sin\varphi + \cos\varphi}. \tag{14.11}$$

By this value, the normals and tangents of both bodies are known (14.6), (14.9), as well as the position of the contact point C_2 on the parabola, given by $\mathbf{r}_{P\Sigma 2}$ in (14.7). Furthermore, it holds that $\mathbf{n}_1 = -\mathbf{n}_2$, $\mathbf{t}_1 = -\mathbf{t}_2$. In a second step one could solve, for example, the equation $\mathbf{r}_D^T \mathbf{t}_1 = 0$ to additionally determine the contact parameter σ_1, and thus to locate the contact point C_1 on the straight line. We will skip this step, since σ_1 will not be needed in the following. To determine the distance g_N of the bodies, one has by (10.9) just to project $\mathbf{r}_D$ on $\mathbf{n}_1$,

$$g_N = \frac{1}{\sqrt{1+a^2}} \Big(b - y + ax - az - (p_1\sigma_2^2 + p_2)(a\,\mathrm{s}\varphi + \mathrm{c}\varphi) \Big). \tag{14.12}$$

The relative velocities of the contact points C_1 and C_2 in the normal and the tangential direction are now simply derived according to (10.10), i.e. $\gamma_N = -\mathbf{n}_1^T \mathbf{v}_{C1} - \mathbf{n}_2^T \mathbf{v}_{C2}$ and $\gamma_T = -\mathbf{t}_1^T \mathbf{v}_{C1} - \mathbf{t}_2^T \mathbf{v}_{C2}$ with $\mathbf{n}_i$, $\mathbf{t}_i$, $\mathbf{v}_{Ci}$ from (14.5), (14.6), (14.8), (14.9) and σ_2 from (14.11). We obtain

$$\gamma_N = \frac{1}{\sqrt{1+a^2}} \Big(-\dot{y} + a\dot{x} - a\dot{z} + (\tfrac{1}{2p_1} + p_1\sigma_2^2 - p_2)(a\,\mathrm{c}\varphi - \mathrm{s}\varphi)\,\dot{\varphi} \Big) \tag{14.13}$$

$$\gamma_T = \frac{1}{\sqrt{1+a^2}} \Big(a\dot{y} + \dot{x} - \dot{z} - (p_1\sigma_2^2 + p_2)(a\,\mathrm{s}\varphi + \mathrm{c}\varphi)\,\dot{\varphi} \Big). \tag{14.14}$$

Finally, we need the temporal changes of (14.14), $\dot{\gamma}_N$ and $\dot{\gamma}_T$. Direct differentiation with respect to time yields

$$\begin{aligned} \dot{\gamma}_N = & \frac{1}{\sqrt{1+a^2}} \Big(-\ddot{y} + a\ddot{x} - a\ddot{z} + (\tfrac{1}{2p_1} + p_1\sigma_2^2 - p_2)(a\,\mathrm{c}\varphi - \mathrm{s}\varphi)\,\ddot{\varphi} \Big) \\ & + \frac{\dot{\varphi}^2}{\sqrt{1+a^2}} \Big(p_2 - \tfrac{1}{2p_1} - p_1\sigma_2^2(3 + 8p_1^2\sigma_2^2) \Big) (a\,\mathrm{s}\varphi + \mathrm{c}\varphi), \end{aligned} \tag{14.15}$$

$$\begin{aligned} \dot{\gamma}_T = & \frac{1}{\sqrt{1+a^2}} \Big(a\ddot{y} + \ddot{x} - \ddot{z} - (p_1\sigma_2^2 + p_2)(a\,\mathrm{s}\varphi + \mathrm{c}\varphi)\,\ddot{\varphi} \Big) \\ & + \frac{\dot{\varphi}^2}{\sqrt{1+a^2}} \Big(2p_1\sigma_2(p_2 - p_1\sigma_2^2) - \sigma_2 \Big) (a\,\mathrm{s}\varphi + \mathrm{c}\varphi), \end{aligned} \tag{14.16}$$

where the changes in the contact parameter, $\dot{\sigma}_2 = \sigma_2'(\varphi)\,\dot{\varphi}$ have to be regarded due to the moving contact points.

Let us now state the constraint terms $\mathbf{w}_i$ and $\bar{w}_i$ of the system's relative velocities and accelerations

$$\begin{aligned} \gamma_i(\mathbf{q}, \mathbf{u}) &= \mathbf{w}_i^T(\mathbf{q})\,\mathbf{u}, \\ \dot{\gamma}_i(\mathbf{q}, \mathbf{u}, \dot{\mathbf{u}}) &= \mathbf{w}_i^T(\mathbf{q})\,\dot{\mathbf{u}} + \bar{w}_i(\mathbf{q}, \mathbf{u}), \end{aligned} \qquad (i = N1, N2, T1, T2) \tag{14.17}$$

for the complete damper model of Figure 14.1. This can be done directly by the results (14.13)–(14.16) when identifying the coordinate tuple $(z\,\varphi\,x\,y)$ with $(q_1\,q_5\,q_6\,q_7)$ from $\mathbf{q}$ in (14.1). With the abbreviation

$$A := \frac{1}{\sqrt{1+a^2}} \tag{14.18}$$

and by writing again s for sin and c for cos, the resulting expressions are

$$\begin{pmatrix}\mathbf{w}_{N1}^T\\ \mathbf{w}_{N2}^T\end{pmatrix} = A\begin{pmatrix}-a & 0 & 0 & 0 & (sq_5-acq_5)(p_2-\frac{1}{2p_1}-p_1\sigma_1^2) & a & -1\\ 0 & 0 & a & 0 & (sq_5+acq_5)(p_2-\frac{1}{2p_1}-p_1\sigma_2^2) & -a & -1\end{pmatrix} \tag{14.19}$$

$$\begin{pmatrix}\mathbf{w}_{T1}^T\\ \mathbf{w}_{T2}^T\end{pmatrix} = A\begin{pmatrix}-1 & 0 & 0 & 0 & (cq_5+asq_5)(-p_2-p_1\sigma_1^2) & 1 & a\\ 0 & 0 & -1 & 0 & (cq_5-asq_5)(-p_2-p_1\sigma_2^2) & 1 & -a\end{pmatrix} \tag{14.20}$$

$$\begin{pmatrix}\bar{w}_{N1}\\ \bar{w}_{N2}\end{pmatrix} = Au_5^2\begin{pmatrix}(cq_5+asq_5)(p_2-\frac{1}{2p_1}-3p_1\sigma_1^2-8p_1^3\sigma_1^4)\\ (cq_5-asq_5)(p_2-\frac{1}{2p_1}-3p_1\sigma_2^2-8p_1^3\sigma_2^4)\end{pmatrix} \tag{14.21}$$

$$\begin{pmatrix}\bar{w}_{T1}\\ \bar{w}_{T2}\end{pmatrix} = Au_5^2\begin{pmatrix}(cq_5+asq_5)((2p_2p_1-1)\sigma_1-2p_1^2\sigma_1^3)\\ (cq_5-asq_5)((2p_2p_1-1)\sigma_2-2p_1^2\sigma_2^3)\end{pmatrix} \tag{14.22}$$

where subscripts 1 and 2 denote the left and right contact points in Figure 14.1, respectively. Furthermore, σ_1 and σ_2 are the values of the parabola contact parameters which are given for the left contact point by equation (14.11), and for the right contact point when substituting the angle of inclination of the plane $a = \cot\gamma$ by $-a$,

$$\sigma_1 = \frac{-1}{2p_1}\,\frac{a\cos q_5-\sin q_5}{a\sin q_5+\cos q_5}, \qquad \sigma_2 = \frac{-1}{2p_1}\,\frac{-a\cos q_5-\sin q_5}{-a\sin q_5+\cos q_5}. \tag{14.23}$$

As a whole the configuration consists of a coupled two-point contact problem, where the normal and tangential forces of the two contact points influence each other and depend on the system's dynamics.

14.3 Numerical Results

To get appropriate initial values for the numerical integration we determine first, with $F_2 = F_4 = 0$, the frictionless equilibria of the damper using equation (10.16) for constrained motion. Frictionless contact ($\mu = 0$) means that the tangential constraints do not exist; thus only the equations of the normal constraints have to be considered (subscript N in equations (14.19)–(14.22)). Therefore we get from (10.16) that

$$\begin{aligned}
&\mathbf{M}\dot{\mathbf{u}} - \mathbf{h} - \mathbf{W}_N\boldsymbol{\lambda}_N = 0 \quad \in \mathbb{R}^7\\
&\dot{\gamma}_N = \mathbf{W}_N^T\dot{\mathbf{u}} + \bar{\mathbf{w}}_N = 0 \quad \in \mathbb{R}^2\\
&\gamma_N = \mathbf{W}_N^T\mathbf{u} = 0\\
&\mathbf{g}_N = 0.
\end{aligned} \tag{14.24}$$

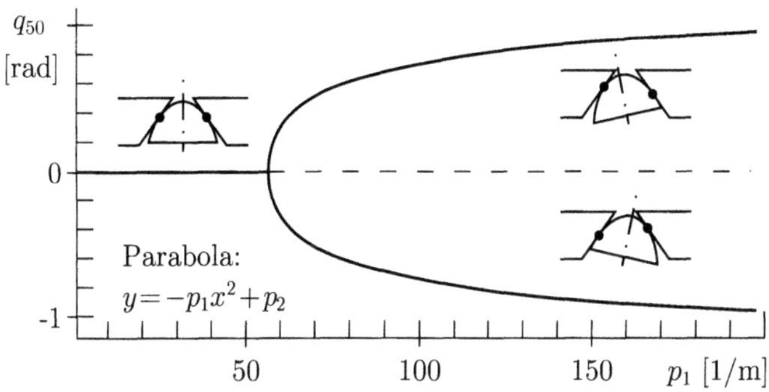

Figure 14.3. Frictionless equilibrium positions of the damper

Equilibria (subscript 0) are characterized by vanishing velocities and accelerations ($\dot{\mathbf{u}}_0 = 0$, $\mathbf{u}_0 = 0$), whereas the displacements $\mathbf{q}_0$ and normal forces $\boldsymbol{\lambda}_{N0}$ are constant. From (14.24) we therefore get nine nonlinear algebraic equations for $\mathbf{q}_0 \in \mathbb{R}^7$ and $\boldsymbol{\lambda}_{N0} \in \mathbb{R}^2$, which read

$$\begin{aligned} \mathbf{h}(\mathbf{q}_0) + \mathbf{W}_N \boldsymbol{\lambda}_{N0} &= 0 \quad && \in \mathbb{R}^7, \\ \mathbf{g}_N(\mathbf{q}_0) &= 0 && \in \mathbb{R}^2. \end{aligned} \tag{14.25}$$

An evaluation of equations (14.25) in terms of the damper's parabola parameter p_1 yields, for example, the angular displacements q_{50} depicted in Figure 14.3. For small values of p_1 only the symmetric equilibrium $q_{50} = 0$ exists. When the parabola parameter p_1 is enlarged, $q_{50} = 0$ becomes unstable and two nonsymmetric equilibria arise. Before starting the numerical integration for a given value of p_1 the equilibrium $\mathbf{q}_0$ is computed and then used as the initial displacement for the simulation of the system under the influence of friction. It turns out that in many cases computer time is reduced by more than two orders of magnitude due to the fact that the stationary situation is known in advance and need not be realized by time-consuming numerical integration.

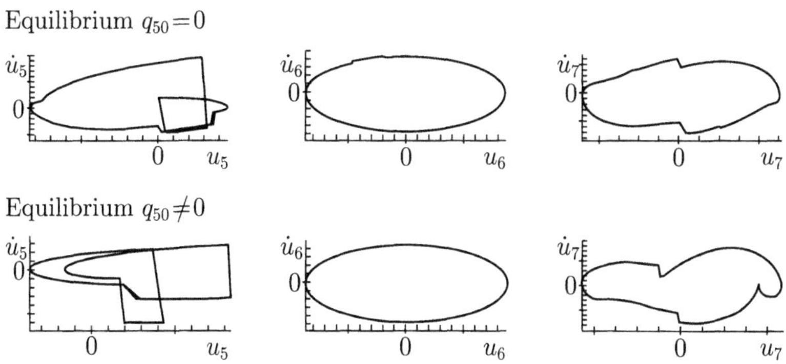

Figure 14.4. Dynamic behavior of the damper

The aim of the design of a friction damper is to obtain the largest possible dissipation of energy by friction. Thus, the damper has to exhibit sufficiently long phases of sliding and show a low tendency to lock up. Using an original data set from industry, Figure 14.4 shows the computed damper accelerations as a function of their velocities for a symmetric and a nonsymmetric equilibrium. The damper shows a different dynamic behavior, depending on the equilibrium, and is working in both cases, which can be seen on the discontinuous changes in accelerations due to stick-slip transitions.

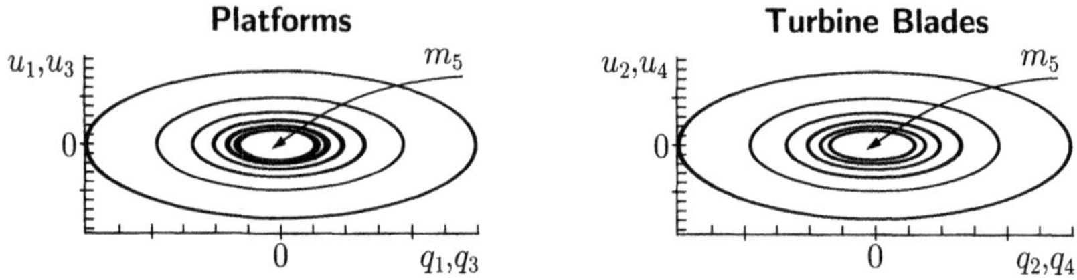

Figure 14.5. Phase portraits of platforms and blades

As one task it had to be examined whether the blade and platform amplitudes can be reduced by installing different dampers. In Figure 14.5 the phase portraits of the blade and platform coordinates are plotted for different damper masses m_5. Obviously, an increased damper mass leads to a reduction of the amplitudes, which unfortunately results mainly, for this data set, from the changes in the system's mass and is not an effect of dry friction produced by the damper. Figure 14.6 shows the platform and blade amplitudes in dependency on the damper mass m_5. The dots correspond to the amplitudes $\hat{q}$ of the coordinates in Figure 14.5. The solid line results from a simulation of an associated linear system which is obtained when the damper is assumed not to work. In this case, the damper together with the two platforms can be regarded as one rigid body, and the model in Figure 14.1 becomes linear. The difference between the dots and the solid line then shows exactly the reduction of the amplitudes by friction. For that special data set, the damper

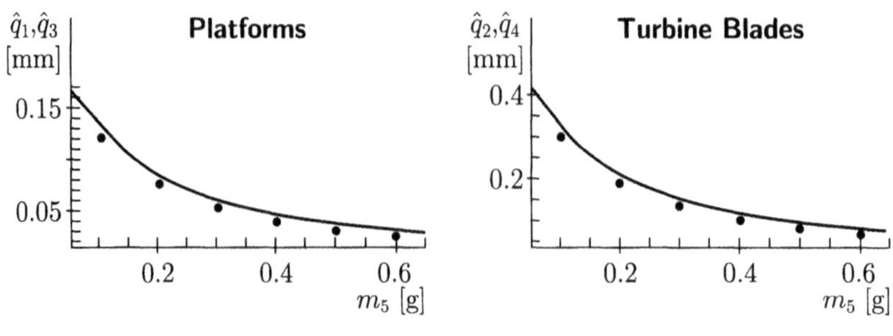

Figure 14.6. Comparison of linear and nonlinear model

shows a nice nonlinear friction-induced behavior, but is not able to influence the amplitudes of the system significantly.

Computations on a different damper, however, have shown that significant optima can be realized by a suitable adaptation of certain parameters like inclination angle γ and damper mass m_5, as reported in [47]. In this report, the theoretical model as presented above was additionally verified by measurements at a laboratory test setup. For the experiments the centrifugal force was realized by a spring force, the excitation was applied electromagnetically, and the blades were replaced by two flexible bars, with all the geometries scaled up from a real device. The amplitudes of the damper and the blades were measured by six inductive displacement transducers, and the spring force by a strain gauge arrangement. For more details we again refer to [47].

15. Concluding Remarks

Traditionally, the dynamics of rigid body systems is attributed to the area of analytical mechanics which connects mathematics with applied technical mechanics. In Germany, analytical mechanics has reached its peak in 1949 with the book of Hamel, and is today not continued in the classical form. Technical mechanics on the one hand is now represented by engineers and deals mainly with application problems, abstract differential geometry on the other hand has to be regarded as an independent branch of mathematics. The interconnection of these disciplines, however, is more and more disregarded. For engineers it is nearly impossible to enter the geometrical branch of mechanics, because the mathematical requirements are very demanding. Thus, much of interdisciplinary and structural knowledge provided classically by analytical mechanics is going to be lost as a consequence. Abstract modeling techniques, for example, such as a direct modeling of physical processes in spaces of higher dimensions, may lead to very compact and powerful models. It requires, however, certain knowledge of both areas, mechanics and mathematics, which should be taught at the universities within the framework of analytical mechanics. The capabilities of commercial software are by far not sufficient to solve these problems adequately. Therefore, it is up to the universities to develop the corresponding theoretical basis and efficient solution techniques. The same holds true for any kind of applications which require rigorous use of interdisciplinary knowledge of physical behavior, mathematical formulation and numerical evaluation.

Classical mechanics is a very well established science and its theory and application are taught in mechanics classes all over the world. One would therefore expect that its theoretical foundations are laid down perfectly in a large number of books. Taking a closer look at literature, it turns out that the opposite is the case: The available books on mechanics are either just meant as an introduction to this subject for undergraduates, or they are already focused on special topics. There are only a very few authors which try to put together the foundations of mechanics in an axiomatic and consistent way, such that the more specific but fundamental equations of all the classical branches of mechanics could be extracted. This situation is not acceptable for a science so closely connected to mathematics, especially as from the mathematical point of view all the required concepts are available; indeed, the development of these mathematical tools was often inspired by mechanical problems. The design of such an axiomatic framework for mechanics is difficult, but necessary to guarantee an effective interdisciplinary education in the classical branches of mechanics. It requires both, mathematicians and engineers, and is carried out in an alternating process: The fundamental equations and working principles of the different branches are provided from the engineers and subsequently put into the right setting as precisely formulated statements by mathematicians. Such an attempt was made in Chapter 2 from which the rigid body dynamics equations have been extracted and further analyzed in Chapter 3. In mechanics it was never intended to compute the processes in nature or technology themselves, but only models of them as sufficiently accurate approximations. An axiomatic treatment of mechanics must therefore regard and contain precisely the models and modeling techniques which have been proven to be suitable and successful for treating and solving practical problems. Besides the continua, this includes as an equal member the rigid body. Thus, forces acting at single points, free moments and point masses, but also inertia and force magnitudes which are only meaningful when stated in integral form have to be accepted. From the beginning, they must fully be taken into account in the theory in a consistent way, and must not be excluded or separately treated as special cases. On the contrary, density functions such as mass densities, stresses and the like are already restrictions which narrow down a general framework of mechanics more than necessary. From the viewpoint of modeling — and this is the only that counts — rigid bodies are as natural or artificial as continua, and must therefore fit in the same axiomatic frame. Axiomatic mechanics is also expected to be of benefit to teaching. There are enough critical points in the comprehension of theoretical mechanics which might be clarified. For instance, the principle of virtual work as one of the most fundamental keys in mechanics can not be grasped without a deep understanding of the duality of linear spaces, a concept from linear algebra which is usually not taught in engineering mathematics. Experience shows that this principle remains obscure and mysterious for most of the students, and that, for the same reason, even the teaching staff has developed in many cases just an intuitive idea about it.

The major part of this book deals with set-valued force laws which are expressed by the generalized gradient of non-convex and non-smooth potential functions. Problems of that kind can basically be reduced to the question of what happens when "half of a classical bilateral constraint is removed." What remains is a unilateral constraint, a concept that is practically not treated in classical mechanics. However, its structure is so rich and is of general importance such that instantaneously an entire new research area opens up, the field of non-smooth dynamics. Again, it is the principle of virtual work which acts as one of the keys to this mystery. In contrast to classical dynamics it holds now as a variational inequality which puts into duality the cones of admissible forces and virtual displacements. This reflects the ambivalent character of set-valued force laws in the sense that they may act as both, impressed forces and constraint forces, depending on the actual dynamic state of the system. One should be aware that these two types of forces are treated completely different in classical mechanics with respect to formulation and evaluation of the equations of motion: Impressed forces are continuous functions of the coordinates and can therefore be directly evaluated when the displacements and velocities of the system are known. Constraint forces originate from bilateral constraints and can not be represented by such functions. Usually certain values of the coordinates are prescribed but the forces are arbitrary. The evaluation of the constraint forces is thus done either by inverting the corresponding force characteristics which yields additional equations depending on the displacements known as constraints. Or one chooses appropriate reduced sets of new coordinates, called minimal coordinates, such that the constraints are automatically satisfied and thus eliminated from the equations of motion. This approach makes sense as long as the above mentioned classification applies to all forces within a multibody system but fails for the mixed-type forces considered here. As already mentioned, this type of forces leads to variational and hemivariational inequalities which are variational expressions for differential inclusions obtained by subdifferentiation or generalized differentiation of certain convex or non-convex superpotential functions. Chapters 4 to 9 treat the formulation and physical interpretation of set-valued force laws in dynamics as well as the evaluation of the system's directional accelerations. The interaction laws are represented by non-smooth potential functions and then derived through generalized differentiation. The resulting multifunctions consider the cases of smooth force characteristics, bilateral constraints, as well as combinations of them like unilateral constraints, dry friction, or prestressed springs. Without making any difference between the nature of these forces three variational inequalities are stated, together with the corresponding differential inclusions. Following classical mechanics they are called the principles of d'Alembert, Jourdain, and Gauss, where the latter provides a strictly convex minimization problem closely related to the classical principle of least constraints. Although these results are expected to be of much more general character, they have been obtained under the assumption that the potentials are composed of one-dimensional functions. This restriction was necessary to avoid second directional (upper sub-)derivatives, quite complicated objects when the functions to be differentiated are not longer lower semicontinuous.

With the use of set-valued force laws a great variety of new questions turns up: From the mathematical point of view, existence and uniqueness problems have to be investigated. When non-existence or non-uniqueness occurs, physical interpretations of these phenomena are required, a task which fortunately always succeeds. An obvious task is to combine non-smooth analysis with the theory of Riemannian manifolds, a step which will further be enlightening for a deep understanding of the geometry behind non-smooth dynamics. Models of mechanical systems containing set-valued force laws require special algorithms for their numerical evaluation. In particular two classes are of exceptional importance: The evaluation of complementarity problems and the design of special routines for numerical integration, better known as time-stepping methods. Recently, there is tremendous progress in all these directions, and one can expect a flood of new results in the near future. The introduction of inequality laws extends considerably the pool of tools for modeling and is, of course, not restricted to systems of rigid bodies: In continuum mechanics, set-valued force laws also occur when formulating unilateral contact problems with friction or adhesion, or in problems of plasticity. In hydraulics, check valves, control valves and the like are modeled in this way; corresponding elements in electric circuits are diodes and transistors. Set-valued interaction laws expressing inequality problems occur in nearly every field of physics, and one might expect them to be implemented in commercial computer software within the next decade.

Mechanics is and will be a science fundamental for the comprehension of processes in nature and technology. The concepts used in mechanics require a high level of abstraction, which is the reason why the subject is judged as difficult in education. It is well known that the ability to deal soundly with abstract concepts can only be gained by a permanent training on the subject, but not via crash courses or the like. Thus, with respect to teaching, mechanics must be looked upon as a long-term subject which accompanies the students for many semesters in order to guarantee a sound and broad education.

References

1. Abraham R., Marsden J.E. (1978) Foundations of Mechanics. The Benjamin Cummings Publishing Company Inc., Reading Massachusetts
2. Abraham R., Marsden J.E., Ratiu T.S. (1988) Manifolds, Tensor Analysis, and Applications. Applied Mathematical Sciences 75, Springer, New York
3. Alart P., Curnier A. (1991) A Mixed Formulation for Frictional Contact Problems Prone to Newton Like Solution Methods. Comp Meth Appl Mech Eng 92(3):353–357
4. Ambrosio L. (1990) Existence Theory for a New Class of Variational Problems. Arch Rat Mech Anal 111:291–322
5. Angeles J. (1988) Rational Kinematics. Springer Tracts in Natural Philosophy Vol. 34, Springer, New York
6. Anitescu M., Potra F.A., Stewart D.E. (1999) Time-Stepping for Three-Dimensional Rigid Body Dynamics. Comp Meth Appl Mech Eng 177(3):183–197
7. Arnold V.I. (1980) Mathematical Methods of Classical Mechanics. Graduate Texts in Mathematics, Springer, New York Heidelberg Berlin
8. Aubin J. (1977) Applied Abstract Analysis. John Wiley & Sons, New York
9. Aubin J., Ekeland I. (1984) Applied Nonlinear Analysis. John Wiley & Sons, New York
10. Aubin J., Frankowska H. (1990) Set-Valued Analysis. Birkhäuser, Boston
11. Ball R.S. (1998) A Treatise on the Theory of Screws. Cambridge University Press, Cambridge
12. Ballard P. (2000) The Dynamics of Discrete Mechanical Systems with Perfect Unilateral Constraints. Arch Rat Mech Anal 154(3):199–274
13. Baraff D. (1994) Issues in Computing Contact Forces for Non-Penetrating Rigid Bodies. Algorithmica 10:292–352.
14. Baraff D. (1994) Fast Contact Force Computation for Nonpenetrating Rigid Bodies. In: Proceedings of the SIG-GRAPH 94 Computer Graphics Proceedings, Orlando, July 24–29, 1994. Annual Conference Series, 23–34
15. Batlle J.A., Condomines A.B. (1991) Rough Collisions in Multibody Systems. Mech Mach Theory 26(6):565–577
16. Blajer W. (1993) Dynamic Analysis of Constrained Multibody Systems in Orthonormalized Tangent Space. In: Schiehlen W. (Ed.) Advanced Multibody System Dynamics. Kluwer Academic Publishers, The Netherlands, 415–420
17. Blajer W. (1997) A Geometric Unification of Constrained System Dynamics, Multibody System Dynamics 1:3–21
18. Boltzmann L. (1922) Vorlesungen über die Prinzipe der Mechanik, I. Teil. Verlag von Johann Ambrosius Barth, Leipzig
19. Brach R.M. (1991) Mechanical Impact Dynamics. John Wiley & Sons, New York
20. Bremer H. (1988) Dynamik und Regelung mechanischer Systeme. Teubner Studienbücher Mechanik, B.G. Teubner, Stuttgart

21. Brogliato B. (1996) Nonsmooth Impact Mechanics. Models, Dynamics and Control. Lect. Notes in Control and Information Sciences Vol. 220, Springer, Heidelberg New York
22. Brogliato B., ten Dam A.A., Paoli L., Génot F., Abadie M. Numerical Simulation of Finite Dimensional Multibody Nonsmooth Mechanical Systems. Submitted to ASME Applied Mechanics Reviews, May 2000
23. Chatterjee A. (1997) Rigid Body Collisions: Some General Considerations, New Collision Laws, and Some Experimental Data. PhD Thesis, Cornell University, Ithaca New York
24. Clarke F.H. (1983) Optimization and Nonsmooth Analysis. John Wiley & Sons, New York
25. Clarke F.H., Ledyaev Y.S., Stern R.J., Wolensky P.R. (1998) Nonsmooth Analysis and Control Theory. Graduate Texts in Mathematics, Springer, New York
26. Cottle R.W., Dantzig G.B. (1968) Complementary Pivot Theory of Mathematical Programming. Lin Alg Appl 1:103–125
27. Cottle R.W., Pang J.S., Stone R.E. (1992) The Linear Complementarity Problem. Computer Science and Scientific Computing, Academic Press, London
28. Curnier A., He Q.C., Klarbring A. (1995) Continuum Mechanics Modelling of Large Deformation Contact with Friction. In: Raous M., Jean M., Moreau J.J. (Eds.) Contact Mechanics. Plenum Press, New York, 145–158
29. ten Dam A.A. (1997) Unilaterally Constrained Dynamical Systems. Doctoral Thesis, Rijksuniversiteit Groningen, The Netherlands
30. Eberhard P. (1999) Collision Detection and Contact Approaches for a Hybrid Multibody System / Finite Element Simulation. In: Pfeiffer F., Glocker Ch. (Eds.) IUTAM Symposium on Unilateral Multibody Contacts, Proc. of the IUTAM Symposium held in Munich, Germany, August 3–7, 1998. Kluwer Acad. Publ., The Netherlands, 193–202
31. Elstrodt J. (1996) Maß- und Integrationstheorie. Springer, Berlin Heidelberg New York
32. Facchinei F., Fischer A., Kanzow Ch. (1997) A Semismooth Newton Method for Variational Inequalities: The Case of Box Constraints. In: Ferris M.M., Pang J.-S. (Eds.) Complementarity and Variational Problems: State of the Art. SIAM, Philadelphia, 76–90.
33. Fischer A. (1992) A Special Newton-Type Optimization Method. Optimization 24:269–284
34. Fischer U., Stephan W. (1972) Prinzipien und Methoden der Dynamik. VEB Fachbuchverlag Leipzig, Leipzig
35. Frémond M. (1995) Rigid Bodies Collisions. Physics Letters A 204:33–41
36. Génot B., Brogliato B. (1999) New Results on Painlevé Paradoxes. European J Mechanics A/Solids 18(4):653–677
37. Glocker Ch. (1995) Dynamik von Starrkörpersystemen mit Reibung und Stößen. VDI-Fortschrittberichte Mechanik/Bruchmechanik, Reihe 18, Nr. 182, VDI-Verlag, Düsseldorf
38. Glocker Ch. (1998) The Principles of d'Alembert, Jourdain, and Gauß in Nonsmooth Dynamics. Part I: Scleronomic Multibody Systems. ZAMM 78(1):21–37
39. Glocker Ch. (1999) Discussion of d'Alembert's Principle for Non-smooth Unilateral Constraints. ZAMM 79(1):S91–S94
40. Glocker Ch. (1999) Decomposition of Scalar Force Interactions. In: Pfeiffer F., Glocker Ch. (Eds.) IUTAM Symposium on Unilateral Multibody Contacts, Proc. of the IUTAM Symposium held in Munich, Germany, August 3–7, 1998. Kluwer Acad. Publ., The Netherlands, 15–24

41. Glocker Ch. (1999) Velocity Jumps Induced by C^o-Constraints. In: Proceedings of the ASME 1999 Design Engineering Technical Conferences, Las Vegas, Nevada, September 12 - 15, 1999. ASME, New York, CD Rom
42. Glocker Ch., Pfeiffer F. (1992) Dynamical Systems with Unilateral Contacts. Nonlinear Dynamics 3:245–259
43. Glocker Ch., Pfeiffer F. (1994) Stick-slip Phenomena and Application. In: Thompson J.M.T., Bishop S.R. (Eds.) Nonlinearity and Chaos in Engineering Dynamics, IUTAM Symposium, UCL, London, July 1993. John Wiley & Sons, New York, 103–113.
44. Glocker Ch., Pfeiffer F. (1995) Multiple Impacts with Friction in Rigid Multibody Systems. Nonlinear Dynamics 7:471–497
45. Hairer E., Norsett S.P., Wanner G. (1987) Solving Ordinary Differential Equations I. Nonstiff Problems. Springer Series in Computational Mathematics, Springer, Berlin Heidelberg
46. Hairer E., Wanner G. (1991) Solving Ordinary Differential Equations II. Stiff and Differential-Algebraic Problems. Springer Series in Computational Mathematics, Springer, Berlin Heidelberg
47. Hajek M. (1990) Reibungsdämpfer für Turbinenschaufeln. VDI-Fortschrittberichte Schwingungstechnik, Reihe 11, Nr. 128, VDI-Verlag, Düsseldorf
48. Haj-Fraj A., Pfeiffer F. (1998) Dynamics of Automatic Vehicle Transmissions. In: Proc. of the 4th Internatinal Conference on Motion and Vibration Control MOVIC'98. Zurich, Switzerland, August 25–28, 3, 839–844
49. Haj-Fraj A., Pfeiffer F. (1999) Dynamics of Gear Shift Operations in Automatic Transmissions. In: Starr, Leung, Wright, Sandoz (Eds.) Proc. 1st Int. Conf. on Integrating Dynamics, Monitoring and Control for the 21st Century , Balkema, Rotterdam, 29–35
50. Haj-Fraj A., Pfeiffer F. (1999) Modelling of Wet Clutches in Automatic Transmissions. In: Proc. 2nd Int. Conf. on Recent Advances in Mechatronics, Istanbul, Turkey, May 24–26, 9–15
51. Hamel G. (1910) Bemerkungen zu den vorstehenden Aufsätzen der Herrn F. Klein und R. v. Mises. Zeitschrift für Mathematik und Physik 58(1-2):195–196
52. Hamel G. (1967) Theoretische Mechanik. Springer, Berlin
53. Hurmuzlu Y., Marghitu D.B. (1994) Rigid Body Collisions of Planar Kinematic Chains with Multiple Contact Points. Int J Robotics Research 13(1):82–92
54. Ivanov A.P. (1992) Energetics of a Collision with Friction. J Appl Maths Mechs 56(4):527–534
55. Ivanov A.P. (1995) On Multiple Impact. J Appl Maths Mechs 59:887–902
56. Ivanov A.P. (1999) On the Problem of Constrained Collisions. In: Pfeiffer F., Glocker Ch. (Eds.) IUTAM Symposium on Unilateral Multibody Contacts, Proc. of the IUTAM Symposium held in Munich, Germany, August 3–7, 1998. Kluwer Acad. Publ., The Netherlands, 107–116
57. Jean M. (1995) Frictional Contact in Collections of Rigid or Deformable Bodies: Numerical Simulation of Geomaterial Motions. In: Selvadurai A.P.S., Boulon M.J. (Eds.) Mechanics of Geomaterial Interfaces. Elsevier Science B.V., 463–486
58. Jean M., Moreau J.J. (1985) Dynamics in the Presence of Unilateral Contacts and Dry Friction; A Numerical Approach. Note Technique 85-5, Lab de Mecanique Generale des Milieux Continues, Univ. des Sci. et Techniques du Languedoc, Montpellier
59. Jean M., Moreau J.J. (1992) Unilaterality and Dry Friction in the Dynamics of Rigid Body Collections. In: Curnier A. (Ed.) Proc. Contact Mechanics International Symposium, Lausanne, Switzerland, October 7–9, 1992. Presses Polytechniques et Universitaires Romandes, Lausanne, 31–48

60. Jourdan F., Jean M., Alart P. (1998) An Alternative Method between Implicit and Explicit Schemes Devoted to Frictional Contact Problems in Deep Drawing Simulation. J Mater Process Technol 80–81:257–262
61. Keller J.B. (1986) Impact with Friction. J Appl Mech 53:1–4
62. Klarbring A. (1986) Quadratic Programs in Frictionless Contact Problems. Int J Engng Sci 24(7):1207–1217
63. Klarbring A. (1986) A Mathematical Programming Approach to Three-Dimensional Contact Problems with Friction. Comp Meth Appl Mech Eng 58:175–200
64. Klarbring A., Björkman G. (1988) A Mathematical Programming Approach to Contact Problems with Friction and Varying Contact Surface. Comp Struct 30(5):1185–1198
65. Klarbring A. (1992) Mathematical Programming and Augmented Lagrangian Methods for Frictional Contact Problems. In: Curnier A. (Ed.) Proc. Contact Mechanics International Symposium, Lausanne, Switzerland, October 7–9, 1992. Presses Polytechniques et Universitaires Romandes, Lausanne, 409–422
66. Klein F. (1910) Zu Painleves Kritik der Coulombschen Reibgesetze. Zeitschrift für Mathematik und Physik 58(1-2):186–191
67. Lötstedt P. (1979) A Numerical Method for the Simulation of Mechanical Systems with Unilateral Constraints. TRITA-NA-7920, Dept. of Numerical Analysis and Computing Science, The Royal Institute of Technology, Sweden
68. Lötstedt P. (1981) Coulomb Friction in Two-Dimensional Rigid Body Systems. ZAMM 61:605–615
69. Lötstedt P. (1982) Mechanical Systems of Rigid Bodies Subject to Unilateral Constraints. SIAM J Appl Math 42(2):281–296
70. Lötstedt P. (1982) Numerical Simulation of Time-Dependent Contact and Friction Problems in Rigid Body Mechanics. TRITA-NA-8214, Dept. of Numerical Analysis and Computing Science, The Royal Institute of Technology, Sweden
71. Lubich Ch., Nowak U., Pöhle U., Engstler Ch. (1992) MEXX – Numerical Software for the Integration of Constrained Multibody Systems. Preprint SC 92-12, Konrad-Zuse-Zentrum für Informationstechnik, Berlin
72. de Luca T., Facchinei F., Kanzow Ch. (1996) A Semismooth Equation Approach to the Solution of Nonlinear Complementarity Problems. Mathematical Programming 75:407–439
73. de Luca T., Facchinei F., Kanzow Ch. (2000) A Theoretical and Numerical Comparison of Some Semismooth Algorithms for Complementarity Problems. Comp Opt Appl 16(2):173–205
74. Mabrouk M. (1998) A Unified Variational Model for the Dynamics of Perfect Unilateral Constraints. European J Mechanics A/Solids 17(5):818–842
75. Maisser P. (1983) Ein Beitrag zur Theorie diskreter elektromechanischer Systeme mit Anwendungen in der Manipulator-/Robotertechnik, Dr. Sc. Nat. Dissertation, Fakultät für Mathematik und Naturwissenschaften, Technische Hochschule Ilmenau, Ilmenau
76. Maisser P. (1988) Analytische Mechanik von Mehrkörpersystemen. ZAMM 68(10):463–481
77. Maisser P. (1997) Differential-Geometric Methods in Multibody Dynamics, Nonlinear Analysis 30(8):5127–5133
78. Marsden J.E., Ratiu T.S. (1999) Introduction to Mechanics and Symmetry. Springer, New York
79. Martins J.A.C., Pinto da Costa A. (2000) Stability of Finite Dimensional Nonlinear Elastic Systems with Unilateral Contact and Friction. Int J Solids Structures 37(18):2519–2564
80. May H., Panagiotopoulos P.D. (1985) F.H. Clarke's Generalized Gradient and Fourier's Principle. ZAMM 65(2):125–126

81. Meitinger Th. (1998) Dynamik automatisierter Montageprozesse. VDI-Fortschrittberichte Fertigungstechnik, Reihe 2, Nr. 476, VDI-Verlag, Düsseldorf
82. Meltzer G., Melzig-Thiel R., Schatte M. (1980) Ein Mathematisches Schwingungsmodell für das Menschliche Hand-Arm-System. Maschinenbautechnik 29(2):54–58
83. v. Mises R. (1910) Zur Kritik der Reibungsgesetze. Zeitschrift für Mathematik und Physik 58(1-2):186–195
84. Monteiro Marques M.D.P. (1993) Differential Inclusions in Nonsmooth Mechanical Problems. Shocks and Dry Friction. Birkhäuser, Basel
85. Moreau J.J. (1963) Les Liaisons Unilatérales et le Principe de Gauß. C.R. Acad Sci Paris 256:871–874
86. Moreau J.J. (1979) Application of Convex Analysis to Some Problems of Dry Friction. Trends in Applications of Pure Mathematics to Mechanics 2:263–280
87. Moreau J.J. (1983) Standard Inelastic Shocks and the Dynamics of Unilateral Constraints. In: Del Piero G., Maceri F. (Eds.) Unilateral Problems in Structural Analysis, CISM Courses and Lectures Vol. 288. Springer, Wien, 173–221
88. Moreau J.J. (1986) Une Formulation du Contact a Frottement Sec; Application au Calcul Numerique. Technical Report 13, Serie II, C.R. Acad. Sci. Paris
89. Moreau J.J. (1987) Une Formulation de la Dynamique Classique. Technical Report 5, Serie II, C.R. Acad. Sci. Paris
90. Moreau J.J. (1988) Bounded Variation in Time. In: Moreau J.J., Panagiotopoulos P.D., Strang G. (Eds.) Topics in Nonsmooth Mechanics. Birkhäuser, Basel, 1–74
91. Moreau J.J. (1988) Unilateral Contact and Dry Friction in Finite Freedom Dynamics. In: Moreau J.J., Panagiotopoulos P.D. (Eds.) Non-Smooth Mechanics and Applications, CISM Courses and Lectures Vol. 302. Springer, Wien, 1–82.
92. Moreau J.J. (1993) New Computation Methods in Granular Dynamics. A.A. Balkema, Rotterdam, Powders & Grains 93:227–232
93. Moreau J.J. (1994) Some Numerical Methods in Multibody Dynamics: Application to Granular Materials. European J Mechanics A/Solids 13(4):93–114
94. Moreau J.J. (1999) Numerical Aspects of the Sweeping Process. Comp Meth Appl Mech Eng 177:329–349
95. Murty K.G. (1988) Linear Complementarity, Linear and Nonlinear Programming. Sigma Series in Applied Mathematics, Vol. 3, Heldermann, Berlin
96. Panagiotopoulos P.D. (1975) A Nonlinear Programming Approach to the Unilateral Contact-, and Friction-Boundary Value Problem in the Theory of Elasticity. Ingenieur Archiv 44:421–432
97. Panagiotopoulos P.D. (1977) Ungleichungsprobleme in der Mechanik. Habilitation, Technische Hochschule Aachen, Aachen
98. Panagiotopoulos P.D. (1982) Ungleichungsprobleme und Differentialinklusionen in der Analytischen Mechanik. Annual Public. School of Technology, Aristotle University Thessaloniki, Greece, Vol. Θ', 100–139
99. Panagiotopoulos P.D. (1983) Hemivariational Inequalities, Existence and Approximation Results. In: Del Piero G., Maceri F. (Eds.) Unilateral Problems in Structural Analysis, CISM Courses and Lectures Vol. 288. Springer, Wien, 223–246
100. Panagiotopoulos P.D. (1985) Inequality Problems in Mechanics and Applications. Birkhäuser, Boston Basel Stuttgart
101. Panagiotopoulos P.D. (1993) Hemivariational Inequalities. Springer, Berlin Heidelberg
102. Pang J.S., Trinkle J.C., Lo G. (1995) A Complementarity Approach to a Quasi-Static Multi-Rigid-Body Contact Problem. Comp Opt Appl 5:139–154

103. Pang J.S., Trinkle J.C. (1996) Complementarity Formulations and Existence of Solutions of Dynamic Multi-Rigid-Body Contact Problems with Coulomb Friction, Mathematical Programming 30:199–226
104. Paoli L., Schatzman M. (1998) Theoretical and Numerical Study for Vibrations with Unilateral Constraints: Case of a Non-convex Set of Constraints. In: Idelsohn S.R., Onate E., Dvorkin E.N. (Eds.) Computational Mechanics – New Trends and Applications, Proc. of the 4th World Congress on Computational Mechanics, Buenos Aires, Argentina, June 29–July 2, 1998. CIMNE, Barcelona, CD Rom
105. Paoli L., Schatzman M. (1999) A Numerical Scheme for Impact Problems. Prepreint de l'Equipe d'Analyse Numérique, UMR 5585, June, submitted to SIAM J. on Numerical Analysis.
106. Paoli L., Schatzman M. (1999) Approximation et Existence en Vibro-Impact. C.R. Acad Sci Paris 329I:1103–1107
107. Papastavridis J.G. (1999) Tensor Calculus and Analytical Dynamics. CRC Press LLC, Boca Raton
108. Del Piero G. (1992) Shock Dynamics for a Rigid Block over a Rigid Plane. In: Curnier A. (Ed.) Proc. Contact Mechanics International Symposium, Lausanne, Switzerland, October 7–9, 1992. Presses Polytechniques et Universitaires Romandes, Lausanne, 3–12
109. Pfeiffer F. (1910) Zur Frage der sog. Coulombschen Reibungsgesetze. Zeitschrift für Mathematik und Physik 58(3):273–311
110. Pfeiffer F. (1984) Mechanische Systeme mit unstetigen Übergängen. Ingenieur Archiv 54(3):232–240
111. Pfeiffer F. (1988) Theorie des Getrieberasselns. VDI Berichte 697:45–65
112. Pfeiffer F. (1991) Dynamical Systems with Time Varying or Unsteady Structure. ZAMM 71(4):T6–T22
113. Pfeiffer F. (1993) Complementarity Problems of Stick-Slip Vibrations. In: Dynamics and Vibrations of Time-Varying Systems and Structures, DE–Vol. 56. ASME, New York, 43–50
114. Pfeiffer F. (1994) Methoden zur Nichtlinearen Antriebstechnik. VDI Berichte 1153:599–624
115. Pfeiffer F. (1999) Unilateral Problems of Dynamics. Archive of Applied Mechanics 69:503–527
116. Pfeiffer F., Glocker Ch. (1996) Multibody Dynamics with Unilateral Contacts. John Wiley & Sons, New York
117. Pfeiffer F., Hajek M. (1992) Stick-slip Motion of Turbine Blade Dampers. Phil Trans Royal Soc A338:503–517
118. Pfeiffer F., Kücükay F. (1985) Eine erweiterte mechanische Stoßtheorie und ihre Anwendung in der Getriebedynamik. VDI-Z 127(9):341–349
119. Popp K. (1999) Examples of Non-smooth Mechanical Systems – An Overview. In: Pfeiffer F., Glocker Ch. (Eds.) IUTAM Symposium on Unilateral Multibody Contacts, Proc. of the IUTAM Symposium held in Munich, Germany, August 3–7, 1998. Kluwer Acad. Publ., The Netherlands, 233–242
120. Prandtl L. (1910) Bemerkungen zu den vorstehenden Aufsätzen der Herrn F. Klein und R. v. Mises und G. Hamel. Zeitschrift für Mathematik und Physik 58(1-2):196–197
121. Rockafellar R.T. (1972) Convex Analysis. Princeton University Press, Princeton New Jersey
122. Rockafellar R.T. (1980) Generalized Directional Derivatives and Subgradients of Nonconvex Functions. Can J Math 32(2):275–280

123. Rockafellar R.T. (1981) The Theory of Subgradients and Its Application to Problems of Optimization. Convex and Nonconvex Functions. R & E research and education in mathematics, Heldermann, Berlin
124. Rockafellar R.T., Wets R.J-B. (1998) Variational Analysis. Springer, Berlin Heidelberg
125. Rossmann Th. (1998) Eine Laufmaschine für Rohre. VDI-Fortschrittberichte Meß-, Steuerungs- und Regelungstechnik, Reihe 8, Nr. 732, VDI-Verlag, Düsseldorf
126. Rossmann Th., Pfeiffer F., Glocker Ch. (1997) Efficient Algorithms for Nonsmooth Dynamics. Proc. of the ASME Int. Mech. Eng. Congress and Exposition, Dallas, Texas, November 16–21, 1997. ASME, New York
127. Rudin W. (1981) Real & Complex Analysis. Tata McGraw-Hill, New Delhi
128. Schatzman M. (1978) A Class of Nonlinear Differential Equations of Second Order in Time. Nonlinear Analysis Theory Methods and Applications 2(3):355–373
129. Schatzman M. (1998) Uniqueness and Continuous Dependence on Data for One-dimensional Impact Problem. Math Comp Modeling 28(4-8):1–18
130. Schiehlen W. (1997) Multibody System Dynamics: Roots and Perspectives. Multibody System Dynamics 1(2):149–188
131. Schiehlen W. (1999) Unilateral Contacts in Machine Dynamics. In: Pfeiffer F., Glocker Ch. (Eds.) IUTAM Symposium on Unilateral Multibody Contacts, Proc. of the IUTAM Symposium held in Munich, Germany, August 3–7, 1998. Kluwer Acad. Publ., The Netherlands, 287–298
132. Seyfferth W., Pfeiffer F. (1994) Modelling of Time-Varying Contact Problems in Multibody Systems. Proc. of 12th Symp. on Eng. Appl. of Mechs., Montreal, Canada, 579–588
133. Simo J.C., Laursen T.A. (1992) An Augmented Lagrangian Treatment of Contact Problems Involving Friction. Comp Struct 42(1):97–116
134. Stavroulakis G.E. (1993) Convex Decomposition for Nonconvex Energy Problems in Elastostatics and Application. European J Mechanics A/Solids 12(1):1–20
135. Stewart D.E. (1998) Convergence of a Time-Stepping Scheme for Rigid Body Dynamics and Resolution of Painlevé's Problem. Arch Rat Mech Anal 145(3):215–260
136. Stewart D.E., Trinkle J.C. (1996) An Implicit Time-Stepping Scheme for Rigid Body Dynamics with Inelastic Collisions and Coulomb Friction. Int J Num Methods Engineering 39(15):2673–2691
137. Stiegelmeyr A., Pfeiffer F. (1999) Chimney Dampers. In: Pfeiffer F., Glocker Ch. (Eds.) IUTAM Symposium on Unilateral Multibody Contacts, Proc. of the IUTAM Symposium held in Munich, Germany, August 3–7, 1998. Kluwer Acad. Publ., The Netherlands, 299–308
138. Stiegelmeyr A., Pfeiffer F. (1999) A Time Stepping Algorithm for Mechanical Systems with Unilateral Contacts. In: Proceedings of the ASME 1999 Design Engineering Technical Conferences, Las Vegas, Nevada, September 12 - 15, 1999. ASME, New York, CD Rom
139. Stronge W.J. (1990) Rigid Body Collisions with Friction. Proc Royal Soc A 421:169–181
140. Stronge W.J. (1999) Mechanics of Impact for Compliant Multi-Body Systems. In: Pfeiffer F., Glocker Ch. (Eds.) IUTAM Symposium on Unilateral Multibody Contacts, Proc. of the IUTAM Symposium held in Munich, Germany, August 3–7, 1998. Kluwer Acad. Publ., The Netherlands, 137-144
141. Tabarrok B., Rimrott F.P.J. (1994) Variational Methods and Complementary Formulations in Dynamics. Kluwer Academic Publishers, The Netherlands

142. Trinkle J.C., Pang J.S., Sudarsky S., Lo G. (1997) On Dynamic Multi-Rigid-Body Contact Problems with Coulomb Friction. ZAMM 77(4):267–279
143. Vol'pert A.I., Hudjaev S.I. (1985) Analysis in Classes of Discontinuous Functions and Equations of Mathematical Physics. Martinus Nijhoff, Dordrecht Boston Lancaster
144. Wang Y., Mason M.T. (1992) Two-Dimensional Rigid-Body Collision with Friction. J Appl Mech 59:635–642
145. Wittenburg J. (1977) Dynamics of Systems of Rigid Bodies. B.G. Teubner, Stuttgart
146. Wolfsteiner P. (1999) Dynamik von Vibrationsförderern. VDI-Fortschrittberichte Fertigungstechnik, Reihe 2, Nr. 511, VDI-Verlag, Düsseldorf

List of Figures

Index

Printing (Computer to Film): Saladruck, Berlin
Binding: Stürtz AG, Würzburg

GPSR Compliance

The European Union's (EU) General Product Safety Regulation (GPSR) is a set of rules that requires consumer products to be safe and our obligations to ensure this.

If you have any concerns about our products, you can contact us on ProductSafety@springernature.com

In case Publisher is established outside the EU, the EU authorized representative is:

Springer Nature Customer Service Center GmbH
Europaplatz 3
69115 Heidelberg, Germany

Batch number: 10584448

Printed by Printforce, the Netherlands